AF864714

Deep Technology for Sustainable Fisheries and Aquaculture

Amaj Rahimi-Midani

Deep Technology for Sustainable Fisheries and Aquaculture

Springer

Amaj Rahimi-Midani
Poseidon-AI PTE Limited
Singapore, Singapore

ISBN 978-981-99-4916-8 ISBN 978-981-99-4917-5 (eBook)
https://doi.org/10.1007/978-981-99-4917-5

© The Editor(s) (if applicable) and The Author(s), under exclusive license to Springer Nature Singapore Pte Ltd. 2023
The book uses the United Nations Sustainable Development Goals (UNSDGs) but does not represent the viewpoints of the UN officials. Data and information gathered in this book are from years of fieldwork conducted by Poseidon-AI PTE LTD in various countries in Asia, Africa, and Latin America.
This work is subject to copyright. All rights are solely and exclusively licensed by the Publisher, whether the whole or part of the material is concerned, specifically the rights of translation, reprinting, reuse of illustrations, recitation, broadcasting, reproduction on microfilms or in any other physical way, and transmission or information storage and retrieval, electronic adaptation, computer software, or by similar or dissimilar methodology now known or hereafter developed.
The use of general descriptive names, registered names, trademarks, service marks, etc. in this publication does not imply, even in the absence of a specific statement, that such names are exempt from the relevant protective laws and regulations and therefore free for general use.
The publisher, the authors, and the editors are safe to assume that the advice and information in this book are believed to be true and accurate at the date of publication. Neither the publisher nor the authors or the editors give a warranty, expressed or implied, with respect to the material contained herein or for any errors or omissions that may have been made. The publisher remains neutral with regard to jurisdictional claims in published maps and institutional affiliations.

This Springer imprint is published by the registered company Springer Nature Singapore Pte Ltd.
The registered company address is: 152 Beach Road, #21-01/04 Gateway East, Singapore 189721, Singapore

To my grandfather, who taught me the fundamentals of business,
To my mother, who works tirelessly to empower women and girls in the primary sector,
To my wife, who is a role model for women and girls in STEM, and
To my father, whose unwavering support gave me the courage to swim against the strongest currents.

Preface

As I walked through the halls of the United Nations (UN), I found myself trying to answer a simple question: Is the UN an implementing body or a coordinating organization? To respond to this question, and because the UN works in every country on the planet, I needed to find criteria that everyone could understand. Although the United Nations Sustainable Development Goals (UNSDGs) are good criteria, bureaucratic procedures at both the national and international levels prevent many approaches from demonstrating their results in a meaningful way.

The truth is that no organization with limited resources, funding, and time can positively impact lives while overcoming time-consuming and exhausting bureaucratic procedures. On the other hand, rising population, increasing climate-related incidents, and ongoing global conflicts highlight the need for innovative and novel approaches to addressing the twenty-first-century challenges.

Deep technologies can address global concerns and demonstrate reasonable results in a shorter amount of time and more effectively, and these results can be simply presented in internationally recognized indexes. Impactful venture capital (VC) investments enable the expansion and development of deep technologies more quickly.

As a result, I decided to establish a company that combines artificial intelligence (AI), machine learning (ML), image processing, sensing, and automation to address the twenty-first-century challenges while also significantly contributing to UNSDGs.

Poseidon-AI is a Singapore-based company that assists the global aquaculture sector in growing sustainably. Poseidon-AI employs a low-cost device and an intelligence software to enable farmers and communities to produce various seafood products in a sustainable manner, with the results displayed in meaningful socioeconomic and environmental metrics.

The goal of incorporating Poseidon-AI is to use deep technology in primary industries, particularly aquaculture, which produces the most protein. I am convinced that deep tech can assist organizations such as the UN in identifying a single, primary role as an efficient coordinating body, while deep tech approaches can efficiently solve many global problems.

That is why I decided to write this book, which explains the major issues confronting the aquaculture industry and how deep technology has aided in overcoming these obstacles. The book explains not only the theoretical aspects but also the environmental and socioeconomic outcomes using real-world examples.

Poseidon-AI did not own these associations, farms, or organizations, but by utilizing deep technology, Poseidon-AI acted as intelligent experts for farmers and vulnerable communities to sustainably produce their own healthy and nutritious food while generating revenue and/or lowering operating costs.

Basically, incorporating Poseidon-AI was in response to increasing global needs, requiring faster and more unconventional methods. Many hurtful comments, unfair judgments, and painful thoughts were made against this novel movement along this path, which is still moving forward because it is based on strong scientific, environmental, and humanitarian beliefs. However, this did not prevent Poseidon-AI from expanding and assisting more people around the world, thanks to the vision best described by Jack London:

> *The proper function of man is to live, not to exist. I shall not waste my days in trying to prolong them. I shall use my time. I would rather be ashes than dust . . .*

New York, NY, USA
February 20, 2023

Amaj Rahimi-Midani

Contents

About the Author

Amaj Rahimi-Midani graduated from Pukyong National University (PKNU) in Busan, South Korea. He pursued his postdoctoral fellowship as a climate change modeler by analyzing the climate change impact on European freshwater species. He is the founder and CEO of Poseidon-AI located in Singapore. He has extensive working experience with international organizations in implementing sustainable practices related to water, soil, aquaculture, and fisheries around the world. He has many peer-reviewed international publications to his credit and delivered numerous oral and poster presentations in numerous international meetings, conferences, and congresses.

Abbreviations

ADP	Adenosine triphosphate
AHPN	Acute hepato-pancreatic necrosis
AI	Artificial intelligence
AIDC	Automatic identification and data capture
AMOC	Atlantic meridional overturning circulation
ANN	Artificial neural network
AWS	Amazon Web Services
BCG	Boston Consulting Group
Br^-	Bromide
Br_2	Bromine
BW	Body weight
$Ca(OH)_2$	Calcium hydroxide
Ca^{2+}	Calcium
$CaCO_3$	Calcium carbonate
CEC	Cation exchange capacity
CH_4	Methane
CNN	Convolutional neural network
CO_2	Carbon dioxide
CO_3^{2-}	Carbonates
CVD	Carp viremia disease
CWT	Curvelet wavelet transform
D.O.	Dissolved oxygen
DBTL	Design-build-test-learn
DCT	Discrete cosine transform
DFT	Discrete Fourier transform
DIP	Digital image processing
DL	Deep learning
DNA	Deoxyribonucleic acid
DPCM	Differential pulse-code modulation
DT	Decision tree
EAP	Electric air pressure
EAs	Evolutionary algorithms
EC	Electric conductivity

EM	Expectation maximization
EMS	Early mortality syndrome
EP	Evolutionary programming
ES	Evolution strategy
ESP	Exchangeable sodium percentage
EUS	Epizootic ulcerative syndrome
FAO	Food and Agriculture Organization
FCR	Feed conversion rates
Fe^+	Iron
GA	Genetic algorithm
GCM	Google cloud messaging
GCRV	Grass carp retrovirus
GDP	Gross domestic product
GHG	Greenhouse gas
GLCM	Gray-level co-occurrence matrix
GP	Genetic programming
GPS	General Problem Solver
H	Height
H_2CO_3	Carbonic acid
HBD	Housing and Development Board
$HBrO_2$	Bromous acid
$HBrO_3$	Bromic acid
HCO_3-	Bicarbonates
HHNV	Hypodermal and hematopoietic necrosis virus
HKTs	High-affinity potassium transporters
HLB	Helminthosporium leaf blotch
HOBr	Hypobromous acid
IAS	Integrated aquaculture systems
ICT	Information and communication
IoTs	Internet of Things
IP	Internet Protocol
IPCC	Intergovernmental Panel on Climate Change
IPNV	Infectious pancreatic necrosis virus
IPv6	Internet Protocol version 6
JPL	Jet propulsion laboratory
K_2CO_3	Potassium carbonate
KHV	Koi herpesvirus
KNN	K-nearest neighbor
KOH	Potassium hydroxide
M2M	Machine-to-machine
MAG	Ministerio de Agricultura y Ganadería
Mg^{2+}	Magnesium
ML	Machine learning
MOC	Meridional overturning circulation

MQTT	Message Queue Telemetry Transport
MT	Methyltestosterone
N	Nitrogen
NASA	The National Aeronautics and Space Administration
NB	Naïve Bayes
NH_3	Ammonia
NO_2-	Nitrite
NO_2	Nitrogen dioxide
NO_3-	Nitrate
NO_3-N	Nitrate nitrogen
NSCCs	Nonselective cation channels
OMZs	Oxygen minimum zones
ORP	Oxidation-reduction potential
OSI	Open systems interconnection
PAE	*P* acquisition efficiency
PCA	Principal component analysis
pCO_2	Pressure of carbon dioxide
PGPR	Plant growth-promoting rhizobacteria
pH	Power of hydrogen
PO_4-P	Phosphate phosphorus
PPP	Public-private partnership
R&D	Research and development
RAS	Recirculating aquaculture systems
RCPs	Representative concentration pathways
ROS	Reactive oxygen species
RWS	Roulette wheel selection
SAR	Synthetic aperture radar
SASs	Salt-affected soils
SGPV	Salmon gill poxvirus
SO_4^{2-}	Sulfate
SQL	Shenzhen quarantine laboratory
STEM	Science, Technology, Engineering, and Mathematics
SVM	Support vector machine
TCDC	Technical Cooperation among Developing Countries
TDSs	Total dissolved solids
TiLV	Tilapia lake virus
TL	Total length
TROs	Total residual oxidants
UAVs	Unmanned aerial vehicles
UN	United Nations
UNDP	United Nations Development Programme
UNSDGs	United Nations Sustainable Development Goals
USD	United States dollar
USDA	United States Department of Agriculture
UV	Ultraviolet

VAMS	Vision-based automatic system
VC	Venture capital
W	Weight
WCA	Weight calculation algorithms
WSD	White spot disease
WWI	World War I
YSFI	Yellow Sea Fisheries Institute

Introduction

1

World aquaculture production is growing much faster than the two other sources of animal protein which are animal husbandry and capture fisheries. Aquaculture is booming in Asia, and it's only getting started. Aquaculture is expected to become the primary source of seafood production, and thus public interest in this sector is growing. However, there is a significant knowledge gap between what is practiced on the ground and what is taught in academia. This book strives to share and disseminate deep technological solutions used in the real world to bridge the gap and convey the excitement of the rapidly growing aquaculture sector.

Aquaculture is defined as the farming of aquatic organisms such as mollusks, crustaceans, fish, and aquatic plants. Farming is a type of production intervention in the rearing process. It also implies that cultivated stocks are owned by corporations or individuals.

Farming was invented during the Neolithic period (around 8000–4000 B.C.). During this time, seven independent origins of farming were documented: the Amazon basin, China, the Middle East, New Guinea, Mexico, West Africa, and the Andes. During this Age, many major cereal crops and root crops were all domesticated, as well as the husbandry of major farm animals. These shifts from hunting-gathering to major agriculture and animal husbandry resulted in increased land productivity for human consumption and human population per unit land area.

Aquaculture began thousands of years after the Neolithic Age in China, where common carp (*Cyprinus carpio*) was cultured. The common carp is a native Chinese species that may have been farmed as early as 2000–1000 B.C. (FAO 1988). The first aquaculture text, dated around 500 B.C., is attributed to Fan Lin, a Chinese politician who believed that part of his wealth was derived from his fishponds. Other continents, on the other hand, only recently began to practice aquaculture. This delay in aquaculture application on some continents may be due to humankind's terrestrial presence. In other words, humans were unable to easily comprehend the critical parameters of aquatic environments. Low content of O_2 compared to high solubility of Carbon Dioxide (CO_2) in water, pH, salinity level, heavy metals, phytoplankton,

© The Author(s), under exclusive license to Springer Nature Singapore Pte Ltd. 2023

A. Rahimi-Midani, *Deep Technology for Sustainable Fisheries and Aquaculture*,
https://doi.org/10.1007/978-981-99-4917-5_1

and zooplankton concentrations are some of the environmental factors that affect aquatic organisms; thus, understanding them can play an important role in the development of the aquaculture sector on various continents. Furthermore, most of the aquatic animals cultured in the aquaculture industry are poikilotherms (cold blooded), and their metabolic rates are greatly influenced by environmental conditions. Today, the difficulties in controlling the influence of environmental factors persist, slowing the development of this sector.

Another issue confronting the aquaculture industry is the lack of genetical selection of the species. Only a few species have been subjected to hybridization selection, including common carp, trout, tilapias, catfish, and salmon. Brood stocking is used in breeding for these species.

Until a few decades ago, capture production outpaced aquaculture production. Natural productivity of fresh, brackish, and marine waters is finite, and thus a finite number of products can be harvested by capture fisheries. The capture fisheries face numerous challenges that can be summarized as stock size unpredictability, exploitation capacity and the difficulties encountered in regulating this capacity, and low productivity. The average harvest from oceans for direct human consumption or through the use of fishmeal is estimated to be 2.5 kg/ha/year of ocean surface. Over time, this has resulted in 77% of marine species being overexploited (FAO 2018).

In addition to this, there are a couple of other factors that boost the capture production. It is estimated that approximately 25% of capture production is used to produce fishmeal. Fishmeal is a good source of omega-3 fatty acids, which are used in animal feed. Another factor is that a significant portion of bycatch and discards are from the capture production.

Aquaculture now accounts for more than half of global seafood production, emphasizing the importance of food supply from the aquatic environment. Seaweed makes a significant contribution to the aquaculture production. Contrary to popular belief, not all seaweeds have low levels of feed protein. Depending on the season and species, red algae (Rhodophyta) contain 33–47% protein (Fleurence 1999). However, a large portion of the catch is used for fishmeal production rather than human consumption. In 2018, roughly 21 million tonnes of capture fishery production (fish, shrimp, and mollusks) were used for fishmeal (FAO 2018).

Food security for the world's population is a major concern, and aquaculture has now surpassed capture production as the primary source of animal food for direct human consumption. Seafood accounts for 25–30% of all animal protein production consumed by humans each year. While livestock production grows at a 1.5% annual rate, aquaculture grows at a 7% annual rate. As a result, aquaculture will play a growing role in global food security.

1.1 Aquaculture for Food Production

One of the most efficient methods of producing protein for human consumption is aquaculture. Bivalve culture, for example, produces protein without the use of any feed. According to Pérez-Camacho et al. (2013), a floating raft of the mussel (*Mytilus*

Fig. 1.1 Common Carp (*Cyprinus carpio*) is now stocked, cultured, and consumed in many countries worldwide. (The photograph was taken in Lake Lipno, which is located in the southern part of the Czech Republic (Czechia))

galloprovincialis) in Galicia rias, located in North-West Spain, can reach a typical yield growth of 200 kg per rope gross weight. The protein content of the flesh accounts for roughly 60% of the total weight of the wet flesh. A hectare contains rafts, and rafts produce 16 tonnes per hectare annually (Duarte et al. 2009). This is more productive than equivalent terrestrial livestock husbandry. The mussels only feed on the particulates found in the water after filtering it. Fish culture is another efficient method of producing edible protein when compared to other types of animal husbandry. Carp (Fig. 1.1), for example, contains 30% protein by weight. When compared to chicken, pork, and beef, this has the highest conversion efficiency. The efficiency of fish culture extends even further to top predators like Atlantic salmon. When three animals are compared (Atlantic salmon, chickens, and pigs), all three have excellent Feed Conversion Ratios (FCRs), but Atlantic salmon has the highest market value. Furthermore, Atlantic Salmon has the highest percentage of energy and protein retention from feed. The main reasons for high energy savings in fish species are buoyancy, poikilothermy, ammonium excretion, and low energy consumption during reproduction. Having said that, based on the FCRs of Atlantic salmon, it cannot be concluded that all marine species have high FCRs. Over decades of research and development (R&D), very efficient feed for Atlantic Salmon has been developed, allowing for high FCRs for this species; however, other valuable species, such as bluefin tuna, lack these types of in-depth studies.

Tuna production is currently based on captured small fish. Due to the lack of suitable artificial diets, small frozen wild caught fish are used as feed for tuna aquaculture. This fish normally consumes small fish like herring, mackerel, and sardines. According to Estess et al. (2014), the FCRs for Pacific Bluefin Tuna (*Thunnus orientalis*) on mixed feed made up of sardines, squid meat, gelatin, and different vitamins are 23:1 and 18:1 with feed made of frozen sardines.

With the boom in aquaculture production in the 1990s, many speculated that the sector was undergoing a blue revolution, transforming the productivity of aquatic environments with new and innovative technologies. Parallels with the agricultural Green Revolution in the decades after World War I led to the development of this vision. The Green Revolution emphasized modern agricultural practices such as mechanization, heavy fertilizer and pesticide usage, genetically improved stocks, and feed formulation, all of which had negative environmental consequences. Learning from the failures of the Green Revolution, Diana et al. (2013) stated that aquaculture must avoid a similar long-term growth path.

The increase in aquaculture production over the last three decades has come from both freshwater and marine production. Most of this near exponential production comes from freshwater production, establishing freshwater as the primary environment for aquatic production. However, oceans are vast, and as technology advances, there will be more movement into the blue oceans (Simpson 2011).

According to Food and Agriculture Organization's (FAO) statistics, nearly 90% of total global aquaculture production comes or is produced by least developing and developing countries. Aquaculture production in these countries increased significantly, with annual rates of 8.4% and 7%, respectively. Asia contributes significantly to this increase in aquaculture production due to massive production in Indonesia and China. Although Asian countries have high-value aquaculture products such as shrimp, grouper, seabass, and scallops, a large portion of their production comes from traditional freshwater pond culture, which produces carp, catfish, and tilapias.

Because freshwater species are herbivores, omnivores, and detritivores, polyculture is a common practice. These practices are frequently carried out in ponds with primitive technologies and low stocking density. These ponds are frequently fertilized with organic fertilizers like animal and human wastes. Freshwater culture is the primary source of animal protein in communities where animal proteins are expensive.

1.1.1 State of Aquaculture Around the World

Total fisheries and aquaculture production reached 214 million tonnes in 2020, a 3% increase from 213 million tonnes in 2018 (FAO 2022). The limited growth is due to a 4.4% decrease in capture production as a result of fewer caught pelagic species, lower Chinese landings, and the effects of the COVID-19 pandemic in 2020 (FAO 2022).

Global aquaculture production reached 122.6 million tonnes in 2020, with 87.5 million tonnes of aquatic animals and 35.1 million tonnes of algae, for a total value of 281.3 billion USD (FAO 2022). According to FAO, all regions, with the exception of Africa, continued to grow their aquaculture sector in 2020, with leading countries including Chile, China, and Norway. The decline in African production was caused by a decrease in aquaculture production in Egypt and Nigeria, Africa's two major producing countries. Inland waters produced approximately 54.4 million tonnes, while marine and coastal waters produced 68.1 million tonnes (FAO 2022).

According to the FAO, marine fishery stocks continued to decline in 2020. According to the report, biologically sustainable marine fishery stock levels fell from 90% in 1974 to 64.6% in 2019, with maximally sustained fish stocks accounting for 57.3% and underfished stocks accounting for 7.2%.

Utilization of fisheries and aquaculture products has increased dramatically in recent decades. Human consumption accounted for 157 million tonnes (approximately 89%) of total global production (excluding algae), up from 67% in the 1960s. Over 20 million tonnes of produced fishmeal and fish oil were used for ornamental fish, in bait production pharmaceutical applications, pet foods, and direct aquaculture and livestock usage.

Aquatic food consumption increased at a 3% annual rate from 1961 to 2019 (almost twice the annual world population growth rate). In 2019, the annual per capita consumption was 20.5 kg (FAO 2022). In 2019, aquatic foods provided approximately 17% of animal proteins and 7% of total proteins. Aquatic food consumption per capita increased in upper-middle-income countries between 1961 and 2019. Aquatic food provided approximately 20% of the average per capita intake of animal protein for planet's 3.3 billion population.

Merino et al. (2012) investigated the feasibility of current rates of fish consumption and population growth until 2050, taking into account relevant scenarios for global and regional climate, capture fishery estimation under different marine ecosystems, global population, fish meal and fish oil prices, and technological advancement in the aquaculture field. The findings demonstrated that meeting rising demand requires sustainable management of fish resources.

1.1.2 South-East Asia and China Aquaculture Production

China, Indonesia, India, Vietnam, the Philippines, Bangladesh, South Korea, Egypt, Norway, and Japan are the top ten aquaculture producers, according to FAO. China is currently in first place, while Indonesia has significantly increased its output and risen to the second place in a relatively short period of time. This increase in output is the result of increased seaweed production. As previously stated, eight out of ten aquaculture producers are from or are located in Asia. Although the Asian region produces the most aquaculture, the European and North American regions import the most. Carnivorous fish with high market value account for more than 60% of fish production in developed countries (Fig. 1.2). These species are cultured in high density and require high-value protein feed inputs. As a result, these producers use low market-value products from capture fisheries as feed. To put it another way, we

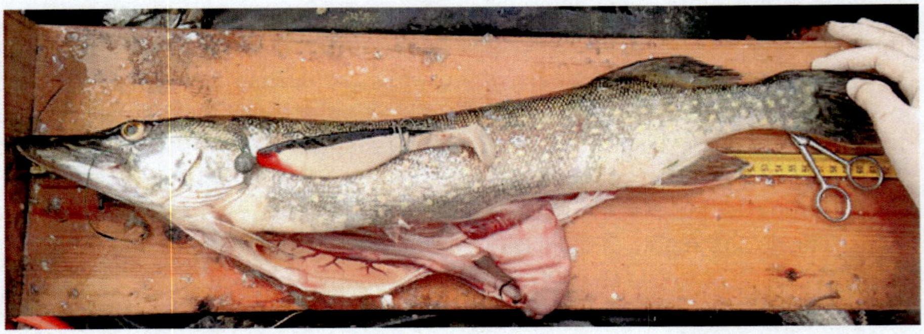

Fig. 1.2 Stomach contents of a Pike Perch (*Sander lucioperca*). This carnivorous fish is extremely popular among European and North American consumers, forcing local governments to artificially stock it in lakes, rivers, and ponds. The aggressive pike perch, as shown, cut freshwater fishing bait and survived until it was captured for scientific analysis

Fig. 1.3 Various carp species are cultured in many areas around the world. Bighead carp (*Aristichthys nobilis*) can grow rapidly in harsh environments with cold winters

use fish to produce edible, high protein, and more expensive fish, which is neither sustainable nor environmentally friendly.

China is an important player in the development of modern aquaculture. In the early 1100 B.C., there is evidence of earthen pond culture of common carp. During Tang Dynasty (618–917 A.D.), five species of carp were cultured (Silver Carp (*Hypophthalmichthys molitrix*), bighead carp (*Aristichthys nobilis*) (Fig. 1.3), grass carp (*Ctenopharyngodon idellus*), black carp (*Mylopharyngodon piceus*), and mud carp (*Cirrhinus molitorella*)). The Yangtze and Pearl rivers were used to collect wild caught fingerlings. There are records of oyster farming from 206 B.C. to 220 A.D. (Han Dynasty), and marine aquaculture and milkfish became popular during the Ming Dynasty (1368–1644 A.D.).

China alone has increased global aquaculture production by 50 million tonnes per year in recent years. This accounts for roughly 60% of global output. But how did China increase its aquaculture production in such a short period of time?

The solution is an integrated system of components with strong government commitments. This system is divided into three layers: first, five-year plans with specific goals established by national and local government bodies in collaboration with the Ministry of Science; second, research and development institutes represented by the Chinese Academy of Fisheries Sciences and three other institutions; and third, implementing national and provincial technology centers in charge of integrating new technologies into farms and businesses.

The outstanding growth of aquaculture in China is due to escalation rather than major technological advancement. Due to the Chinese government's ongoing assistance, this nation has emerged as the only significant producer whose output exceeds that of the capture fisheries.

Aquaculture development, on the other hand, was environmentally destructive and not sustainable during the rapid growth period. In South-East Asia and China, these developments were under the mentality of "move fast and get rich faster" causing enterprises not considering long-term environmental impacts. Under these conditions, environmentally friendly and sustainable aquaculture may be the best solution for global seafood production and planet's food security.

1.2 Recirculating Aquaculture Systems and Integrated Aquaculture Systems

Recirculating Aquaculture Systems (RAS) have been used to produce fish in land-based facilities for several decades. RAS systems are those in which less than 10% of the system volume is exchanged while more than 90% of the system volume is recirculated (Timmons et al. 2001). RAS can be used to catch fish at various stages of life and in various environmental conditions. RAS can produce fish in a controlled environment with minimal water usage that are independent and separate from the surrounding environment. Furthermore, RAS determines and optimizes fish growth rates and health by monitoring water quality and controlling incoming water (Attramadal et al. 2012; Blancheton 2000; Blancheton et al. 2009; Martins et al. 2009, 2010; Timmons and Ebeling 2007). RAS also permits point-source discharge, which permits decreased water discharge and waste collection for later use.

One of the major limiting factors in making RAS production economically viable is economies of scale (De Ionno et al. 2006). However, RAS production and technology have recently advanced to the point of sophistication and scale that allows for cost-effective and commercial fish production. RAS is used on a variety of scales, from small-scale farming systems to land development facilities. The principles of engineering and management remain the same regardless of production scale. RAS cannot be used in these areas due to the growing population and the need for rapid progress, particularly in rural areas and for vulnerable communities. As a result, our strategy must prioritize conserving natural resources, improving resource efficiency, increasing productivity and profitability, and improving quality through lower unit cost of production.

Integrated Aquaculture System (IAS) is an example of mixed farming mostly practiced in South-East Asia and East Asian countries. The technology involves a combination of fish integrated with crop. Farm waste is important in IAS because it improves the economy of production while reducing the negative environmental impact of farming. IAS provides quality food, resource utilization, farm waste recycling, job creation, and economic development.

Inspired by IAS and RAS, aquaponic systems came into picture enabling communities and urban population to produce fish and cultivate plants in small areas and with cheaper costs (fraction of the cost needed for RAS). In aquaponic systems, water circulates from the fish tank through filters, plant grow beds, and back to the fish. The sustainability of aquaponics considers environmental and socioeconomic dynamics. Socioeconomically, these systems have low recurring costs as well as offering quality of life improvement. Environmentally, aquaponic systems prevent similar aquaculture wastes from entering waterbodies and polluting the watersheds.

1.3 Deep Tech in Twenty-First Century and the Importance of Sustainability

Many research institutions, international organizations, governmental and non-governmental organizations believe that climate change is the most serious environmental threat that modern society faces. Climate change has had numerous effects, including sea level rise, desertification, and changes in weather patterns. According to Mach et al. (2019), socioeconomic problems like flooding, water scarcity, climate refugees, and armed conflict are already being brought on by climate change-related floods and droughts, the disappearance of entire Pacific islands, and an increase in destructive hurricanes. Social inequality or "climate apartheid" makes matters worse when comparing global levels of mitigation and adaptation to climate change (Mach et al. 2019).

Since the 1950s, there have been a number of parallel accelerations in industrial and socioeconomic processes, GHG, energy, water, scarcity, and raw material use, as well as changes in land use. In addition to these quick changes, there has been an increase in global inequality in terms of wealth, freedom, and health over the past century. Under these circumstances, it is hoped that technological innovation and entrepreneurship will aid in finding solutions to these environmental changes.

The United Nations (UN) has defined the Sustainable Development Goals (SDGs) as a process of combining efforts at the market, society, and country levels to innovate in areas such as social equality, economic progress, and the environment. The central idea behind the SDGs approach is the need for global collaboration in responding to climate change and other environmental issues while also considering poverty eradication, increased well-being, and global economic progress. The SDGs approach recognizes that sustainable progress is no longer solely a goal of developing countries, but also includes responsibility and challenges shared by the developed countries.

Technology plays a fundamental role in facing today's global challenges. Usually, situations that need innovative solutions are often complex and challenging, thus, it involves high risks, unique and unexplored business models as well as entrepreneurial mindsets. Consequently, when the ambition is to increase sustainable technological innovation to reach global SDGs, it is essential to eliminate any disruption that sustainable tech might encounter (Boesten 2021).

A tech startup can be a viable solution to the rapidly changing world. However, the fact that startups fit better in the role of innovating markets doesn't mean that they are free of any troubles. According to Castellas and Ormiston (2018) startups typically have no company legacy, thus, the first phase of their existence is dedicated to finding investments. Since there is no track record of the technology in the market, these investments happen to be under high risk. Venture Capital (VC) investors are the ones who mainly get involved in this specific level of risk (Bocken 2015).

It is critical to shift away from incidental investments in sustainable technology and toward more systematic investments in the sustainable tech sector in order to make a significant impact through technological innovation. For example, the SDGs require an estimated 1.4 trillion USD, which is not expected to be met solely through philanthropic or governmental investments, but rather with at least half coming from private sector investments (Castellas and Ormiston 2018).

The SDGs specifically mention the mainstreaming of sustainability and sustainable development in companies, identifiable in goal "12/6 Encourage companies, especially large and transnational companies, to adopt sustainable practices and to integrate sustainability information into their reporting cycle" (Inter-Agency and Expert Group on Sustainable Development Goal Indicators 2016).

It is referred to as "impact investment" when investors invest their money for reasons other than financial gain. This book demonstrates the impact of investment stance not only in terms of relevant SDGs, but also provides success stories of these investments on the aquaculture sector, environment, and communities worldwide.

1.4 The Effect of Environmental Variables and the Impact of Climate Change

Aquaculture provides half of all fish production for human consumption (FAO 2016). However, a largest portion of world aquaculture is produced in tropical and subtropical climate regions, and geographically in the Asian region. For this reason, impacts from climate change are likely to have greater consequences, with direct impacts on global food fish supply. It is also predicted that global warming and the consequent increase in water temperature could impact negatively on aquaculture in temperate climate zones, because it could exceed the optimal temperature range of cultured organisms, as opposed to potential optimal (positive) impacts through enhanced growth in tropical and subtropical zones. On the other hand, higher temperatures can cause other impacts such as eutrophication in inland waters along with possible outbreaks of virulent pathogens.

Another important but indirect impact of climate change on aquaculture is limitation on fish meal and fish oil available for feeds through reduction in raw material supplies. It is understood that these limitations will be mainly felt by aquaculture in temperate regions, where finfish culture is based on carnivorous species.

1.4.1 Observed Changes in Global Climate

Knowledge and information on the climate are based on multiple pieces of evidence, including direct and indirect observations and historical reconstructions. Based on this information, Intergovernmental Panel on Climate Change (IPCC) concluded that global warming was unequivocal, and many of the observed changes since the 1950s are unprecedented compared with proceeding decades to millennia. Until the nineteenth century, the earth's average surface temperature has increased by more than 0.8 °C, and now is warming at a rate of more than 0.1 °C every decade (Hansen et al. 2010). Heat waves are more common globally, though data's degree of certainty and dependability varies by continent (Hartmann et al. 2013).

It is believed that atmospheric concentration of Green House Gases (GHGs) such as CO_2, methane (CH_4), and nitrogen dioxide (NO_2), are the largest contributors to the global warming, by acting like a thermal blanket (IPCC 2014). The world's oceans absorb 93% of the heat caused by anthropogenic climate change, and ice and snowmelt absorb another 3–4%. For this reason, any small change in the balance of heat between ocean and atmosphere would have huge impacts on the global air temperature (Reid 2016). Additionally, oceans sequester about 25% of the CO_2 released as a result of anthropogenic activities (Le Quéré et al. 2018).

For forecasting and assessing future changes in climate system, the IPCC uses simulation models based on scenarios of anthropogenic forcing. In comparison to pre-industrial values, Representative Concentration Pathways (RCPs) are the scenarios that simulate potential heat or radiative forcing values in the year 2100 (+2.6 W/m^2, +4.5 W/m^2, +6.0 W/m^2 and +8.5 W/m^2). The four RCPs are based on possible socioeconomic trends such as population trends, economic activity, lifestyle, energy use, land use patterns, climate, and technology policy. RCP 2.6 is an emission pathway that leads to very low GHGs, RCP 4.5 and RCP 6.0 reflect two stabilization scenarios in which total radiative forcing is stabilized after 2100 with differential speed, while RCP 8.5 representative of scenarios that lead to high GHG levels (van Vuuren et al. 2011).

With the exception of RCP 2.6, it is estimated that under the other three scenarios, the average global atmospheric temperature change for the end of the twenty-first century will likely exceed 1.5 °C (relative to the average of the 1850–1900 period). Many aspects of anthropogenic climate change are thought to be irreversible for centuries to come (IPCC 2014).

1.4.2 Impact of Climate Change on Aquatic Systems

Global warming has serious consequences for the hydrological cycle. Temperature, climate patterns, and precipitation all have an impact on the quality and seasonality of water resources, resulting in unavoidable changes in aquatic ecosystems. According to the IPCC report, climate change is already causing permafrost warming and thawing in high latitude regions, and in high-elevation regions it is driving glacier shrinkage, with consequences for downstream water resources. The melting of Arctic Sea ice is disrupting or slowing the global ocean conveyor belt, which has consequences for marine ecosystems (Liu et al. 2017). Precipitation has varied across regions since 1901, increasing in mid-latitude land areas of the Northern Hemisphere while decreasing in others (Hartmann et al. 2013).

Global river discharges have not demonstrated changes associated with global warming. However, it is expected that in the near future, the contribution of snowmelt to river flow will increase (Jha et al. 2006; Pervez and Henebry 2015; Siderius et al. 2013). Additionally, in many rivers and wetlands, change in precipitation will substantially alter ecologically important attributes of flow regimes. In the short term, the intensity and frequency of heavy precipitation events over land are likely to increase. Kirtman et al. (2013) showed there is low confidence in projections of changes in the frequency of tropical cyclones.

Since the 1960s, anthropogenic forcing has made substantial contribution to the warming of above 700 m of the ocean (Cheng et al. 2017), with an average increase of 0.7 °C in the surface area from 1900 to 2016 (Huang et al. 2015). Ocean temperature trends are positive in most regions and warming is more prominent in the Northern Hemisphere, especially the North Atlantic. Approximately 64% of the extra anthropogenic energy accumulated in the ocean's and sea's upper layers.

Freshwater systems will experience an increase in water temperature as a result of an increase in air temperature. This is due to the shallow nature of surface water and its sensitivity to changes in atmospheric temperature.

Another important component of aquatic systems is Dissolved Oxygen (D.O). According to IPCC (2014) changes in the concentrations of D.O., global carbon and nitrogen cycles will be heavily impacted. Average concentrated D.O. in the oceans varies, ranging from super saturated in Antarctic waters to none in the coastal sediments. There are variety of Oxygen Minimum Zones (OMZs) exist in open oceans, deep basins of semi-enclosed seas, deep fjords, coastal upwelling zones, and other areas with restricted circulation, however, up until recent decades there were no clear evidence in decrease in oxygen concentrations (IPCC 2014).

Global warming is the ultimate cause of the ongoing deoxygenation in many parts of the open sea (Breitburg et al. 2018). It is estimated that ocean warming accounts for app15% of total global oxygen loss and more than 50% of the oxygen less in the upper 1000 m of the ocean. Reducing ventilation caused by intensified stratification accounts for the remaining 85% global ocean oxygen loss. Depending on the oxygen tolerance levels of different species and taxonomic groups, oxygen influences biogeochemical and biological processes.

The average annual extent of Arctic Sea ice decreased at the rate of 3.5–4.1% per decade over the period of 1979–2012, while over the same period, the extent of perennial sea ice decreased by $11.5 \pm 2.1\%$ per decade. According to Vaughan et al. (2013) the average annual extent of Antarctic Sea ice increased by 1.2–1.8% per decade over the period of 1979–2012. Additionally, between 2003 and 2009 most ice loss was from glaciers in Alaska, the Canadian Arctic, the Southern Andes, the periphery of the Greenland ice sheet, and the Asian Mountains (Vaughan et al. 2013). Due to melting of ice and snow, reduction of glaciers, changes in water levels and flows occur in aquatic systems. Reduction in mountain glaciers will have an impact on the lake levels and river flows in the short term until likely disappearance in the mid-term, while it is expected that melting ice will impact the sea level.

According to Dangendorf et al. (2017), sea level has risen by an average of 31 mm per year due to climatic and non-climatic factors. This rate varies by region, being three times the global average in the Western Pacific and zero or negative in the Eastern Pacific. Between 1901 and 2010, sea level rose by a global average of 0.19 m. According to Kopp et al. (2014), the projected global mean level rise (SLR) between 2000 and 2010 will be between 0.5 and 1.2 m under RCP 8.5, 0.4 and 0.9 m under RCP 4.5, and 0.3 and 0.8 m under RCP 2.6.

Ocean circulation influences local climates by globally redistributing heat and freshwater. Meridional Overturning Circulation (MOC) plays an important role in heat redistribution by transporting heat from the tropics to middle and high latitudes, as well as carbon sequestration. The Atlantic Meridional Overturning Circulation (AMOC) is gradually weakening, causing a cooling of sea surface temperature in the subpolar Atlantic Ocean and a warming and northward shift of the Gulf stream (Caesar et al. 2018; Thornalley et al. 2018). Upwelling may benefit primary production and nutrient inputs while increasing the presence of low oxygen in shelf habitats (Bakun et al. 2015).

Up-taking atmospheric CO_2 is the primary cause of reduction in the pH of the ocean called ocean acidification. Increase in atmospheric CO_2 means ocean absorbs more CO_2. This causes the partial pressure of CO_2 (pCO_2) at the surface and a decrease in water pH and saturation of calcium carbonate ($CaCO_3$), which is important for all shell forming aquatic animals (Pörtner et al. 2014).

Ocean acidification caused by ocean uptake of CO_2 started since the beginning of the industrial era; it is estimated that pH of ocean surface water has dropped by an average of 0.1, corresponding to a 26% increase in acidity (IPCC 2014; Jewett and Romanou 2017). Acidification is already 50% higher in the Northern Atlantic than in the subtropical Atlantic and since the cold water can absorb more CO_2, the Arctic water are acidifying faster than the global average. According to Harris et al. (2013), corrosive condition events in the California current have increased in frequency, severity, duration, and spatial extent.

Near freshwater sources and heavily tidal environments, the oceanic pH is subject to natural variability; and observed trends in global ocean pH exceed the range in natural seasonal variability in most of the oceans (Henson et al. 2017) and according to Gattuso et al. (2015) expected to exceed even further in years to come.

Impact of ocean acidification on marine organisms vary with their physiological abilities to adapt. However, research on these effects is still in its early stages, and so far, only short-term responses in laboratory conditions have been studied (McElhany 2017). Scientific research on the responses and adaptive capacity of marine organisms is expanding (Munday 2017).

At the base of marine food web, phytoplankton production is located which is in charge of controlling the energy and food available to higher trophic levels and eventually to fish. Projections of global marine primary production under climate change scenarios show both increase and decline by ±20% by 2100 (Taucher and Oschlies 2011). This uncertainty is partly because primary production is an integrator of changes in light, temperature, and nutrient as well as uncertainty in sensitivity of tropical ocean primary production to climate change. It is estimated that global marine primary production will decline by 6 ± 3% by 2100 (Kwiatkowski et al. 2017). In some Arctic and boreal freshwater lakes, the primary production has been increased (Michelutti et al. 2005) but declined in the tropical ones (O'Reilly et al. 2003).

References

Attramadal KJK, Salvesen I, Xue RY, Oie G, Storseth TR, Vadstein O, Olsen Y (2012) Recirculation as a possible microbial control strategy in the production of marine larvae. Aquac Eng 46: 27–39

Bakun A, Black BA, Bograd SJ, García-Reyes M, Miller AJ, Rykaczewski RR, Sydeman WJ (2015) Anticipated effects of climate change on coastal upwelling ecosystems. Curr Clim Change Rep 1(2):85–93

Blancheton JP (2000) Developments in recirculating systems for Mediterranean fish species. Aquac Eng 22:17–31

Blancheton JP, Bosc P, Hussenot JME, D'Orbcastel ER, Romain D (2009) The 'new' European fish culture systems: recirculating systems, offshore cages, integrated systems. Cahiers Agric 18: 227–234

Bocken NMP (2015) Sustainable venture capital - catalyst for sustainable start-up success? J Clean Prod 108:647–658

Boesten Y (2021) Mainstreaming sustainable deep tech venture capital. Master thesis, Radboud University

Breitburg D, Levin LA, Oschlies A, Grégoire M, Chavez FP, Conley DJ, Garçon V et al (2018) Declining oxygen in the global ocean and coastal waters. Science 359(6371):eaam7240

Caesar L, Rahmstorf S, Robinson A, Feulner G, Saba V (2018) Observed fingerprint of a weakening Atlantic Ocean overturning circulation. Nature 556:191–196

Castellas EI, Ormiston J (2018) Impact investment and the sustainable development goals: embedding field-level frames in organisational practice. Contemp Issues Entrepreneurship Res 8:87–101

Cheng L, Trenberth KE, Fasullo J, Boyer T, Abraham J, Zhu J (2017) Improved estimates of ocean heat content from 1960 to 2015. Sci Adv 3(3):e1601545

Dangendorf S, Marcos M, Woppelmann G, Conrad CP, Frederikse T, Riva R (2017) Reassessment of 20th century global mean sea level rise. Proc Natl Acad Sci U S A 114(23):5946–5951

De Ionno PN, Wines GL, Jones PL, Collins RO (2006) A bioeconomic evaluation of a commercial scale recirculating finfish growout system—an Australian perspective. Aquaculture 259:315–327

Diana JS, Egna HS, Chopin T et al (2013) Responsible aquaculture in 2050: valuing local conditions and human innovations will be key to success. Bioscience 63(4):255–262

Duarte CM, Holmer M, Olsen Y et al (2009) Will the oceans help feed humanity? Bioscience 59(11):967–976

Estess EE, Coffey DM, Shimose T et al (2014) Bioenergetics of captive Pacific bluefin tuna (Thunnus orientalis). Aquaculture 434:137–144

Fleurence J (1999) Seaweed proteins: biochemical, nutritional aspects and potential uses. Trends Food Sci Technol 10:25–28

Food and Agriculture Organization (FAO) (1988) History of aquaculture. FAO, Rome, Italy

Food and Agriculture Organization (FAO) (2016) State of world fisheries and aquaculture. FAO, Rome, Italy

Food and Agriculture Organization (FAO) (2018) State of world fisheries and aquaculture. FAO, Rome, Italy

Food and Agriculture Organization (FAO) (2022) State of world fisheries and aquaculture. FAO, Rome, Italy

Gattuso J-P, Magnan A, Billé R, Cheung WWL, Howes EL, Joos F, Allemand D et al (2015) Contrasting futures for ocean and society from different anthropogenic CO_2 emissions scenarios. Science 349(6243):aac4722

Hansen J, Ruedy R, Sato M, Lo K (2010) Global surface temperature change. Rev Geophys 48(4): RG4004

Harris KE, DeGrandpre MD, Hales B (2013) Aragonite saturation state dynamics in a coastal upwelling zone. Geophys Res Lett 40(11):2720–2725

Hartmann DL, Klein Tank AMG, Rusticucci M, Alexander LV, Brönnimann S, Charabi Y, Dentener FJ et al (2013) Observations: atmosphere and surface. In: Stocker TF, Qin D, Plattner G-K, Tignor M, Allen SK, Boschung J, Nauels A, Xia Y, Bex V, Midgley PM (eds) Climate change 2013: the physical science basis. Contribution of Working Group I to the fifth assessment report of the Intergovernmental Panel on Climate Change. Cambridge University Press, Cambridge, UK, pp 159–254

Henson SA, Beaulieu C, Ilyina T, John JG, Long M, Séférian R, Tjiputra J, Sarmiento JL (2017) Rapid emergence of climate change in environmental drivers of marine ecosystems. Nat Commun 8:art: 14682

Huang B, Banzon VF, Freeman E, Lawrimore J, Liu W, Peterson TC, Smith TM, Thorne PW, Woodruff SD, Zhang H-M (2015) Extended reconstructed sea surface temperature Version 4 (ERSST. v4). Part I: Upgrades and intercomparisons. J Clim 28:911–930

Inter-Agency and Expert Group on Sustainable Development Goal Indicators (2016) Report of the Inter-Agency and Expert Group on Sustainable Development Goal Indicators (E/CN.3/2016/2/Rev.1), Annex IV. Report of the Inter-Agency and Expert Group on Sustainable Development Goal Indicators, Annex IV

Intergovernmental Panel on Climate Change (IPCC) (2014) In: Barros VR, Field CB, Dokken DJ, Mastrandrea MD, Mach KJ, Bilir TE, Chatterjee M et al (eds) Climate change 2014: impacts, adaptation, and vulnerability. Part B: Regional aspects. Contribution of Working Group II to the fifth assessment report of the Intergovernmental Panel on Climate Change. Cambridge University Press, Cambridge, UK

Jewett L, Romanou A (2017) Ocean acidification and other ocean changes. In: Wuebbles DJ, Fahey DW, Hibbard KA, Dokken DJ, Stewart BC, Maycock TK (eds) Climate science special report: fourth National Climate Assessment, vol I. U.S. Global Change Research Program, Washington, DC, pp 364–392

Jha M, Arnold JG, Gassman PW, Giorgi F, Gu RR (2006) Climate change sensitivity assessment on Upper Mississippi River Basin stream flows using SWAT. J Am Water Resour Assoc 42(4): 997–1016

Kirtman B, Power SB, Adedoyin JA, Boer GJ, Bojariu R, Camilloni I, Doblas-Reyes FJ et al (2013) Near-term climate change: projections and predictability. In: Stocker TF, Qin D, Plattner G-K, Tignor M, Allen SK, Boschung J, Nauels A, Xia Y, Bex V, Midgley PM (eds) Climate change 2013: the physical science basis. Contribution of Working Group I to the fifth assessment report of the Intergovernmental Panel on Climate Change. Cambridge University Press, Cambridge, UK, pp 953–1028

Kopp RE, Horton RM, Little CM, Mitrovica JX, Oppenheimer M, Rasmussen DJ, Strauss BH, Tebaldi C (2014) Probabilistic 21st and 22nd century sea-level projections at a global network of tide-gauge sites. Earth's Future 2(8):383–406

Kwiatkowski L, Bopp L, Aumont O, Ciais P, Cox PM, Laufkötter C, Li Y, Séférian R (2017) Emergent constraints on projections of declining primary production in the tropical oceans. Nat Clim Chang 7:355–358

Le Quéré C, Andrew RM, Friedlingstein P, Sitch S, Pongratz J, Manning AC, Korsbakken JI et al (2018) Global carbon budget 2017. Earth Syst Sci Data 10(1):405–448

Liu W, Xie SP, Liu Z, Zhu J (2017) Overlooked possibility of a collapsed Atlantic meridional overturning circulation in warming climate. Sci Adv 3(1):e1601666

Mach KJ, Kraann CM, Adger WN et al (2019) Climate as a risk factor for armed conflict. Nature 571:193–197

Martins CIM, Ochola D, Ende SSW, Eding EH, Verreth JAJ (2009) Is growth retardation present in Nile tilapia Oreochromis niloticus cultured in low water exchange recirculating aquaculture systems? Aquaculture 298:43–50

Martins CIM, Eding EH, Verdegem MCJ, Heinsbroek LTN, Schneider O, Blancheton JP, D'Orbcastel ER, Verreth JAJ (2010) New developments in recirculating aquaculture systems in Europe: a perspective on environmental sustainability. Aquac Eng 43:83–93

McElhany P (2017) CO_2 sensitivity experiments are not sufficient to show an effect of ocean acidification. ICES J Mar Sci 74(4):926–928

Merino G, Barange M, Blachard JL, Harle J, Holmes R, Allen I, Allison EH, Badjeck MC, Dulvy NK, Holt J, Jennings S, Mullon C, Rodwell LD (2012) Can marine fisheries and aquaculture meet fish demand from a growing human population in a changing climate? Glob Environ Chang 22(4):795–806

Michelutti N, Wolfe AP, Vinebrooke RD, Rivard B, Briner JP (2005) Recent primary production increases in arctic lakes. Geophys Res Lett 32(19):L19715

Munday P (2017) New perspectives in ocean acidification research: editor's introduction to the special feature on ocean acidification. Biol Lett 13(9):art: 20170438

O'Reilly CM, Alin SR, Plisnier P-D, Cohen AS, McKee BA (2003) Climate change decreases aquatic ecosystem productivity of Lake Tanganyika, Africa. Nature 424(6950):766–768

Pérez-Camacho A, Labarta U, Vinseiro V et al (2013) Mussel production management: raft culture without thinning-out. Aquaculture 406–407:172–179

Pervez MS, Henebry GM (2015) Assessing the impacts of climate and land use and land cover change on the freshwater availability in the Brahmaputra River basin. J Hydrol Regional Stud 3: 285–311

Pörtner H-O, Karl DM, Boyd PW, Cheung WWL, Lluch-Cota SE, Nojiri Y, Schmidt DN, Zavialov PO (2014) Ocean systems. In: Field CB, Barros VR, Dokken DJ, Mach KJ, Mastrandrea MD, Bilir TE, Chatterjee M et al (eds) Climate change 2014: impacts, adaptation, and vulnerability. Part A: Global and sectoral aspects. Contribution of Working Group II to the fifth assessment report of the Intergovernmental Panel on Climate Change. Cambridge University Press, Cambridge, UK, pp 411–484

Reid PC (2016) Ocean warming: setting the scene. In: Laffoley D, Baxter JM (eds) Explaining ocean warming: causes, scale, effects and consequences. IUCN, Gland, Switzerland, pp 17–45

Siderius C, Biemans H, Wiltshire A, Rao S, Franssen WHP, Kumard P, Gosain AK, van Vliet MTH, Collins DN (2013) Snowmelt contributions to discharge of the Ganges. Sci Total Environ 468–469(Supplement):S93–S101

Simpson S (2011) The blue food revolution: making aquaculture a sustainable food source. Sci Am 304:54–61

Taucher J, Oschlies A (2011) Can we predict the direction of marine primary production change under global warming? Geophys Res Lett 38(2):L02603

Thornalley DJR, Oppo DW, Ortega P, Robson JI, Brierley C, Davis R, Hall IR et al (2018) Anomalously weak Labrador Sea convection and Atlantic overturning during the past 150 years. Nature 556:227–230

Timmons MB, Ebeling JM (2007) Recirculating aquaculture. Cayuga Aqua Ventures, Ithaca, NY

Timmons MB, Ebeling JM, Wheaton FW, Summerfelt ST, Vinci BJ (2001) Recirculating aquaculture systems. NRAC. Cayuga Aqua Ventures, Ithaca, NY van Vuuren DP, Edmonds J, Kainuma M, Riahi K, Thomson A, Hibbard K, Hurtt GC et al (2011) The representative concentration pathways: an overview. Clim Chang 109:5–31

Vaughan DG, Comiso JC, Allison I, Carrasco J, Kaser G, Kwok R, Mote P et al (2013) Observations: cryosphere. In: Stocker TF, Qin D, Plattner G-K, Tignor M, Allen SK, Boschung J, Nauels A, Xia Y, Bex V, Midgley PM (eds) Climate change 2013: the physical science basis. Contribution of working group I to the fifth assessment report of the Intergovernmental Panel on Climate Change. Cambridge University Press, Cambridge, UK, pp 317–382

2 Deep Tech Practices in Aquaculture

Flying into space, developing virtual currencies, developing efficient electric vehicles, and many other new technologies are improving humankind's life on Earth. But, in today's world, how to define disruptive technologies? Deep tech or hard tech refers to the type of organization, typically a startup, that develops these disruptive technologies.

According to Techwork (2021), definition of deep tech is "*technology that is based on tangible engineering innovation or scientific advances and discoveries.*" The fundamental focus of these organizations is to pioneer new solutions that solve society's biggest problems, including but not limited to climate change, clean energy, food safety, and food security. These deep tech firms are not similar to traditional companies since in deep tech organizations scientists and tech experts collaborate toward a common goal. Deep tech harnesses cutting-edge technologies to tackle twenty-first century challenges and great tangible social shifts.

Deep tech covers areas such as Artificial Intelligence (AI) and Machine Learning (ML); advanced material sciences; big data; biotech; blockchain; language processing; photonics and electronics; robotics; vision and speech algorithms; and quantum computing.

Several different industries, including the ones listed below, can profit from deep technology capabilities:

In the energy market, deep tech companies are using software packages and applying technologies to investigate the efficiency of wind and solar energy with Internet of Things (IoTs). Storage capacity is now one of the largest obstacles for renewable energy, and deep tech companies are working to enhance batteries and improve cloud-based solutions.

Humanity is facing food insecurity as one of the single biggest issues in the twenty-first century. Deep tech farming solutions are trying to move us from traditional farming practices toward more sustainable methods of production. Integrated farming systems that use blockchains, big data analysis, and biotechnologies to forecast outcomes with a high degree of accuracy.

© The Author(s), under exclusive license to Springer Nature Singapore Pte Ltd. 2023

A. Rahimi-Midani, *Deep Technology for Sustainable Fisheries and Aquaculture*,
https://doi.org/10.1007/978-981-99-4917-5_2

As the population grows, it is necessary to develop reliable infrastructure to support it. However, it is becoming increasingly important for societies and their governing bodies to reduce carbon emissions and foster sustainable solutions. According to Brooks Shehata (2020), pandemics like COVID-19 will become more frequent and thus societies need to develop environmentally sensitive infrastructure. According to the report by G20, it is recommended that nations employ Public-Private Partnership (PPP) as well as deep tech and green finances.

Deep technology applications in healthcare and life sciences include AI diagnosis, health tracking and diagnostics, electronic health records, and a variety of other applications. Precision medicine has made significant progress as a result of the field's rapid development, allowing scientists to collect accurate healthcare data and present tailor-made treatments.

Quantum computing, as an important aspect of deep technology, has aided in the advancement of computing and processing capabilities. Because computers are in charge of today's computing needs, deep tech can assist computers in tackling specific areas of problems.

According to the Boston Consulting Group (BCG), deep tech innovations are affecting all industries and appear to be at a crossroads of massive demand shifts driven by megatrends and scientific progress. The deep tech approach assists today's world in solving or addressing problems that were left unsolved or only partially solved in previous waves. Deep tech ventures use the Design-Build-Test-Learn (DBTL) cycle and rely on three approaches (BCG 2021): problem-oriented, convergence approaches, and techs (Fig. 2.1).

Stokes (1997) defined two dimensions that distinguished between basic and applied research (Fig. 2.2). The first dimension is whether the research is motivated by a desire for fundamental understanding, and the second dimension is whether the research is motivated by a desire for practical application. Stokes (1997) distinguished three quadrants: Bohr's, Edison's, and Pasteur's. Bohr's quadrant describes pure basic research and is distinguished by the pursuit of fundamental understanding. Edison's quadrant demonstrates a strong interest in utility and little interest in knowledge. Finally, and perhaps most intriguingly, there is the Pasteur's quadrant, named after Louis Pasteur, who managed to advance science while always keeping the end user in mind (vaccinations, fermentations, and pasteurization).

As shown in Fig. 2.1, problem orientation is one of the deep tech dimensions. Under Pasteur's quadrant, problem orientation generates optionality meaning it leverages the deep understanding of science and technology to address the vast possible sets of problems (Fig. 2.3).

Design thinking is directly related to deep tech problem statement. A good problem statement is dependent on three factors: It must be human-centered; it must be broad enough to allow for creative freedom while remaining narrow enough to be manageable (BCG 2021).

In deep tech, human-centered characteristic focuses on possible impact of answering to the critical needs. With the use of deep technology, ventures can delve further and find more fundamental solutions. Yet using science and technology to go deeper is not just the most crucial step, it's also the most challenging. It is

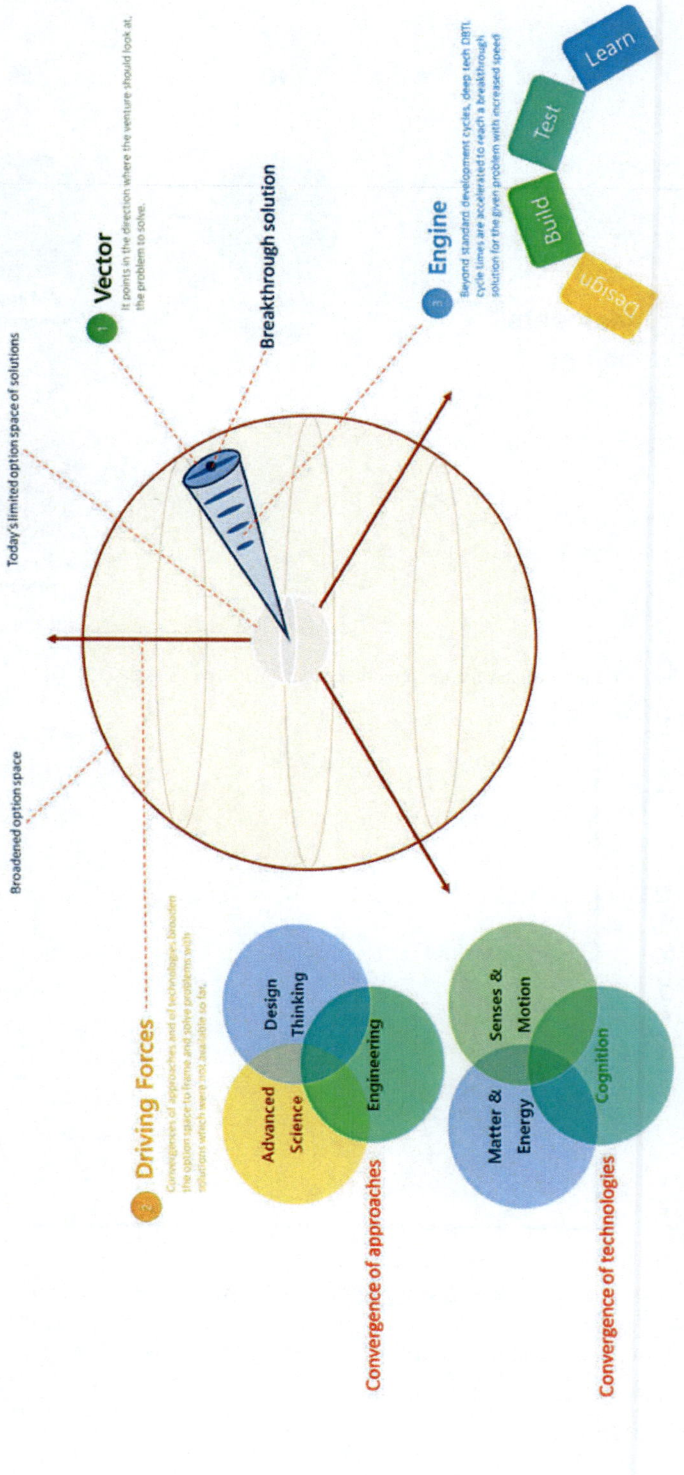

Fig. 2.1 Deep tech approach. (Source: BCG and Hello Tomorrow analysis)

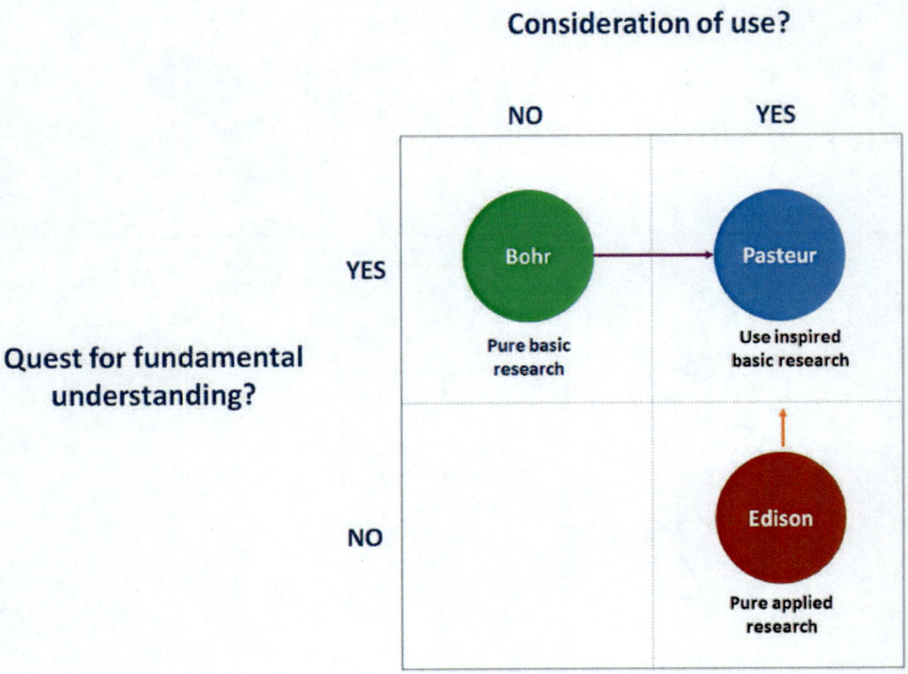

Fig. 2.2 Pasteur's quadrant used for deep tech ventures. (Source: Stokes (1997))

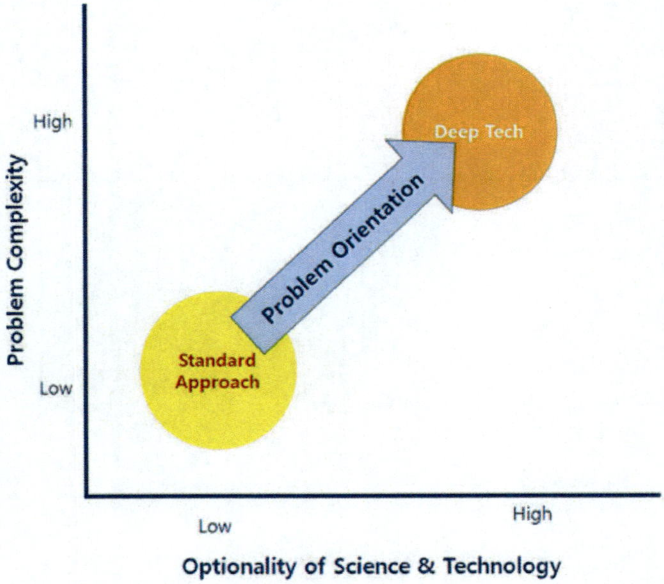

Fig. 2.3 Problem orientation and the role of optionality. (Source: BCG and Hello Tomorrow analysis)

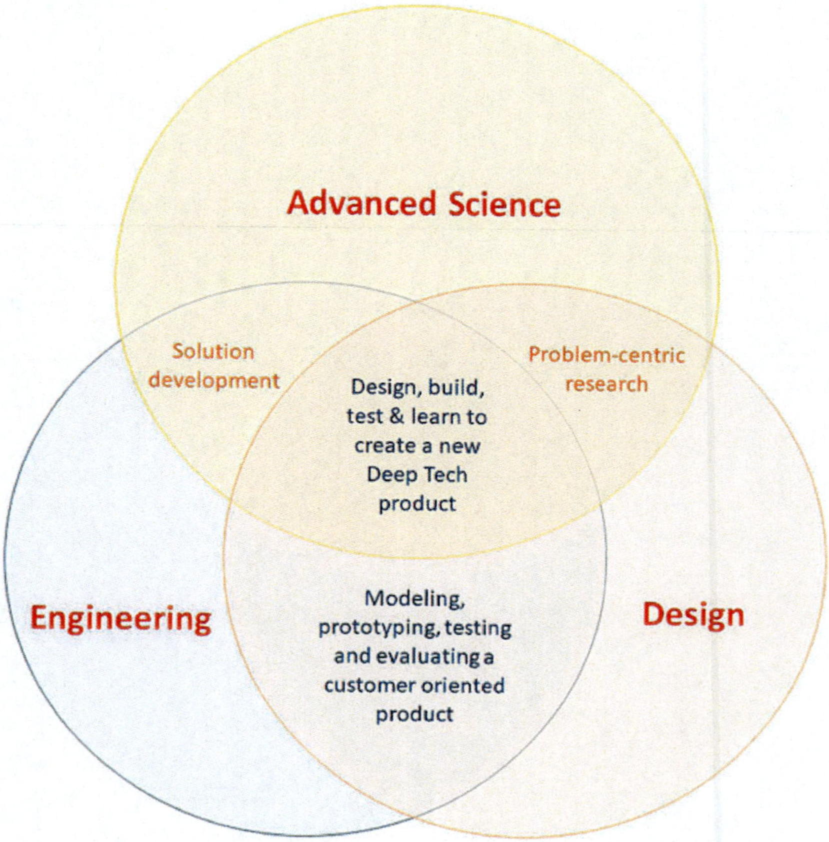

Fig. 2.4 Convergence of three approaches for deep tech ventures. (Source: BCG and Hello Tomorrow analysis)

believed that going deeper needs to be counterbalanced with other two characteristics of deep tech problem orientation. Therefore, problem orientation can impact deep tech along three main dimensions: first, being product/market fit; second, being able to define their strategies based on values; third; the correct problem orientation should serve as technical goal of the venture, driving all aspects of venture's operation and organization and not only just the market strategy.

The convergence of approaches is a key enabler of moving to Pasteur's quadrant. It will begin with design, which will allow for interdisciplinary co-creation of content analysis, framing, problem solving, and ideation (BCG 2021). Following that, it will progress into more advanced science by providing a thorough understanding of matter, computation, cognition, sensing, and motion, as well as the theory to solve the problem (BCG 2021). The procedure must be carried out in parallel (Fig. 2.4).

Convergence of the technologies is another key enabler of deep tech that works beyond the convergence of approaches (Fig. 2.5). In Fig. 2.1's Venn diagram, there

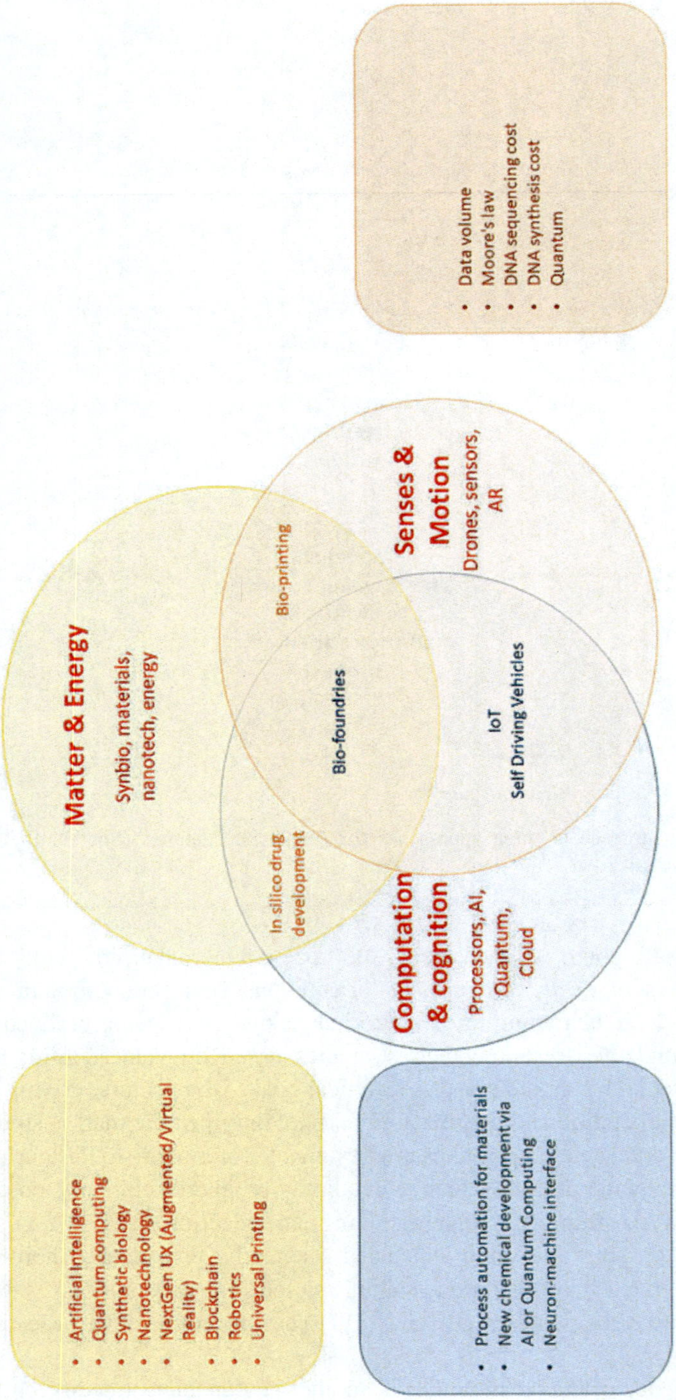

Fig. 2.5 Convergence of technologies. (Source: BCG and Hello Tomorrow analysis)

was one overlap between computing and cognition as well as sensing and motion, allowing for the development of IoTs, robots, and self-driving cars. Adding “Matter and Energy” will add two additional overlaps which enables a whole new different approach to innovation.

It is critical that everyone involved in a deep tech venture feels at ease with multiple topics and areas while mastering one. This is referred to as a “T-Shape,” and it is essential for people with a strong scientific background and deep specialization (BCG 2021). This is because people must communicate in a common language and comprehend the arguments of others. It is also necessary, but only at the individual and team levels, to have multidisciplinary skills because it is extremely rare for someone to be able to master all relevant aspects.

2.1 Artificial Intelligence and Machine Learning

2.1.1 Introduction to Artificial Intelligence

To determine whether an entity is intelligent or not, the Turing test was developed by Alan Turing (1950) as an empirical test of artificial intelligence. The test got three players, human interrogator in room number 1, another human in room number 2, and an artificial entity in the room number 3. The human interrogator can only use textual device to communicate between the human in room number 2 and artificial entity in room number 3. Then it is asked from the human interrogator to distinguish the other human from the artificial entity. The Turing test is only passed if the interrogator fails to distinguish between the human and the artificial entity. In other words, the artificial entity is so advanced in responding to the questions asked by the human interrogator that it will not demonstrate any difference from the human in room 2 and thus, it is intelligent. Note that in Turing test, the assumption is that physical interaction is not necessary for intelligence.

In the 1980s, Searle (1980) presented the Chinese room thought experiment, which defied the Turing test. The Chinese room experiment assumes that a computer program can write Chinese sentences using characters as inputs, then process them and output written sentences in Chinese characters. The Turing test will be passed if it can process and convince the Chinese interrogator that it is a human. “Does the program literally understand Chinese or is it only simulation ability to understand Chinese?” asked Searle.

To answer this question, Searle sat in a closed room holding a book with an English version of the program, and adequate paper and pencils to carry out the instructions of the program. The Chinese interrogator provided Chinese sentences through a slot in the door, and Searle could process them using the program’s instructions and send Chinese sentences back through the same slot. He mentioned that he did the same task as the computer that passed the Turing test. He argued that computer and him both doesn’t understand Chinese and hence, if the computer isn’t understanding the conversation, then it is not thinking, and therefore, it does not have an intelligent mind. He formulated the strong AI which according to him is:

"The appropriately programmed computer really is a mind, in the sense that computers given the right programs can be literally said to understand and have other cognitive states" (Searle 1980). Based on the Chinese room experiment, he concluded that strong AI is not possible.

The main question is whether a computer can have a mind (strong AI) or can only simulate a mind (weak AI). According to Chalmers (1996), the distinction between the weak AI and the strong AI is a great concern to the philosophers who are discussing the notion of consciousness.

Premier attempts at AI involved modeling the neurons in the brain. An artificial neuron is a binary variable. McCulloch and Pitts (1943) proposed the binary variable that is switched to either on or off and Hebb (1949) further developed it by proposing Hebbian learning for neural networks. An artificial neural network consists of a large collection of neural units (artificial neurons), whose behavior is roughly based on how real neurons communicate with each other in the brain. Each neural unit is connected with many other neural units, and links can enhance or inhibit the activation state of adjoining units. The network architecture consists of multiple layers of neural units. A signal initiates at the input layer, traverses through hidden layers, and finally culminates at the output layer. Since the 1950s, neural networks have waned in popularity, but deep learning applications have made use of them thanks to improved algorithms for training them and substantially faster computer processing (Goodfellow et al. 2016).

In 1951, Minsky and Edmonds created the first neural network computer, and AI emerged as a new discipline with the goal of creating computer systems that could learn, react, and make decisions in a complex, changing environment.

A discipline known as cognitive science seeks to answer the question of how the mind presents and processes information. It combines philosophy, psychology, artificial intelligence, neuroscience, linguistics, and anthropology, and emerged as a separate discipline around the same time as AI. AI is concerned with the creation of an artificial mind, whereas cognitive science is concerned with empirical studies of the human mind.

Newell and Simon (1961) in 1955–1956 developed a program called the Logic Theorist that mimic the problem-solving skills of a human. The Logic Theorist is considered the first AI program. According to McCorduck (2004), Logic Theorist was able to prove 38 of the first 52 theorems in White and Russell's Principia Mathematica and found shorter proofs for some of them.

General Problem Solver (GPS) was a program intended to work as a universal problem solver machine, developed by Newell and Simon. GPS solved simple problems such as the Towers of Hanoi but did not scale up. Gelernter (1959) developed the Geometry Theorem Prover, which proved the theorems in elementary Euclidean plane geometry.

McCarthy (1958) designed a hypothetical program to accept the next axioms about the environment, and reason with them without being reprogrammed, called the Advice Taker. The Advice Taker advanced the notion of separating the representation of the world from the manipulation of the representation.

To understand and obtain a manageable grasp on an entity that can reason intelligently with all aspects of the world, researchers developed microworlds which the most well-known of these is the blocks world (Winograd 1972; Winston 1973). This world consists of a set of blocks placed on a table. A robot then has the task of manipulating the blocks in various ways. However, systems that could prove theorems containing limited number of facts that behaved well in microworld could not be scaled up to systems that could prove theorems involving many facts with complex worlds.

As previously stated, all early AI efforts worked in limited domains and solved simple problems; however, their programs were incapable of handling complex problems and failed to scale up. These methods are referred to as weak methods due to their inability to scale up. As a result, many researchers concentrated their efforts on creating systems that solved difficult problems in specialized domains. Because they used domain-oriented knowledge, these systems were dubbed knowledge-based systems. These systems are also known as expert systems because they frequently perform expert-level tasks. ACRONYM (Brooks 1981), DENDRAL (Lindsay et al. 1980), and XCON are examples of successful knowledge-based systems (McDermott 1982). Knowledge-based systems were based on logic, performed exact inference, and arrived at categorical conclusions, however, in many domains especially medicine, the conclusions cannot be certain (Szolovits and Pauker 1978). The best attempt has been made to incorporate certainty elements into knowledge-based systems. According to Buchanan and Shortliffe (1984), the certainty factors incorporated in a medical system called MYCIN which is used for diagnosing bacterial infections and prescribing treatment for them.

Since it is so complex and does not accurately reflect how people reason, rule-based representations of uncertain information and reasoning are rarely used (Neapolitan 1989). According to Pearl (1986), humans identify local probabilistic causal relationships between individual propositions and reason with these relationships. For managing uncertain inference in AI, Bayesian networks are the texts that synthesize many of the findings from Probabilistic Reasoning in Expert Systems (Neapolitan 1989) and Probabilistic Reasoning in Intelligent Systems (Pearl 1988).

Optimization problems involved in natural selection as its paradigm, use the evolutionary mechanism to obtain approximate solutions to these problems and thus, it is called evolutionary computation (Fraser 1958; Holland 1975; Koza 1992; Fogel 1994). On the other hand, when some group of autonomous and non-intelligent entities interact, swarm intelligence emerges. Swarm intelligence is used for algorithms that solve many practical problems (Kennedy and Eberhart 2001; Dorigo and Gambardella 1997).

Edelman (2006) explains the organization of higher brain functions in terms of a process known as neuronal group selection also called neural Darwinism. He developed a number of robot-like Brain-Based Devices (BBDs) that interact with the real-world environments (Edelman 2007).

2.1.2 Introduction to Machine Learning

ML is the knowledge that shows us how computer algorithms learn to do things; for example, learning to complete a task, make accurate predictions, or to behave intelligently. The learning is always based on some sort of observation and/or data, direct experience, or instruction. Thus, ML is about learning to do better in the future, from the experienced of the past.

The main idea behind ML is automatic methods. The goal is for the device to learn with algorithms and do the learning automatically without human intervention or assistance. The ML paradigm is as "programming by example," meaning there is a specific task that the computer can come up by its own program based on previous examples.

ML is a subarea of AI, meaning that it is nearly impossible to build any intelligent system without using learning to get there. Furthermore, a system is intelligent if it has the capacity to learn. There are many examples of ML problems such as customer segmentation, face detection, fraud detection, medical diagnosis, optical character recognition, spam filtering, spoken language understanding, topic spotting, and weather prediction.

However, the main question is when ML should be used. There are two aspects of a given problem that necessitate the use of machine learning. The first aspect is when the tasks are overly complex, and the second is the task's adaptability.

There are numerous tasks that humans perform on a daily basis, but the ability to perform them routinely is insufficient to extract a well-defined program, such as driving and speech. Another set of tasks to which ML can significantly contribute is the analysis of large and complex sets of data, such as astronomical data, converting medical archives into medical knowledge, weather prediction, genomic data analysis, web search engines, and E-commerce. The availability of larger amounts of digitally recorded data demonstrated that machines could find meaningful information that is too complex for humans to comprehend.

One of the problems of programmed tools is that once it is installed and it cannot be changed. However, through time many tasks change or change from one user to another. ML tools by nature adapt to changes in the environment they interact and thus, they offer a solution to such problems. Programs that decode written texts—where a fixed program can adjust to differences in the handwriting of different users—adopt automatically to changes in spam e-mails; programs that detect spam; and speech recognition programs are typical examples of successful applications of machine learning to such problems.

ML has branched into several subfields for dealing with different tasks. ML paradigms can be categorized according to four criteria: supervised and unsupervised learning, active and passive learning styles, teacher helpfulness, and online and batch learning protocols.

2.1.2.1 Supervised and Unsupervised

Learning involves an interaction between environment and the learner, and so, learning tasks can divide according to the nature of that interaction. The first classification is the difference between supervised and unsupervised learning. Consider the task of spam email versus the task anomaly detection. There will be a setting in which the learner receives training emails with labels spam/not-spam. The learner will Figure out a rule for labeling a new email. In contrast, for the anomaly detection task, each learner receives a vast collection of unlabeled emails as training, and their objective is to find any unusual messages.

A scenario of supervised learning is one in which the experience contains important information that is omitted from the hypothetical test cases. In this setting, the acquired expertise is aimed to predict the missing information for the test data. In this condition, the environment is the teacher that supervises the learner by providing the extra information. In unsupervised learning, there is no distinction between training and test data. The learner processes input data with the goal of coming up with a compressed version of data.

The intermediate learning setting also exists for the situation in which training examples contain more information than the test examples. In this situation, the learner is required to predict even more information for the test examples. Such learning frameworks are mainly investigated under the title of reinforcement learning.

2.1.2.2 Active and Passive Learners

The paradigms can vary by the role played by the learner. An active learner interacts with the environment at training time while a passive learner only observes the information provided by the environment without posing any influence. The learner of the spam filter is waiting for users to mark the email coming; hence, the learner is usually passive. If the users label emails chosen by learner, to enhance their understanding of what spam is, then learner is an active one.

2.1.2.3 Helpfulness of the Teacher

When a scientist learns about nature, the environment is thought of as passive without any need of the learner. Such learning can be modeled by postulating the training data generated by a random process. The is the base block in the branch of "statistical learning." Additionally, learning occurs when the learner's input is generated by an adversarial "teacher." The adversarial teacher model can be used as a worst-case scenario when no milder setup can be assumed.

2.1.2.4 Online and Batch Learning Protocol

Finally, the distinction between situations in which the learner must respond online, throughout the learning process compared to the situation that learner must gain expertise only after processing a large amount of data. The example is when a stockbroker makes decisions based on the experience gained over time while in data mining settings, the data miner has large amounts of training data before concluding.

2.1.2.5 How Machine Learning Can Be Applied

Big data emphasizes a change in the scale of data. However, there has been a change in the nature of the data as well. As mentioned, ML can deal with unconventional data that is high dimensional for standard estimation methods.

Satellites have been taking images of the earth for many years which thanks to ML, it can be used as an economically meaningful input. Donaldson and Storeygard (2016) showed an overview of the growing literature in economics using satellite data, including how luminosity at night correlates with economic output (Henderson et al. 2012) or estimating future harvest size (Lobell 2013). Satellite images provide a large number x vector of image-based data for measuring crop yield. These images are then matched from the yield data from the y variable. ML is the essential tool that extracts and scales economically meaningful information from this data.

Where economic consequences are absent, such as in areas of poverty in developing nations, the new sources of data are trustworthy (Blumenstock 2016). Jean et al. (2016) used the outcome from the satellite data in five African countries to train a neural net for predicting the local economies of these countries. ML was also used to forecast economic outcomes from massive networks of data. For instance, cell phone data was used to estimate Rwandan wealth, enabling researchers to calculate individual-level poverty (Blumenstock et al. 2015). In New York City and Boston, images from Google Street View were used by Glaeser et al. (2016) to measure block-level income.

Language provides powerful sources of data similar to satellite images. Kang et al. (2013) used restaurant reviews from Yelp.com to predict the outcome of hygiene inspections. Antweiler and Frank (2004) built an algorithm to use online messages to explain the market volatility and economically modest effects on stock return.

Corporate financial information is a reliable source for financial economists. Companies release detailed reports of their financial positions. Kogan et al. (2009) used the information provided from these sources and showed significant predictive information.

ML can also be useful in preprocessing in traditional datasets. ML was applied to match individuals in historical records linking father and sons across censuses, allowing researchers to quantify social mobility during the Great Depression (Feigenbaum 2015). Bernheim et al. (2013) link survey responses to observable behavior, giving economists a tool to infer actual from reported behavior. A subset of survey respondents take part in a laboratory experiment; an ML algorithm trained on this data predicts actual choices from survey responses.

2.1.3 Image Processing

In the 20s, the first interest in transmitting and handling images in digital forms was documented. However, it was only until the mid-60s and after the successful Apollo mission that the interest in this area was fully explored. This delay was due to the lack of strong computer systems and vast required storage. The need for processing

Table 2.1 DIP application in different fields

Field name	Applications
Medical	Automatic detection, classification of tumors in X-ray images, magnetic resonance image (MRI), processing of CAT scan, ultrasound images, chromosome identification, and blood test.
Computer vision	Identification of parts in an assembly line, robotics, tele-operation, autonomous system, and bin-picking.
Remote sensing	Meteorology and climatology, tracking of earth resources, geographical mapping, prediction of agriculture corps, urban growth and weather, flood, and fire control.
Radar and sonar	Detection and recognition of various targets, guidance and maneuvering of aircraft and missiles.
Image transmission	HDTV and 3DTV, teleconferencing, communications over computer networks/satellite, military communication, and space missions.
Office automation	Document storage, retrieval, and reproduction.
Identification systems	Facial, Iris, finger printing ID systems, airport, and bank security.

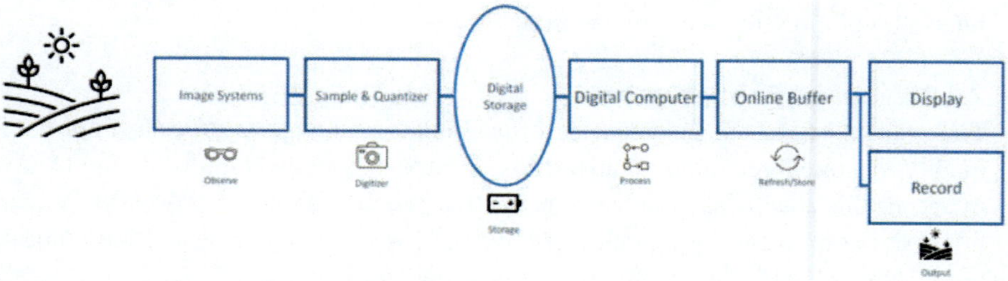

Fig. 2.6 Process of image processing system

lunar images after the Apollo mission forced The National Aeronautics and Space Administration (NASA) Jet Propulsion Laboratory (JPL) to work in the area of Digital Image Processing (DIP).

DIP was initially established to analyze and enhance lunar images, but due to the incredibly quick advancements in algorithm development and computer engineering, it quickly expanded into a variety of new applications. The most important growth appeared to emerge in the areas of medical image processing, data communication and compression, remote sensing, and computer vision. Nowadays, the emphasis is being shifted toward real-time digital image processing. Table 2.1 shows the application DIP in various fields.

A typical image-processing system begins by observing an object and then processes it so that it can be displayed and used for a specific purpose (Fig. 2.6). A digital image is a sampled and quantized version of a 2D function acquired optically, sampled at equally spaced rectangular patterns.

The task of digital image processing entails using computers to handle, transmit, enhance, and analyze digital images. There are three types of processing: high-level processing, intermediate-level processing, and low-level processing.

2.1.3.1 Image Representation and Modeling

Images can be represented in either the spatial or transform domains. Image representation relies heavily on fidelity or intelligibility criteria for measuring image quality. These measures include contrast, spatial frequencies, color, and edge sharpness.

In the spatial domain, the images represented directly indicate the type and the physical nature of the imaging sensors, such as luminance of object in the scene for pictures taken by camera, absorption characteristics of the body tissue for X-ray images, radar cross section of a target for radar imaging, and temperature profile of a region for infrared imaging and gravitational field in an area in geophysical imaging.

Linear statistical models can also be used to model images in the spatial domain. They enable the creation of algorithms that are applicable to an entire ensemble of images rather than a single image. Orthogonal images can be used to extract frequency, content, bandwidth, and power spectra from digital images for use in filtering, compression, and object recognition.

2.1.3.2 Image Enhancement

The goal of image enhancement is to manipulate an image in order to improve its quality. An intelligent human must recognize and extract useful information from an image in this case. Image enhancements that require low-level processing include improved contrast and grayscale, spatial frequency enhancement, pseudo coloring, noise removal, edge sharpening, magnification, and zooming.

2.1.3.3 Image Restoration

The ultimate goal of image restoration is to improve an image's quality. It entails restoring an image that has been corrupted by stochastic or deterministic phenomena. In satellite imaging, deterministic phenomena such as blur are caused by relative motion between the camera and the object, such as atmospheric turbulence. A stochastic phenomenon is one that corrupts the image both additively and multiplicatively. The additive noises are caused by sensor imperfections, thermal noise, and channel noise. The multiplicative noise in Synthetic Aperture Radar (SAR), lasers, and ultrasound images, which are coherent imaging systems, is speckle.

In order to recover the original image, the restoration techniques aim at modeling the degradation and then applying an appropriate scheme (Fig. 2.7). Some of the

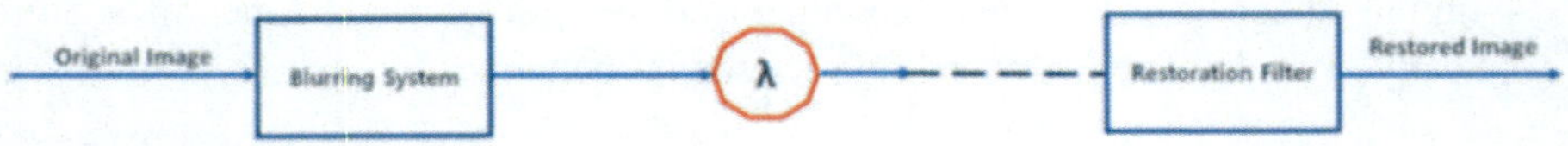

Fig. 2.7 How image restoration system works

methods are image estimation and noise smoothing, deblurring, inverse filtering, and 2D Wiener and Kalman filers.

Image construction is a subset of image restoration in which two or more dimensional objects are gathered from multiple projections. It is commonly found in CT scanners, astronomy, and radar imaging. Radon transform, projection theorem, and reconstruction algorithms are common methods.

2.1.3.4 Image Transforms

Image transformation brings the mapping digital images to the transform domain while using a unitary transform. Examples can be 2D Discrete Fourier Transform (DFT), 2D Discrete Cosine Transform (DCT), and 2D Discrete Wavelet Transform (DWT). In the transform domain, certain characteristics of the images are revealed. Image transformation uses both feature extraction and dimensionality reduction and is considered intermediate level because images are mapped to reduce dimensional feature vectors.

2.1.3.5 Image Data Compression and Coding

It is important to transmit and store images in digital forms and in a reliable source. However, due to the size of the images, it is essential to compress or code the data. For example, LANDSAT sends approximately 3.7×10^{15} bits annually. For this reason, storage and transmission of data requires a large capacity and bandwidth which is expensive.

Eight bits per picture element (pixel) is required for straight digitalization. Using Differential Pulse-Code Modulation (DPCM) or transform coding can reduce this amount to one or two bits per pixel which preserves the quality of images. With the help of frame-to-frame coding, this amount can be reduced even further. For motion parameters from video image sequences, motion-compensated coding can detect and estimate these parameters. For example, the area of research in stereo-video sequence compression and coding for 3D TV and virtual reality applications. Some of the schemes are pixel-by-pixel coding, predictive coding, transform coding, hybrid coding, frame-to-frame coding, and vector quantization.

2.1.3.6 Image Analysis and Computer Vision

Computer vision and image analysis involve segmentation, feature extraction, and classification/recognition. Segmentation techniques are used to isolate the desired object from the scene so that the features can be measured accurately. The useful features are then extracted from the segmented objects (targets).

The goal of image analysis is to develop an automatic interactive system capable of extracting symbolic descriptions from a scene. Pattern recognition is thought to be the inverse of computer graphics because it takes a scene or an image and converts it into an abstract description, numbers, symbols, or a graph. Figure 2.8 depicts the parallels and distinctions between computer graphics, image processing, and pattern recognition.

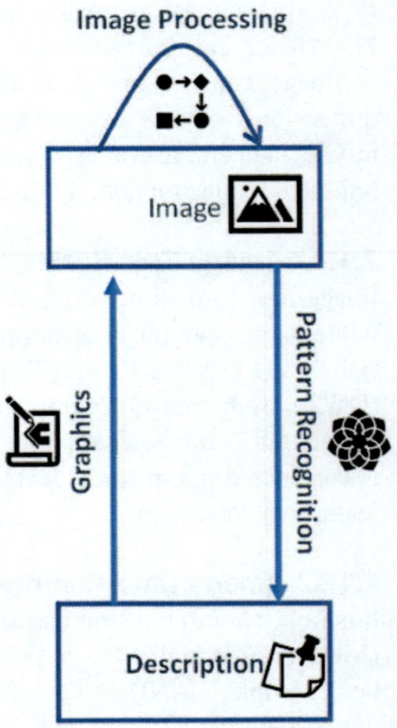

Fig. 2.8 Simple diagram for distinguishing computer graphics, image processing, and pattern recognition

Finally, image processing (Fig. 2.9) is made up of sensors that collect image data, processors that compress data and remove noise, segmentations that isolate objects of interest using edges, feature extraction that extracts a representative set of features from segmented objects, classifiers that classify each object, and structural analyzers that determine the relationships between the classified primitives.

2.1.4 Internet of Things

In today's world, the internet has touched every corner of the globe and is affecting human life in unimaginable ways. Humankind is now entering an era of even more pervasive connectivity where a wide variety of appliances will be connected to the web. Nowadays, IoTs has reached many different players and gained global recognition. As part of IoT ecosystem, the main areas of IoT applications are smart cities, smart cars and mobility, smart home and assisted living, smart industries, public safety, energy and environmental protection, agriculture, and tourism (Fig. 2.10).

There are two definitions of IoT, the first definition simply portraits IoT as an interaction between the physical and digital worlds. According to Vermesan et al. (2013), the digital world interacts with the physical world using Plethora of sensors and actuators. The second definition is a paradigm in which computing and networking capabilities are embedded in any kind of conceivable object (Peña and Fernández 2019).

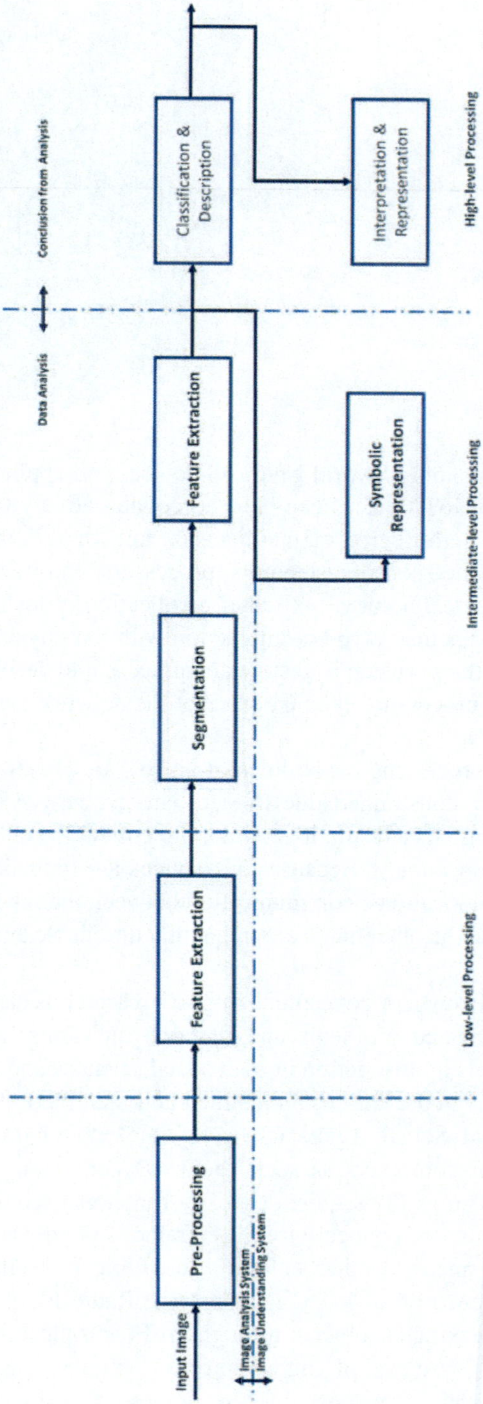

Fig. 2.9 Diagram of image analysis system

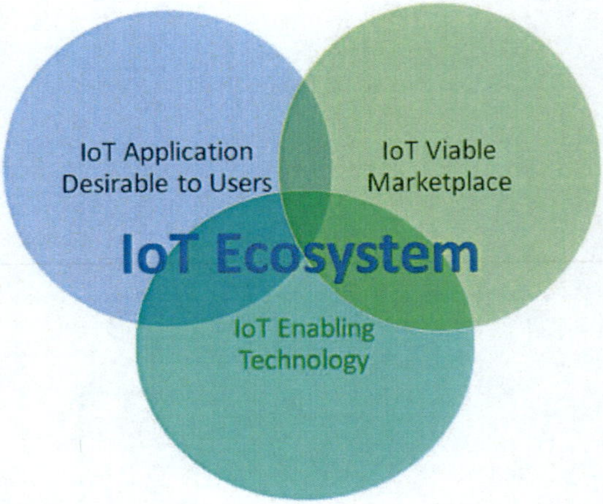

Fig. 2.10 Ecosystem of Internet of Things

IoT refers to a new kind of world where all devices and appliances are connected to a network. All the IoT devices can be used collaboratively to achieve complex tasks that require a high degree of intelligence, and for this the IoT devices are equipped with embedded sensors, actuators, processors, and transceivers.

IoT is not a single technology, but rather a collection of technologies that work together. Sensors and actuators aid in interaction with the physical world. After the sensors collect data, the next step is to store and process it to derive useful inferences from it. Storage and processing can take place at the network's edge or on a remote server.

The storage and processing capabilities of an IoT are constrained by resources, which are frequently constrained due to size, energy, power, and computational capabilities limitations. As a result, it is even more critical to collect the right kind of data with maximum accuracy. Because IoT devices are typically installed in geographically dispersed locations, communication between them is primarily wireless. Wireless channels, on the other hand, are frequently unreliable and have high rates of distortion.

The general framework for conceptualizing IoT research is capable of being used to understand IoT related problems and research questions in conjunction with widely accepted levels of abstraction in both social sciences and computer sciences. In this framework, five core entities are defined and identified.

The framework of Social Actors (S) is flexible enough to accommodate the emerging concept of computers as social actors (Lynn et al. 2015; Zhao 2003). The framework of Things (T) requires two key functional requirements which are data sensing and network connectivity. The framework of Data (D) are discrete artifacts that can connect to other entities. According to Haller et al. (2009), it recognizes the existence of an IoT data chain like Radio Frequency Identification (RFID), enabling tracking of objects through an Electronic Product Code (EPC). Networks (N) are systems of interconnected entities, and this framework accommodates networks between different types of IoTs. Processes (P) are important to how entities interoperate and interact in the IoT. The framework of Processes is essential on how value is created, captured, and delivered in the IoT.

A number of key concepts are surrounding IoT including object identification, information sensing, communication technologies for data exchange and network integration technologies (Shin 2014). The following are definitions for key IoT technologies:

- Cloud computing: According to Mell and Grance (2011), cloud computing is a model for enabling ubiquitous, convenient, on-demand network access to a shared pool of configurable computing resources that can be rapidly provisioned and released with minimal management effort or service provider interaction.
- Dew computing: This concept is an on-premises computer software-hardware organization paradigm in the cloud computing environment where the on-premises computer provides functionality that is independent of cloud services and is also collaborative with cloud services (Wang and Leblanc 2016).
- Edge computing: Iorga et al. (2017) defines this concept as the network layer encompassing the end devices and their users, to provide, for example, local computing capability on a sensor, metering or some other devices that are network accessible.
- Fog computing: This model enables the deployment of distributed, latency-aware applications and services by utilizing fog nodes (physical or virtual) located between smart end devices and centralized (cloud) services (Iorga et al. 2017).
- IPv6: Internet Protocol version 6 (IPv6) is the most recent version of the Internet Protocol (IP). It is an identification and location system for computers on networks and routes traffic across the Internet.
- Machine-to-Machine Communication: M2M communication technologies provide capabilities for devices to communicate with each other through wired and wireless systems (Tsai et al. 2012).
- Radio frequency identification: RFID is a form of Automatic Identification and Data Capture (AIDC) technology that uses electric or magnetic fields at radio frequencies to transmit information. This communication can occur without optical line of sight and over greater distances than other AIDC technologies (Karygiannis et al. 2007).
- Wireless Sensor and Actuator Networks: WSANs are networks of large numbers of minimal capacity sensing, computing, and communicating devices and various types of actuators (Stankovic 2008).

The Things used in the IoT can be characterized by their heterogeneity in terms of computing resources, network connectivity, and software development. This heterogeneity enables depth and breadth of applications and use cases, it introduces complexity, particularly with respect to expected service level requirements. However, as Cavalcante et al. (2015) mentioned, lack of standardization means lack of interoperability. For this reason, reference architecture can help software developers to understand, compare, and evaluate different IoT solutions. For standardizing the concepts and implementation of IoT systems in different domains, several reference architectures have been proposed. For instance, a comparative study with 11 different reference architectures was conducted by Breivold (2017).

Networking technologies enable IoT devices to communicate with other cloud-based devices, applications, and services. To ensure device communication, the Internet uses standardized protocols. These protocols define the rules and formats that devices use to manage networks and send data across them.

The Open Systems Interconnections (OSI) protocol stack is made up of seven layers: application, presentation, session, transport, network, data link, and physical. TCP/IP is composed of four layers: network access and physical layer, Internet, transport, and application.

2.2 Technology Use in Agriculture Sectors

The dveloping world has witnessed growth in food production. For example, in Asia the farmers managed to produce enough food to reduce hunger and malnutrition in the region. Technology advancements and widespread use of high-yielding varieties of staple crops have reduced poverty and malnutrition while simultaneously having a positive impact on the environment by preventing the overuse of marginal lands and delaying deforestation.

In some areas, such as Sub-Saharan Africa, there are still questions regarding how well these technologies will be able to combat hunger. About a billion people reside in areas where agriculture is predominantly rain-fed, and water scarcity prevents irrigated agriculture from spreading. Certain locations with high development potential have declining marginal intensification returns, which is in bad contrast to the possible return from developing vulnerable land (Hazell and Fan 2000).

The development of technologies and practices that enable agricultural growth to match rising food demand will be a major challenge in the coming decades. To reduce hunger and poverty, needs must be designed and processed in such a way that the natural resource base is preserved, and pollution is minimized. Agriculture development in today's world must be broad-based, market-oriented, participatory, decentralized, and driven by innovative approaches that increase factor productivity while conserving resources (Hazell and Lutz 1998). Agroecological approaches, which focus on growing conditions for plants and animals as part of a larger ecosystem, are gaining popularity (Altieri 1995). Major aspects include control of soil erosion and nutrient depletion, biological control of pests and diseases, diversification of activities, interaction between cropping, livestock, and forestry activities.

In the agriculture sector, at low-income societies, a rapid increase in productivity is required to improve the rural incomes and maintain the food supply for the urban population as well as raw materials supply for agro-industrial development, and crop production for exporting purposes. For this level of development in the agriculture sector, policy frameworks with proper incentives for farmers are required.

Despite an increase in agricultural output, the rate of expansion in this sector is typically lower than that of other economic activities. As a result, agriculture's contribution to the microeconomy is diminishing, which highlights the urgency of swiftly implementing more productive technologies. Policymakers are frequently perplexed by the agricultural sector's declining size and rising productivity. Supply analysis is typically used to respond to policy measures such as agricultural input and output prices, trade and exchange rate policies, and the availability of public goods and services. Extension, technological change, and crop substitution can all be supplying responses, each with different implications for resource allocation and the environment. Rural households' responses differ due to differences in expectations and costs.

According to some studies, price reforms will encourage soil depletion, while others claim that they will have a positive effect on farmers' investment in soil conservation and land use (Barrett 1991). The disagreement here is caused by differences in discount rates and risk aversion. Extensification, intensification, input substitution, and output substitution are the four potential responses to changes in relative prices on agricultural resource allocation.

Typically, supply response reactions to changing prices are analyzed through the lens of substituting fertilizers, for reduced availability of nutrients from natural sources due to soil loss. Treating natural soil fertility as a function of capital investment is considered an alternative approach in conservation measures. Strategies for selective intensification and productivity improvement are needed to support sustainable land use.

There are many unknown aspects related to sustainable land use, farmers supply response and agriculture policies. Increase in agricultural production originating from area expansion causes severe environmental impacts such as overgrazing, erosion and sedimentation, and deforestation. According to Binswanger et al. (1987), an increase in output prices leads to a corresponding increase in area but a small increase in yields.

Many agroecological approaches focus on land productivity as a major indicator, with less attention given to returns to labor (Low 1993). Farmers tend to consider yield increase based on agroecological principles from profitability prospect, input efficiency prospect, labor use prospect, risk management, and sustainability prospect.

Over the last decade and with increasing concern over the reduction of cultivation areas, the volume of investments in agriculture is increasing to meet the 50% cultivation growth goal by 2050 (Reardon et al. 2019; FAO 2009; Tsouros et al. 2019). It is believed that modern Information and Communication (ICT) and IoT, data analysis and many modern technologies can improve precision agriculture and smart farming. To grow better and more sustainable food, smart agriculture integrated with ICT is becoming more popular. Unmanned Aerial Vehicles (UAVs) help in irrigation processes to map and monitor the crops, IoT devices can monitor the condition of the fields and ICT can provide automation in agriculture practices. Smart agriculture helps in sustainable development of the food sector as well as promotes recent trends like organic farming (Khatri-Chhetri et al. 2017; Maddikunta et al. 2007).

2.2.1 Smart Agriculture Systems

When combining ICT with agriculture, it is called smart agriculture. Smart agriculture means managing farms with the help of modern ICT to achieve higher productivity while optimizing the human workforce (Gondchawar et al. 2016; Ayaz et al. 2019). In modern agriculture, there are various ICTs used to improve this sector (Table 2.2).

The Internet of Things (IoT) is a collection of computing devices having digital and mechanical parts that may transport data over a network without requiring human or computer intervention (Nizetic et al. 2020). The pillars of IoT are sensors, connectivity, data processing, and user interface. Through sensors, IoT devices gain environmental conditions, send data directly to the controller and the cloud, and receive responses from cloud servers' controller. UAVs are another type of IoT devices that carry cameras and GPS and are controlled by a user or a server using Internet connection.

Drones are recoverable or disposable flying objects that can fly on their own without a pilot using remote control or by pre-arrival programs. Drones have sensors that make them intelligent, such as LIDARs, RADARs, RGB, and infrared cameras (Lee and Choi 2016). Drones are divided into groups based on their size, range, speed, and other characteristics. Drones, for instance, can have rotating or fixed wings. Large fields are a good fit for fixed-wing drones since they are less wind sensitive and can fly farther with higher attitudes while the most suitable drones for small takeoff and landing surfaces are those with rotary wings.

Deep learning models have the capacity to approach human accuracy by learning how to execute pattern analysis directly from images, sounds, texts, and videos. A large set of labeled data is used to train deep learning models, and neural network architecture can have many layers of them (Goodfellow et al. 2016). Deep learning is very effective in agriculture domains, contributing to the spread of Smart Agriculture worldwide (Kamilaris and Prenafeta-Boldu 2018). Deep learning is used in smart agriculture in various domains such as plant detection and classification, plant health assessment, smart pest and herb control, and field analysis and yield estimation.

Large fields make it particularly challenging to identify, categorize, and distinguish between grass and weeds and the real crop. Because of this, it can be difficult

Table 2.2 ICTs used in modern agricultural environments

ICT	Application
Robotics	UAV, UGV, automated manufacturing machines
Sensors	For sensing water, temperature, soil, humidity, light
Data analytics	Modern data analysis, image processing algorithms
Software	Dedicated software solutions that target specific farm types or use platforms that can manage various agricultural processes and systems without knowing the details of the system
Connectivity	Wi-Fi, 5G, LTE, cellular networks, satellites, GPS

to distinguish between grass and weeds, which can reduce revenues and harm the environment owing to overuse of herbicides, insecticides, and fertilizers. In the field of plant classification and identification, many studies have been conducted. In a study conducted by Lu and He (2017), drone images were used to identify grass in grasslands. With the help of drone-based dual camera, Wan et al. (2018) managed to estimate the flower numbers oilseed rape. By use of drone imagery, Hung et al. (2014) developed an algorithm based on the learning method for identifying weeds in the field with 90% accuracy. Sandino et al. (2018) developed an integrated pipeline methodology and gained 96% accuracy using gradient boosted decision trees to map grasses and vegetation in arid lands. With the help of ML methods, Peña et al. (2014) improved classifying summer crops. Priya et al. (2012) proposed an algorithm for plant leaf recognition using a vector machine, classifying up to 32 types of plants.

Food consumption rises steadily along with population expansion, necessitating the use of cutting-edge technologies to ensure the agriculture sector's success. The main goal is to create automated agricultural systems that are resource-efficient, affordable, and productive while utilizing existing resources. With development it is necessary to monitor the crop health, control the spread of diseases, and grade the products before entering the market. Do et al. (2018) evaluated the health of citrus plants using digital photos obtained from drones. Ground-based sensors including a water potential meter, chlorophyll meter, and spectroradiometer are used to monitor the plant conditions. Zermas et al. (2015) built an automated system for the evolution of plant health in cornfields in 2015.

In Henan Province, China, Huang et al. (2019) conducted an experiment for detecting Helminthosporium Leaf Blotch (HLB) disease in wheat farms. The infection turned the leaves color at the initial stage and so with the help of Convolutional Neural Network (CNN) the infected leaves were detected with the accuracy of 91.34%. Dang et al. (2020) used drones equipped with RGB sensors to detect fusarium wilt with 90% accuracy. In order to find plant diseases in remote areas, Bashir and Sharma (2012) employed automated image processing. With the use of 87,848 leaf images from 25 different plants, Ferentinos (2018) created a CNN model for the detection and diagnosis of plant illness, attaining a 99.53% success rate. By creating a web-based image processing system and extracting the features based on color, morphology, and CCV clustering algorithms with k-means, Bhange and Hingoliwala (2015) studied Telya (bacterial blight disease) that affects fruit. A method for diagnosing diseases in potato plants was created by Islam et al. (2017) with the use of image processing and machine learning.

For classifying and sorting of fruits, Kumar et al. (2015) reviewed different classification techniques based on image processing. Costa et al. (2011) reviewed articles for shape analysis of agriculture products. Post-harvest vegetable and fruit grading systems for packaging were developed by Cubero et al. (2014). Al-Marakeby et al. (2013) used a data set of 1000 and developed a vision-based model for sorting food products with 97% accuracy.

Pests are the main issues faced in farming. The use of pesticides is known as a solution for facing the pest problem, however, inefficient, and excess use of pesticides can harm both the crops and the environment and is not economical. Drone usage can expedite the process and stop pesticides from being used excessively or leaching into the soil. But there are number of issues using drones for spreading pesticides such as the wind that cause drift of pesticides to the surrounding fields. For this reason, increased precision can help to limit the loss of pesticides while spraying from above. Faical et al. (2014) presented a drone with system of coupled spray that can communicate with a wireless sensor network. Water-sensitive paper and rhodamine B tracer were compared with the droplet parameters' deposition of Electric Air Pressure (EAP) sprayers and drones under different spray volumes (Wang et al. 2019).

For management and marketing purposes, estimating the crop is important. Yield forecasting assists farmers in improving the quality of their products. The two computer-based methods for estimating performance are regional and counting methods. Based on these methods, Wang et al. (2013) developed an automatic system to estimate apple crop yield. Linker (2017) managed to count the number of apples in orchards under natural illumination with 85% accuracy by using color images. By spectral mixture analysis, Gong et al. (2018) developed a technique to estimate rapeseed yield with remotely sensed data.

2.2.2 Smart Aquaculture Systems

Aquaculture as it is practiced now has numerous negative effects on the environment, limited yield, and labor-intensive processes. Aquaculture industry development that is more ecologically friendly and productive is what smart aquaculture strives to achieve. There are many obstacles in implementing aquaculture systems such as maintaining the quality of the water. Due to the lengthy and laborious procedure, it is frequently hard to treat the water in time for unexpected changes in the water quality of ponds and/or tanks. Disease control is not detectable at an early stage of the spread of diseases. Feed wastage is another problem in aquaculture systems especially when leftovers affect the water quality. Counting fingerling and seeds is another challenge faced by farmers. Smart aquaculture systems can be a great solution to answering these challenges and solve the problems faced in traditional aquaculture systems.

In smart aquaculture systems, several smart devices are combined to track the environment being cultivated in real time and make automated decisions using the data gathered (Sharma and Kumar 2021). It can be improved and controlled remotely by adding IoT devices, AI, 4G/5G connectivity, cloud computing, and robotics. Robots are capable of operating and achieving effective output by managing smart aquaculture systems, facilities, machinery, and equipment (Kassem et al. 2021). The components of smart aquaculture may be summed up as the monitoring of water quality in the systems using water quality sensors (temperature, DO, humidity, transparency, light, pH), and the transmission of the gathered data to the control center with the aid of communication nodes. A cloud-based platform is used to store data, make decisions, analyze feedback from those decisions, and operate the system

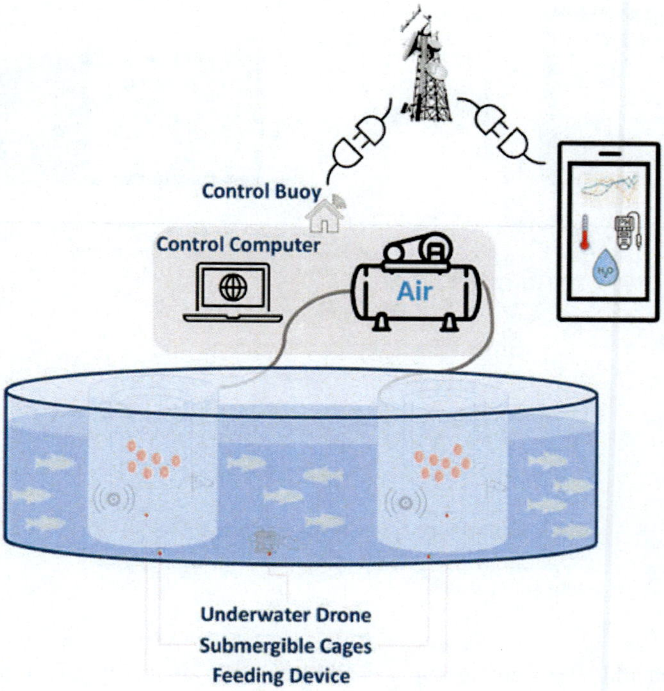

Fig. 2.11 Diagram for smart aquaculture systems. (Source: Chrispin et al. 2020)

intelligently in order to create an aquaculture system that is environmentally friendly and sustainable.

There are various examples of AI and IoT usage in traditional aquaculture systems (Imai et al. 2019). These methods are used in a variety of cultured systems, including cage, pond, hatcheries, and breeding facilities, with a range of objectives, including management of water quality (Hamid et al. 2019; Dzulqornain et al. 2017; Sivabalan et al. 2020; Shubhaker et al. 2020), observation and monitoring, feed supply, and optimization (Rashid et al. 2021; Wang et al. 2021).

AI is utilized in aquaculture for the supply chain of seafood, feeding machinery, drones, seed disease control, regular stock monitoring, and shrimp farming (Chrispin et al. 2020). Figure 2.11 shows the process of deployment of smart devices in aquaculture.

The primary purpose of ML is to solve problems by using computer algorithms and data (Jordan and Mitchell 2015). The main models being used in the aquaculture systems are Artificial Neural Network (ANN) (Zhakov et al. 2020), Decision Tree (DT) (Zhang et al. 2020c), Deep Learning (DL) (LeCun et al. 2015), Ensemble Learning (EL) (Ma 2012), K-Nearest Neighbor (KNN) (Jia and Zhang 2020), Naïve Bayes (NB) (Xu et al. 2020), and Support Vector Machine (SVM) (Tang et al. 2020). The most popular among four types of ML (Supervised Learning, Unsupervised Learning, Semi-Supervised Learning, and Reinforcement learning) which is used in

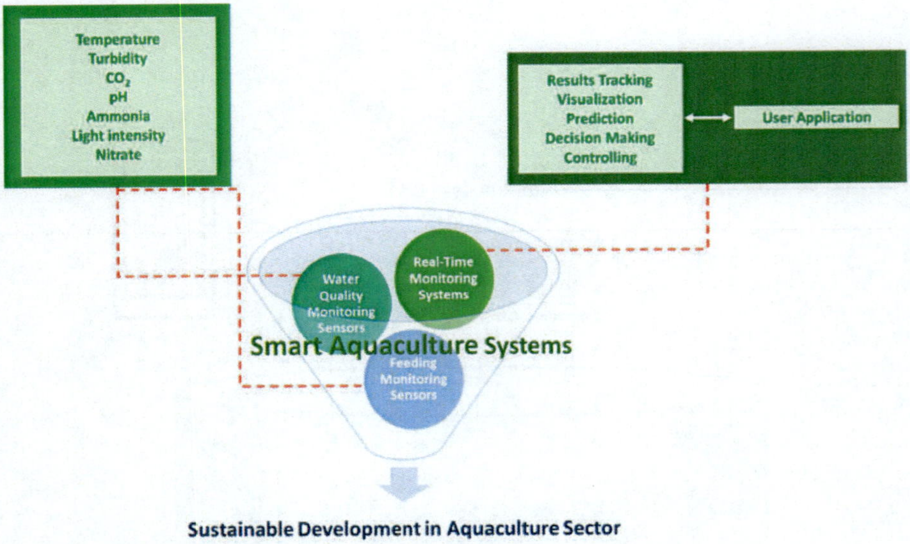

Fig. 2.12 Overview of smart aquaculture systems for sustainable development

smart aquaculture systems is supervised learning. This learning is used for classification and regression by using data for training ML (Kotsiantis 2007). ML implementation in aquaculture can be seen in biomass fish detection (Yang et al. 2020a, b), size and weight estimation (Monkman et al. 2019; Garcia et al. 2019; Li et al. 2020; Fernandes et al. 2020; Zhang et al. 2020a; Petrellis 2021), fingerling counting (França Albuquerque et al. 2019; Le and Xu 2017; Liu et al. 2018; Siddiqui et al. 2018), fish recognition (Xu and Matzner 2018; Cai et al. 2020; Villon et al. 2018; Rauf et al. 2019; Hu et al. 2020; Cao et al. 2020), age detection (Moen et al. 2018; Ordonez et al. 2020), sex identification (Cai et al. 2020; Webb et al. 2019; Barulin 2017, 2019), species classification (Yang et al. 2020a, b; Tharwat et al. 2018; Jalal et al. 2020; Pramunendar et al. 2019; Rum and Nawawi 2021; Deep and Dash 2019; Chhabra et al. 2020), feeding behavior (Zhou et al. 2018; Adegboye et al. 2020), group behavior and abnormal behavior (Han et al. 2020; Zhao et al. 2018; Morimoto et al. 2018), and univariate and multivariate prediction (Cao et al. 2020; Keshtegar and Heddam 2018; Ren et al. 2020; Ta and Wei 2018; Kim et al. 2020; Lu and Ma 2020; Fijani et al. 2019; Barzegar et al. 2020). As it is shown in Fig. 2.12, smart aquaculture can improve all the stages from breeding, nursery to growing stage as well as managing and quality control of water resources, feeding, grading, and counting.

2.2.2.1 Water Quality

The success of aquaculture farms depends on the condition of the water. Fish infections are highly common in many farms across the world and are caused by poor water quality, which affects productivity (Lafferty et al. 2015). The optimal fish production is dependent on the physical, chemical, and biological qualities of the environment (Bhatnagar and Devi 2013). These qualities are determined by variables such as temperature, turbidity, CO_2, pH, alkalinity, ammonia, nitrate, D. O., etc. which the most critical are temperature, D.O., and pH (Rahman et al. 2020).

IoT devices are used in various areas related to smart aquaculture. These devices have opened new trends in the aquaculture sector and guarantee the sustainability of this sector with real-time water monitoring (Rahman et al. 2020). There are four layers in every IoT device working in aquaculture field: physical layer, monitoring layer, virtual layer, and connection protocol (Fig. 2.12). Smart and intelligent aquaculture helps in decreasing labor costs, increasing operational efficiency and higher productivity. Due to the various nature of aquaculture activities, labor management is high in risk and there is still a need for a certain amount of observation, information analyzation, and decision making during the fish farming process. However, intelligent equipment responsible for monitoring fish farming environments; robotic devices for production; data and information sorting; and energy-saving processing equipment will continue to greatly automate different stages of fish farming operations. There are several studies about IoT devices deployed in different areas of aquaculture. Raspberry Pi was used by Chavan et al. (2018) for checking real-time monitoring systems in aquaculture with ammonia, D. O, pH, and temperature sensors. Kim and Shin (2018) set up a RAS with water monitoring sensors, Message Queue Telemetry Transport (MQTT) protocol, and MICOM controller. Low-cost systems with Raspberry Pi were used by Al-Hussaini et al. (2018) for automatic data acquisition, monitoring systems, and fog computing for RAS. Monirul et al. (2019) built an IoT system using sensors (temperature, dissolved oxygen, turbidity, pH, water level, and CO_2 gas) and Arduino.

Nocheski and Naumoski (2018) demonstrated a functional and upgraded IoT system with water quality monitoring sensors and small computer board for data collection and analysis with ability to send notification to the users. Krishna et al. (2019) also set up an IoT system in fishpond to manage fish health as well as the water quality monitoring with an Arduino Uno board, Atmega328 micro controller, Wi-Fi module, Buzzer, LCD (liquid crystal display), and MIT application, providing data that are retrieved from the cloud by the farmer, along with the environmental parameters.

Prabhu (2019) proposed an IoT system that can obtain and analyze data as messages through mobile phone and several actions in managing the environmental conditions with an Arduino Nano Board and ESP8266 Wi-Fi module. Nguyen et al. (2020) deployed an IoT system to monitor water quality and created a model for forecasting quality indicators. Hsu et al. (2020) collected water quality parameter's data through sensors such as Oxidation-Reduction potential (ORP), pH, and temperature. They also employed a monitoring data map in real time to find out the pond conditions. The IoT-based system included Raspberry Pi, Arduino UNO, Bluetooth module, two or more sensor modules (ORP, temperature, pH meter), Google Cloud Messaging (GCM), mobile App device's REGID, prediction analysis, and Web Page Management Mode (HTML combined with PHP). Darus et al. (2020) proposed water monitoring for catfish farms and used Expectation Maximization (EM) clustering. Figure 2.13 shows the IoT water quality system block diagram.

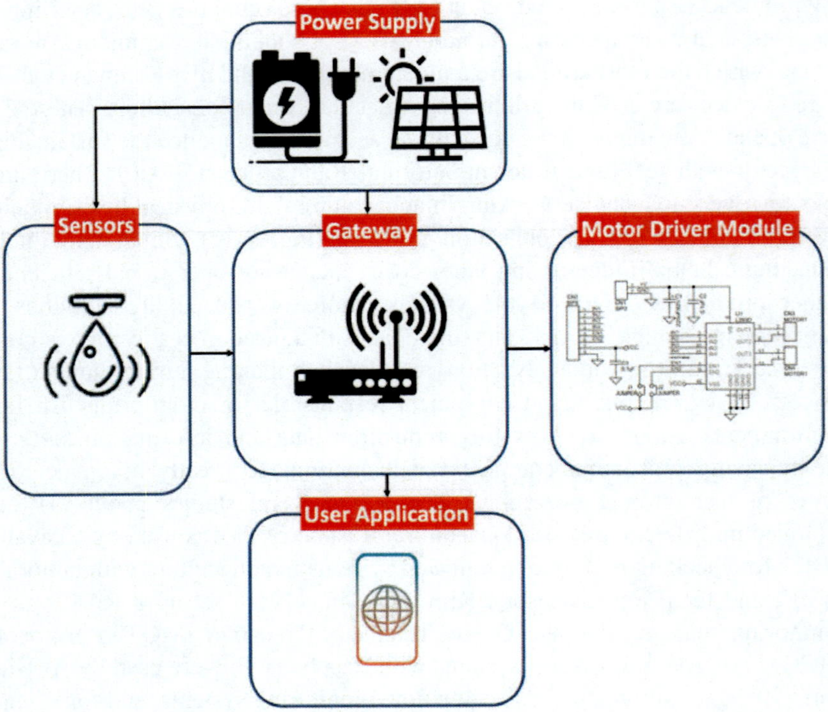

Fig. 2.13 IoT water quality monitoring block diagram. (Source: Darmalim et al. 2019)

2.2.2.2 Feeding Control in Aquaculture Farms

In traditional aquaculture, farmers spread the feed around the pond, cage, or tank depending on the eating characteristics of the species. This will cause multiple problems on the farms such as pollution and feed wastage. Applying IoT system to the feeding system not only controls the feeding amount but also brings many benefits such as saving in manpower and reducing leftovers, and reducing water pollution in aquaculture.

Cao and Xu (2018) implemented machine vision to count feed in order to manage feed residuals in aquaculture farms. The system counts the pellets in a heap. As mentioned, excess feed causes many issues such as decreasing profit, increasing water pollution, and affecting the health of cultured species. Therefore, Cao and Xu (2018) used an algorithm to count the exact amount of feed under different water turbidities. These algorithms can be applied in automatic feeding systems with high accuracy rate.

Harish et al. (2018) built an efficient semi-automatic system that facilitated healthy growth of aquatic organisms in aquaculture systems. This study was conducted not only for a feeding system but also for water quality parameters in a cultured system through sensors. The GSM module is used to alert the cultivator whenever the quality parameters violate the normal range. Daud et al. (2020) studied the suitable condition in aquariums with monitoring the feeding condition and water quality. This system is designed with MEGA and NodeMCU controllers and is controlled by a smartphone in its operation. Arduino is used in the designed system. Wi-Fi communication on the NodeMCU is connected between the smartphone and the controller to control the operation. All collected data of water quality is displayed on LCD.

2.2.2.3 ML in Smart Aquaculture Systems

As previously stated, the ML process can be classified as supervised, semi-supervised, unsupervised, or reinforcement learning, with supervised being the most popular method (Fig. 2.14) (Erickson et al. 2017).

Supervised learning needs to use an algorithm system to obtain experience through training. The algorithms will test new data images to classify either the target images (Moore et al. 2019). Unsupervised learning is the output data with no training dataset. It is very similar to supervised learning. The goal of this type of learning is to classify the input resources into different types. Unsupervised learning uses data without labels as input, therefore classifying the input data is the objective of the model. Jeong et al. (2018) claim that this is the primary distinction between supervised and unsupervised learning. The most crucial machine learning activities are illustrated in Fig. 2.15.

In the case of aquaculture systems, computer vision is the main tool to measure the size of the fish. Images are taken through a camera that is attached in the systems while collected images are sent to software and are analyzed.

Measuring the fish size and grading is one of the necessary tasks during culture period because of the fish characteristics and the need for reaching to the market size. Size grading is the most frequent grading, being done in the rearing stage. Automation with the help of ML can reduce the operation and management costs as well as enhance the quality of production.

White et al. (2006) proposed a system using image processing algorithms to determine and identify the size of different fish species. Images of fish were obtained

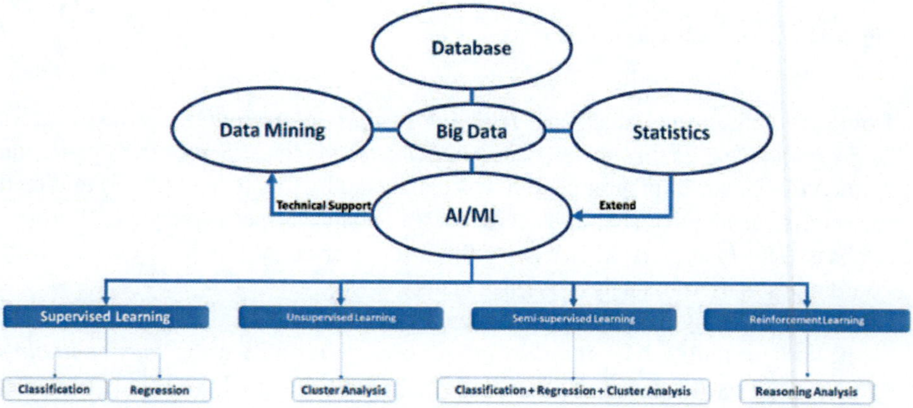

Fig. 2.14 Four types of ML process

Fig. 2.15 ML tasks and main algorithms

from several species such as *Hippoglossoides platessoides*, *Solea vulgaris*, *Microstomus kitt*, *Pleuronectes platessa*, *Sebastes marinus*, *Sebastes mentella,* and *Platichthys flesus*. Similar approach was conducted by Costa et al. (2013) to directly measure size, identify sex, and recognize the abnormal performance of European Seabass (*Dicentrarchus labrax* L.). Elliptic Fourier analysis (EFA) on the outline coordinates is used to analyze the fish shape.

Instead of using expensive instruments, Mustafa et al. (2013) used the FLUDI framework to study fish length. *Rastrelliger kanagurta* and *Selar crumenophthalmus* species were used for this experiment. The experiment used two kinds of camera such as Pentax camera (8.0 megapixel) and Sony (5.0 megapixel). These cameras have 0.74% and 0.19% error, respectively. This allowed FLUDI Software to be developed with high accuracy.

Vision-Based Automatic System (VAMS) was used by Jeong et al. (2018) to determine Total Length (TL), Body Width (BW), Height (H), and Weight (W) without any relation to morphometric characteristics of flatfish. A laser displacement and a load cell were used to measure H and W. The TL and BW were estimated using an algorithm that was created based on the morphological image processing technique.

Weight of the rainbow trout (*Oncorhynchus mykiss*) was assessed by Gerami et al. (2016). The left half of the samples were photographed with a digital Canon IXUS 960IS (12 mega pixels (3000 × 4000); in the red, green, and blue channels). At a 45-cm height above the sample, the camera was placed. The computer was used to transfer all of the captured fish photographs, and MATLAB (Matrix laboratory) version R2009x was used to analyze the data. The study obtained a good result with clear images and assessed weight with high accuracy. From this result, it was concluded that machine vision can be an efficient tool for measuring weight as well as evaluating the visual features of the fish.

Based on the existing information, Sanchez-Torres et al. (2018) implemented a new concept. It is a system with one camera in a controlled set up so it can better manage the sample size and create images with higher resolution. Result showed efficiency, and the method was also applied to feed optimization based on the length and weight of fish.

Sung et al. (2020) aimed to grade flatfish based on the size using machine vision. The designed grader has three components including a conveyor belt, machine vision, and sorter. As fish and their length are detected by image processing, the location of the grader is managed by the length classification. The machine vision part is composed of a camera, LED lights, and a darkroom with minimum costs. Image processing is the main tool for measuring the size of fish. Images are taken through a camera which is attached to a fish measurement system. Afterward, all images are sent to software for analysis, and results are compiled in a Cloud.

2.2.2.4 Fish Disease Control

Diseases in aquaculture farms have negative impact both on the quality of the products and the production levels. Spread of diseases in aquaculture farms is the result of imbalance among many factors coming from host, pathogens, and the environment. There are infectious (parasitic, fungal, bacterial, viral) and non-infectious diseases (environmental, nutritional, and genetical) (Rahman et al. 2019), which are hard to distinguish and hence, to find proper treatment on time. As the result of the spread of fish disease, fish can die in large numbers causing a huge loss to the farmers (Divinely et al. 2019). ML can help detect and alert farmers about the spread of diseases on time.

In Australia, India, Japan, Pakistan, Thailand, and United Kingdom, a disease called Epizootic Ulcerative Syndrome (EUS) with high mortality rate annually causes millions in loss. EUS is a fungal disease caused by *Aphanomyces invadans*. By combing Principal Component Analysis (PCA) and ANN, Malik et al. (2017) developed a method to identify this fish disease. In this method, fish disease images were collected and then morphology operations were applied. The result showed that

FAST-PCA-ANN has better classification accuracy and efficiency than the existing combine technique HOG-PCA-ANN.

In Barak Vally, Assam, Chakravorty et al. (2015) identified four species (*Clarias batrachus*, *Puntius chola*, *Labeo bata*, and *Labeo gonius*) infected by EUS. Chakravorty et al. (2015) proposed two kinds of ML algorithms (Principal Component Analysis (PCA) and K-means clustering). The result showed 90% accuracy for PCA.

Divinely et al. (2019) proposed detecting fungal disease using PNN in a timely and more effective manner. Images are gathered from various internet sources, then put through the preprocessing to prevent unwanted distortions. The Curvelet Wavelet Transform (CWT) has been utilized to boost efficiency, and as a result, the targeted fish diseases such as dropsy, camallanus worm, and ammonia poisoning are categorized. Gray Level Co-Occurrence Matrix (GLCM) was also used to reduce the dimensions and preserve useful information. The result shows that the proposed combination of CWT-GLCM-PNN is an efficient and accurate way to detect fish disease.

Salmon diseases in farms were identified by Ahmeda et al. (2021) using image preprocessing and segmentation. The segmentation extracted features and classified diseases from images with the support of SVM algorithms. The results showed that using SVM is an efficient method to identify the fish diseases with high accuracy rate.

In conclusion, it is very common to use ML to identify fish diseases. There are various ML algorithms used and combined to increase the accuracy for detecting fish diseases, especially fungal diseases.

2.2.2.5 Seed and Fingerling Counting

Counting the seeds and fingerlings is one of the most difficult but important tasks in aquaculture farms. Computer vision technology helps this process to be more efficient and easier by using image processing and video analysis (Li et al. 2021). For counting larvae and juvenile fish, Raman et al. (2016) used an image processing system. In this process, the system detects images of larvae fish and counts them by separating each one. The results showed 82% accuracy in counting larvae and 87% accuracy for juveniles.

Zhang et al. (2020b) used a system for counting fish in cages in an offshore environment. A hybrid neural network model based on multi-column CNN and DCNN was used to observe, realize, and count the number fish in real time. The result showed 95.06% accuracy, thanks to the use of hybrid neural networks.

2.2.2.6 Identification and Classification

Identifying species in farms and open waters is costly by using genetics and inefficient by using the traditional methods with naked eyes. ML and computer vision are efficient and low-cost methods to classify and identify fish in a short time with high accuracy. Coz-Rakovac et al. (2009) applied biomedical data using ML to identify three aquaculture affected species seabass (*Dicentrarchus labrax*), sea bream (*Sparus aurata L.*), and mullets (*Mugil* spp.). The decision tree was used and gave the best results among the machine-learning methods with 210 samples classified (85.71%) correctly and 35 (14.29%) incorrectly.

Underwater cameras can also be used for monitoring marine stocks. The images were used in deep learning neural network for automatic classification of species. Allken et al. (2019) applied this technique for blue whiting, Atlantic herring, and Atlantic mackerel and achieved 94% accuracy in classifying these species.

2.3 Twenty-First Century Businesses Working in Smart Aquaculture

In recent years, smart aquaculture has expanded significantly around the world. However, even though applications of AI and ML in aquaculture are progressing, there are many challenges in fully operating automated systems in this sector. The applications of ML and image processing in aquaculture contribute significantly to applying automatic trends in the aquaculture sector as well as improving farming productivity. However, the cost of this implementation is high and only feasible for large-scale farms.

It is believed that the application of ML and computer vision should be more accessible in smart aquaculture not only for hatcheries, nurseries but also in offshore aquaculture systems. The applications can be applied to offshore cage farms to detect diseases and monitor the feeding as well as the weight and the size of the cultured fish.

In most cases, the farmers try to apply traditional methods such as manually monitoring the fish schools and look for fish with weak swimming pattern. However, submerged cameras or automatic feeding systems can help farmers to reliably face the challenges ahead. There are various private deep tech businesses that help farmers in facing these challenges as well as sustainably develop their farms. In this section, some of these deep tech companies are mentioned.

2.3.1 Aquabyte

Aquabyte is a Norwegian-American company that works to solve problems in Norwegian salmon farms. The company employs ML-based cameras to monitor fish welfare and assist salmon farmers in managing their farms in a sustainable manner. Aquabyte assists global salmon farmers in better understanding fish populations, detecting sea lice infestations, and calculating fish weights. The product consists of on-site image capture hardware, cloud computing for data processing, and a user web application.

The Aquabyte platform provides a comprehensive view of salmon fish welfare and growth monitoring. The data collected from each cage provides insights into growth and feeding performance, as well as alerts farmers to the spread of lice infestation in their farms. The data helps farmers better understand daily trends and events, allowing them to act more quickly and precisely.

The camera gathers accurate information about salmon growth, welfare, performance, and environmental conditions. The camera takes images continuously and is installed in the best possible location to allow as many fish as possible to swim past it. The software then purges the unused images while analyzing the high-quality ones.

The system continuously delivers the daily weight via images, and smart algorithms aggregate these measurements to deliver a high accuracy average weight per day. In addition to daily fish weight, K-factor (reflecting physical and biological circumstances interacting with feeding condition, parasitic infections, and physiological factors (Le Cren 1951)) and Specific Growth Rate (SGR) are available, allowing monitoring of individual fish growth and weight development. The system also includes a Welfare Indicator Score (WISE), which monitors and identifies all welfare indicators. WISE continuously monitors the welfare of the cages. Controlling welfare status reduces both welfare risk and the risk of downgrading. WISE is based on the LaksVel standard, which allows for scoring of fish welfare based on image documentation. Aquabyte can also be used to monitor and count lice levels without requiring physical contact. It provides information about daily sea lice development while reducing handling and fish stress.

Although Aquabyte's methods are unique, they can only be used in oligotrophic and mesotrophic waters with high transparency. This approach is only applicable to high-value species such as salmon.

2.3.2 eFishery

eFishery is an Indonesian company that began with feed automation for Indonesian fish farms. The eFishery feed dispenser is an IoT device that uses a mobile app to automatically shoot feed to the area where the fish schools are located. Because there are no cameras in the system, it is difficult to confirm process accuracy, but water quality monitoring sensors can be connected. The app provides farmers with real-time information and assists them in efficiently managing their farms. The feed dispenser is suitable for both fish and shrimp cultured in Indonesia. eFishery also produces feed, assists with farm e-financing in Indonesia, and provides farmers with go-to-market solutions.

2.3.3 Innovasea

Innovasea addresses a variety of issues in the aquaculture industry. The company created submersible Net Pens, which can be used in offshore farming and easily lowered to safe depths, keeping the cages safe from heavy waves and storms. Submersible net pens not only reduce equipment wear and tear, but they also reduce stress on fish stocks. The benefits of using these net pens include avoiding oxygen-depleting algal blooms, locating optimal currents for keeping fish active and healthy, locating ideal water temperatures for optimal fish growth without thermal stress, and avoiding parasites that congregate near the surface.

Realfish Pro is an IoT device developed by Innovasea that combines real-time data monitoring from environmental sensors and AI cameras. Realfish Pro offers powerful tools for monitoring, managing, and controlling key aspects of the fish farm. The system provides critical information at remote farm locations. It allows for continuous monitoring of water quality and the making of data-driven decisions to protect fish stocks. The underwater sensors wirelessly report variables such as D.O., temperature, and salinity using sound waves. Tidal conditions, wind, barometric pressure, and other weather-related parameters are monitored using surface monitoring tools. Physical sensors are also used to monitor mooring line tension and pen infrastructure conditions. Due to the high cost and installation requirements, it is not suitable for all species worldwide, and it is unclear whether the AI component of the system can be used in other conditions such as RAS and inland aquaculture.

2.3.4 Tidal

The tidal system is heavily reliant on underwater image processing and sensing. The system can track fish underwater in extreme conditions such as cold water, salt water, and strong currents. The tidal system can interpret and model fish behaviors and habits over time, allowing fish farmers to make better decisions about the feeding, welfare, and health of their cultured species.

This system, which consists of an underwater camera and environmental sensors, is most commonly used in waters with high transparency. The camera takes pictures, which are then analyzed using computer vision software. This enables farmers to better understand and monitor fish physiology, biology, and feeding habits. The cameras can rotate 360° in order to capture images of the underwater environment. Applied ML will then allow for a simple and quick image interpretation process in remote farms in Norway. The system, on the other hand, collects environmental data such as temperature and salinity to create patterns between fish behavior and health and their living environment.

The system aids in day-to-day farm decisions, such as how much feed is required, while reducing wastes that are normally undetectable with human senses. Furthermore, the system works best in waters with high transparency, making them ideal for species such as salmon.

2.3.5 Umitron

Umitron uses IoT, satellite remote sensing, and AI to create a user-friendly data platform for aquaculture. Farmers can use the technology to improve farm efficiency, manage environmental risks, and increase revenue. It employs computer models and IoT devices to optimize feeding while focusing on fish population.

The Fish Appetite Index (FAI) is a real-time ocean-based fish appetite detection system developed by UMITRON. To calculate fish appetite, advanced machine learning algorithms and analyzed video data are collected directly from farm sites.

Farmers can make data-driven decisions about feeding. The FAI algorithm processes the same visual information as humans, scores fish appetite, and displays it in an easy-to-understand chart. When used in conjunction with a smart feeder, such as the UMITRON CELL, the feed time intervals and amounts can be adjusted automatically with minimal human intervention. Farm owners can use FAI to fine-tune the feeding schedules, ensuring that stocks are always satisfied.

Umitron Pulse is an ocean data service from Umitron that provides a high-resolution map of critical environmental parameters such as water temperature, chlorophyll, dissolved oxygen, salinity, and wave height. UMITRON CELL® is a smart automated feeder for aquaculture. The device can be managed remotely through use of a smartphone or desktop computer via the Cloud. UMITRON CELL® was developed with the aim of solving major issues for aquaculture.

Umitron devices require installation and a large team to operate, while image processing in the background is only applicable to ocean waters with high transparency. On the other hand, the high cost of the devices limits their market to high-value species and with government support.

References

Adegboye MA, Aibinu AM, Kolo JG, Aliyu I, Folorunso TA, Lee S (2020) Incorporating intelligence in fish feeding system for dispensing feed based on fish feeding intensity. IEEE Access 8: 91948–91960

Ahmeda S, Aurpaa TT, Azada AK (2021) Fish disease detection using image-based machine learning technique in aquaculture. J King Saud Univ Comput Inf Sci:1–15

Al-Hussaini K, Zainol SM, Ahmed RB, Daud S (2018) IoT monitoring and automation data acquisition for recirculating aquaculture system using fog computing. J Comput Hardw Eng. https://doi.org/10.63019/jche.v1i2.610

Allken V, Handegard NO, Rosen S, Schreyeck T, Mahiout T, Malde K (2019) Fish species identification using a convolutional neural network trained on synthetic data. ICES J Mar Sci 76:342–349

Al-Marakeby A, Aly AA, Salem FA (2013) Fast quality inspection of food products using computer vision. Int J Adv Res Comput Commun Eng 1(2013):2

Altieri MA (1995) Agroecology: the science of sustainable agriculture. Westview Press, Boulder, Co

Antweiler W, Frank MZ (2004) Is all that talk just noise? The information content of Internet stock message boards. J Financ 59(3):1259–1294

Ayaz M, Ammad-Uddin M, Sharif Z, Mansour A, Aggoune E-HM (2019) Internet-of things (IoT)-based smart agriculture: toward making the fields talk. IEEE Access 7:129551–129583

Barrett S (1991) Optimal soil conservation and the reform of agricultural pricing policies. J Dev Econ 36:167–187

Barulin NV (2017) External sex specific signs in the structure of derivatives of sterlet (Acipenser ruthenus, Linnaeus, 1758) corium. Acta Biol Univ Daugavp 17:9–17

Barulin NV (2019) Using machine learning algorithms to analyse the scute structure and sex identification of sterlet Acipenser ruthenus (Acipenseridae). Aquac Res 50:2810–2825

Barzegar R, Aalami MT, Adamowski J (2020) Short-term water quality variable prediction using a hybrid CNN—LSTM deep learning model. Stoch Environ Res Risk Assess 34:415–433

Bashir S, Sharma N (2012) Remote area plant disease detection using image processing. IOSR J Electron Commun Eng 2(6):31–34

Bernheim D, Bjorkegren D, Naecker J, Rangel A (2013) Non-choice evaluations predict behavioral responses to changes in economic conditions. NBER Working Paper 19269

Bhange M, Hingoliwala H (2015) Smart farming: pomegranate disease detection using image processing. Proc Comput Sci 58:280–288

Bhatnagar A, Devi P (2013) Water quality guidelines for the management of pond fish culture. Int J Environ Sci 3(6):1980

Binswanger HP, Yang Y, Bowers A, Mundlak Y (1987) On the determinants of cross country aggregate agricultural supply. J Econometrics 36:111–131

Blumenstock JE (2016) Fighting poverty with data. Science 353(6301):753–754

Blumenstock JE, Cadamuro G, On R (2015) Predicting poverty and wealth from mobile phone metadata. Science 350(6264):1073–1076

Breivold HP (2017) A survey and analysis of reference architectures for the Internet-of-things. ICSEA 2017:143

Brooks RA (1981) Symbolic reasoning among 3-D models and 3-D images. Artif Intell 17(1–3): 285

Brooks Shehata E (2020) Harnessing Innovation in sustainable infrastructure and investment. T20 Saudi Arabia

Buchanan BG, Shortliffe EH (1984) Rule-based expert system. Addison-Wesley

Cai K, Miao X, Wang W, Pang H, Liu Y, Song JA (2020) Modified YOLOv3 model for fish detection based on MobileNetv1 as backbone. Aquac Eng 91:102117

Cao J, Xu L (2018) Research on counting algorithm of residual feeds in aquaculture based on machine vision. In: Proceedings of the 2018 IEEE 3rd international conference on image, vision and computing (ICIVC), Chongqing, China

Cao X, Liu Y, Wang J, Liu C, Duan Q (2020) Prediction of dissolved oxygen in pond culture water based on K-means clustering and gated recurrent unit neural network. Aquac Eng 91:102122

Cavalcante E, Alves MP, Batista T, Delicato FC, Pires PF (2015) An analysis of reference architectures for the Internet of things. In: International workshop on exploring component-based techniques for constructing reference architectures (CobRA)

Chakravorty H, Paul R, Das P (2015) Image processing technique to detect fish disease. Int J Comput Sci Secur 9:121–131

Chalmers D (1996) The conscious mind: in search of a fundamental theory. Oxford University Press

Chavan MS, Patil VP, Chavan S, Sana S, Shinde C (2018) Design and implementation of IoT based real time monitoring system for aquaculture using Raspberry Pi. Int J Recent Innov Trends Comput Commun 6:159–161

Chhabra HS, Srivastava AK, Nijhawan R (2020) A hybrid deep learning approach for automatic fish classification. In: Singh P, Panigrahi B, Suryadevara N, Sharma S, Singh A (eds) Proceedings of ICETIT 2019, Lecture notes in electrical engineering. Springer, Cham, Switzerland

Chrispin CL, Jothiswaran VV, Velumani T, Agnes Daney Angela S, Jayaraman R (2020) Application of artificial intelligence in fisheries and aquaculture. Biot Res Today 2:499–502

Costa C, Antonucci F, Pallottino F, Aguzzi J, Sun D-W, Menesatti P (2011) Shape analysis of agricultural products: a review of recent research advances and potential application to computer vision. Food Bioprocess Technol 4(5):673–692

Costa C, Antonucci F, Boglione C, Menesatti P, Vandeputtec M, Chatain B (2013) Automated sorting for size, sex and skeletal anomalies of cultured seabass using external shape analysis. Aquac Eng 52:58–64

Coz-Rakovac R, Topic Popovic N, Smuc T, Strunjak-Perovic I, Jadan M (2009) Classification accuracy of algorithms for blood chemistry data for three aquaculture-affected marine fish species. Fish Physiol Biochem 35:641–647

Cubero S, Aleixos N, Albert F, Torregrosa A, Ortiz C, Garciıa-Navarrete O, Blasco J (2014) Optimized computer vision system for automatic pre-grading of citrus fruit in the field using a mobile platform. Precis Agric 15(1):80–94

Dang LM, Hassan SI, Suhyeon I, Kumar Sangaiah A, Mehmood I, Rho S, Seo S, Moon H (2020) UAV based wilt detection system via convolutional neural networks. Sustain Computing Informatics Syst 28:100250

Darmalim U, Darmalim F, Darmalim S, Ahmad A, Budiarto A, Mahesworo B, Pardamean B (2019) IoT solution for intelligent pond monitoring. In: Proceedings of the 3rd international conference on eco engineering development, Solo, Indonesia

Darus F, Ayob MZ, Malaysia G, Hassan HA, Ridzuan IS, Razak MRA, Baharuddin H, Rahman AHA (2020) Potential of using Internet of things (IoT) for water quality monitoring in aquaculture: a case study in freshwater catfish culture in Rawang, Selangor, Malaysia. Test Eng Manag 83:2163–2169

Daud AKPM, Sulaiman NA, Yusof YWM, Kassim M (2020) An IoT-based smart aquarium monitoring system. In: Proceedings of the 2020 IEEE 10th symposium on computer applications & industrial electronics (ISCAIE), Penang, Malaysia

Deep B, Dash R (2019) Underwater fish species recognition using deep learning techniques. In: Proceedings of the 2019 6th international conference on signal processing and integrated networks (SPIN), Noida, India

Divinely SJ, Sivakami K, Jayaraj V (2019) Fish diseases identification and classification using Machine Learning. Int J Adv Res Basic Eng Sci Technol 5:46–51

Do D, Pham F, Raheja A, Bhandari S (2018) Machine learning techniques for the assessment of citrus plant health using UAV-based digital images. In: Autonomous air and ground sensing systems for agricultural optimization and phenotyping III, vol 10664. International Society for Optics and Photonics, p 1066400

Donaldson D, Storeygard A (2016) The view from above: applications of satellite data in economics. J Econ Perspect 30(4):171–198

Dorigo M, Gambardella LM (1997) Ant colonies for the traveling salesperson problem. Biosystems 43:73

Dzulqornain MI, Rasyid MUHA, Sukaridhoto S (2017) Design and development of smart aquaculture system based on IFTTT Model and Cloud Integration. In: Proceedings of the 3rd international conference on electrical systems, technology and information (ICESTI 2017), Bali, Indonesia, 26–29 September 2017, vol 164

Edelman GM (2006) Second nature: brain science and human knowledge. Yale University Press

Edelman GM (2007) Learning in and from brain-based devices. Science 318(5853):1103

Erickson BJ, Korfiatis P, Akkus Z, Kline TL (2017) Machine learning for medical imaging. Radiographics 37:505–515

Faical BS, Pessin G, Geraldo Filho P, Carvalho AC, Furquim G, Ueyama J (2014) Finetuning of UAV control rules for spraying pesticides on crop fields. In: 2014 IEEE 26th international conference on tools with artificial intelligence. IEEE, pp 527–533

Feigenbaum JJ (2015) Intergenerational mobility during the great depression. http://scholar.harvard.edu/jfeigenbaum/publications/jmp

Ferentinos KP (2018) Deep learning models for plant disease detection and diagnosis. Comput Electron Agric 145:311–318

Fernandes AFA, Turra EM, de Alvarenga ER, Passafaro TL, Lopes FB, Alves GFO, Singh V, Rosa GJM (2020) Deep learning image segmentation for extraction of fish body measurements and prediction of body weight and carcass traits in Nile tilapia. Comput Electron Agric 170:105274

Fijani E, Barzegar R, Deo R, Tziritis E, Skordas K (2019) Design and implementation of a hybrid model based on two-layer decomposition method coupled with extreme learning machines to support real-time environmental monitoring of water quality parameters. Sci Total Environ 648: 839–853

Fogel DB (1994) Evolutionary programming in perspective: the top-down view. In: Zurada JM, Marks RJ II, Robinson CJ (eds) Computational intelligence: imitating life. IEEE Press

Food and Agricultural Organization (FAO) (2009) Declaration of the world summit on food security. FAO, Rome, Italy

França Albuquerque PL, Garcia V, Da Silva Oliveira A, Lewandowski T, Detweiler C, Gonçalves AB, Costa CS, Naka MH, Pistori H (2019) Automatic live fingerlings counting using computer vision. Comput Electron Agric 167:105015

Fraser AS (1958) Monte Carlo analyses of genetic models. Nature 181:208

Garcia R, Prados R, Quintana J, Tempelaar A, Gracias N, Rosen S, Vågstøl H, Løvall K (2019) Automatic segmentation of fish using deep learning with application to fish size measurement. ICES J Mar Sci 77:1354–1366

Gelernter H (1959) Realization of a geometry theorem proving machine. In: Proceedings of international conference on information processing. UNESCO

Gerami MH, Safiyari H, Jafari A, Mousavi-Nasab M (2016) Application of Machine-vision to assess weight of fish (Case study: Oncorhynchus mykiss). Iran J Fish Sci 15:575–584

Glaeser EL, Kominers SD, Luca M, Naik N (2016) Big data and big cities: the promises and limitations of improved measures of urban life. Economic Inquiry

Gondchawar N, Kawitkar R et al (2016) IoT-based smart agriculture. Int J Adv Res Comput Commun Eng 5(6):838–842

Gong Y, Duan B, Fang S, Zhu R, Wu X, Ma Y, Peng Y (2018) Remote estimation of rapeseed yield with unmanned aerial vehicle (UAV) imaging and spectral mixture analysis. Plant Methods 14(1):70

Goodfellow I, Bengio Y, Courville A (2016) Deep learning. MIT Press

Gourévitch A, Portincaso M, Legris A, de la Tour A, Hammoud T, Salzgeber T (2021) Challenges of investing into deep tech. https://www.bcg.com/publications/2021/overcoming-challenges-investing-in-digital-technology. The Boston Consulting Group (BCG). May 10th, 2021

Haller S, Karnouskos S, Schroth C (2009) The Internet of things in an enterprise context. In: Future Internet symposium. Springer, Berlin, pp 14–28

Hamid MS, Wahab MAA, Abdullah R, Gani SFBA, Hamzah RA (2019) Development of water quality for smart aquaculture system. J Eng Appl Sci 14:2840–2847

Han F, Zhu J, Liu B, Zhang B, Xie F (2020) Fish schools' behavior detection based on convolutional neural network and spatiotemporal information. IEEE Access 8:126907–126926

Harish B, Manigandan K, Harishankar N, Ram T (2018) Smart aquaculture fish feeding and water quality monitoring. In: Proceedings of the conference on electronics, information and communication Systems (CELICS'18), Tamil Nadu, India

Hazell P, Fan S (2000) Balancing regional development priorities to achieve sustainable and equitable agricultural growth. In: Lee DR, Barrett CB (eds) Critical tradeoffs: agricultural intensification, economic development and the environment in developing countries. CAB International, London

Hazell P, Lutz E (1998) Integrating environmental and sustainability concerns into rural development policies. In: Lutz E (ed) Agriculture and the environment: perspectives on sustainable rural development. World Bank, Washington, DC

Hebb D (1949) The organization of behavior. Wiley

Henderson JV, Storeygard A, Weil DN (2012) Measuring economic growth from outer space. Am Econ Rev 102(2):994–1028

Hsu WC, Chao PY, Wang CS, Hsieh JC, Huang W (2020) Application of regression analysis to achieve a smart monitoring system for aquaculture. Information 11:387

Hu W, Wu H, Zhang Y, Zhang S, Lo C (2020) Shrimp recognition using ShrimpNet based on convolutional neural network. J Ambient Intell Humaniz Comput:1–8

Huang H, Deng J, Lan Y, Yang A, Zhang L, Wen S, Zhang H, Zhang Y, Deng Y (2019) Detection of helminthosporium leaf blotch disease based on UAV imagery. Appl Sci 9:558

Hung C, Xu Z, Sukkarieh S (2014) Feature learning-based approach for weed classification using high resolution aerial images from a digital camera mounted on a UAV. Remote Sens 6(12): 12037–12054

Imai T, Arai K, Kobayashi T (2019) Smart aquaculture system: a remote feeding system with smartphones. In: Proceedings of the 2019 IEEE 23rd international symposium on consumer technologies (ISCT), Ancona, Italy, 19–21 June 2019, pp 93–96

Iorga, M., Feldman, L., Barton, R., Martin, M., Goren, N., and Mahmoudi, C., (2017). The NIST definition of fog computing. NIST special publication (SP) 800-191 (draft). National Institute of Standards and Technology

Islam M, Dinh A, Wahid K, Bhowmik P (2017) Detection of potato diseases using image segmentation and multiclass support vector machine. In: 2017 IEEE 30th Canadian conference on electrical and computer engineering (CCECE). IEEE, pp 1–4

Jalal A, Salman A, Mian A, Shortis M, Shafait F (2020) Fish detection and species classification in underwater environments using deep learning with temporal information. Ecol Inform 57: 101088

Jean N, Burke M, Xie M, Davis WM, Lobell DB, Ermon S (2016) Combining satellite imagery and machine learning to predict poverty. Science 353(6301):790–794

Jeong SJ, Yang YS, Lee KH, Kang JG, Lee DG (2018) Vision-based automatic system for non-contact measurement of morphometric characteristics of flatfish. J Electr Eng Technol 8: 1194–1201

Jia B, Zhang M (2020) Multi-dimensional classification via kNN feature augmentation. Pattern Recogn 106:107423

Jordan M, Mitchell T (2015) Machine learning: trends, perspectives, and prospects. Science 349: 255–260

Kamilaris A, Prenafeta-Boldu FX (2018) Deep learning in agriculture: a survey. Comput Electron Agric 147:70–90

Kang JS, Kuznetsova P, Luca M, Choi Y (2013) Where not to eat? Improving public policy by predicting hygiene inspections using online reviews. In: EMNLP 2013: 2013 conference on empirical methods in natural language

Karygiannis TT, Eydt B, Barber G, Bunn L, Phillips T (2007) Guidelines for securing radio frequency identification (RFID) systems: recommendations of the National Institute of Standards and Technology. NIST Special Publication. NIST, pp 800–898

Kassem T, Shahrour I, El Khattabi J, Raslan A (2021) Smart and sustainable aquaculture farms. Sustainability 13:1–16

Kennedy J, Eberhart RC (2001) Swarm intelligence. Morgan Kaufmann

Keshtegar B, Heddam S (2018) Modeling daily dissolved oxygen concentration using modified response surface method and artificial neural network: a comparative study. Neural Comput Appl 30:2995–3006

Khatri-Chhetri A, Aggarwal PK, Joshi PK, Vyas S (2017) Farmers' prioritization of climate smart agriculture (CSA) technologies. Agric Syst 151:184–191

Kim BJ, Shin KJ (2018) Development of smart fish farming system based on IoT using wasted warm water energy. J Inst Control Robot Syst 24:155–163

Kim S, Alizamir M, Zounemat-Kermani M, Kisi O, Singh VP (2020) Assessing the biochemical oxygen demand using neural networks and ensemble tree approaches in South Korea. J Environ Manag 270:110834

Kogan S, Levin D, Routledge BR, Sagi JS, Smith NA (2009) Predicting risk from financial reports with regression. In: Proceedings of human language technologies: the 2009 annual conference of the North American chapter of the Association for Computational Linguistics. ACM, pp 272–280

Kotsiantis SB (2007) Supervised machine learning: a review of classification techniques. Informatica 31:249–268

Koza J (1992) Genetic programming. MIT Press

Krishna PC, Rao MV, Reddy VR (2019) An IoT based fish health monitoring system in aquaculture farming. Int J Innov Technol Explor Eng (IJITEE) 8:1578–1584

Kumar A, Gill G et al (2015) Automatic fruit grading and classification system using computer vision: a review. In: 2015 Second international conference on advances in computing and communication engineering. IEEE, pp 598–603

Lafferty KD, Harvell CD, Conrad JM, Friedman CS, Kuris AM, Powell EN, Rondeau D, Saksida SM (2015) Infectious diseases affect marine fisheries and aquaculture economics. Annu Rev Mar Sci 7(1):471–496

Le Cren ED (1951) The length-weight relationships and seasonal cycle in gonad weight and condition in the perch (Perca fluviatilis). J Anim Ecol 20:201–219

Le J, Xu L (2017) An automated fish counting algorithm in aquaculture based on image processing. In: Proceedings of the 2016 international forum on mechanical, control and automation (IFMCA 2016). Atlantis Press, Paris, France

LeCun Y, Bengio Y, Hinton G (2015) Deep learning. Nature 521:436–444

Lee S, Choi Y (2016) Reviews of unmanned aerial vehicle (drone) technology trends and its applications in the mining industry. Geosyst Eng 19(4):197–204

Li P, Hua P, Gui D, Niu J, Pei P, Zhang J, Krebs P (2020) A comparative analysis of artificial neural networks and wavelet hybrid approaches to long-term toxic heavy metal prediction. Sci Rep 10: 13439

Li D, Miao Z, Peng F, Wang L, Hao Y, Wang Z, Chen T, Li H, Zheng Y (2021) Automatic counting methods in aquaculture: a review. J World Aquac Soc 52:269–283

Lindsay RK, Buchanan BG, Feigenbaum EA, Lederberg J (1980) Applications of artificial intelligence for organic chemistry: the Dendral Project. McGraw-Hill

Linker R (2017) A procedure for estimating the number of green mature apples in nighttime orchard images using light distribution and its application to yield estimation. Precis Agric 18(1):59–75

Liu L, Lu H, Cao Z, Xiao Y (2018) Counting fish in sonar images. In: Proceedings of the 2018 25th IEEE international conference on image processing (ICIP), Athens, Greece

Lobell DB (2013) The use of satellite data for crop yield gap analysis. Field Crop Res 143:56–64

Low ARC (1993) The low-input, sustainable agriculture (LISA) prescription: a bitter pill for farm households in southern Africa. Project Appraisal 8(2):97–101

Lu B, He Y (2017) Species classification using unmanned aerial vehicle (UAV)-acquired high spatial resolution imagery in a heterogeneous grassland. ISPRS J Photogramm Remote Sens 128:73–85

Lu H, Ma X (2020) Hybrid decision tree-based machine learning models for short-term water quality prediction. Chemosphere 249:126–169

Lynn T, Healy P, Kilroy S, Hunt G, van der Werff L, Venkatagiri S, Morrison J (2015) Towards a general research framework for social media research using big data. In: In 2015 IEEE international professional communication conference (IPCC). IEEE, pp 1–8

Ma Y (2012) Ensemble machine learning: methods and applications. Springer Science & Business Media, New York

Maddikunta PKR, Hakak S, Alazab M, Bhattacharya S, Gadekallu TR, Khan WZ, Pham Q-V (2007) Unmanned aerial vehicles in smart agriculture: applications, requirements, and challenges. arXiv Preprint. arXiv:2007.12874

Malik S, Kumar T, Sahoo AK (2017) A novel approach to fish disease diagnostic system based on machine learning. Adv Image Video Process 5:49–57

McCarthy J (1958) Programs with common sense. Symposium on mechanization of thought processes. National Physical Laboratory, Teddington, England

McCorduck P (2004) Machines who think. A. K. Peters, Ltd

McCulloch W, Pitts W (1943) A logical calculus of the ideas immanent in nervous activity. Bull Math Biophys 5:115

McDermott J (1982) A rule-based configure of computer systems. Artif Intell 19(1):39–88

Mell P, Grance T (2011) The NIST definition of cloud computing. NIST Special Publication, 800 (145), p 7

Moen E, Handegard NO, Allken V, Albert OT, Harbitz A, Malde K (2018) Automatic interpretation of otoliths using deep learning. PLoS One 13:e204713

Monirul I, Mohammod AK, Farzana IJ (2019) Aqua fishing monitoring system using IoT devices. Int J Innov Sci Eng Technol 6:108–114

Monkman GG, Hyder K, Kaiser MJ, Vidal FP (2019) Using machine vision to estimate fish length from images using regional convolutional neural networks. Methods Ecol Evol 10:2045–2056

Moore MM, Slonimsky E, Long AD, Sze RW, Iyer RS (2019) Machine learning concepts, concerns, and opportunities for a pediatric radiologist. Pediatr Radiol 49:509–516

Morimoto T, Zin TT, Itami T (2018) A study on abnormal behavior detection of infected shrimp. In: Proceedings of the 2018 IEEE 7th global conference on consumer electronics (GCCE), Nara, Japan

Mustafa M, Zaidi MZ, Shafry MMR, Ismail MA, Norhaida A (2013) FLUDI: using digital images for measuring fish length. Galaxea J Coral Reef Stud 15:101–106

Neapolitan RE (1989) Probabilistic reasoning in expert systems. Wiley

Newell A, Simon H (1961) GPS, a program that simulates human thought. In: Building H (ed) Lerenede Automaten. R. Oldenbourg

Nguyen TN, Tran TH, Nguyen CN (2020) A forecasting model for monitoring water quality in aquaculture and fisheries IoT systems. In: Proceedings of the 2020 international conference on advanced computing and applications (ACOMP), Quy Nhon, Vietnam

Nizetic S, Solic P, Gonzalez-de DL-D-I, Patrono L et al (2020) Internet of things (IoT): opportunities, issues and challenges towards a smart and sustainable future. J Clean Prod 274: 122877

Nocheski S, Naumoski A (2018) Water monitoring IoT system for fish farming ponds. Industry 4.0 3:77–79

Ordonez A, Eikvil L, Salberg A, Harbitz A, Murray SM, Kampffmeyer MC (2020) Explaining decisions of deep neural networks used for fish age prediction. PLoS One 15:e235013

Pearl J (1986) Fusion, propagation, and structuring in belief networks. Artif Intell 29

Pearl J (1988) Probabilistic reasoning in intelligent systems. Morgan Kaufmann

Peña MAL, Fernández IM (2019) SAT-IoT: an architectural model for a high-performance fog/edge/cloud IoT platform. In: IEEE world forum on Internet of things (WF-IoT). IEEE, pp 633–638

Peña JM, Gutierrez PA, Hervas-Martinez C, Six J, Plant RE, Lopez Granados F (2014) Object-based image classification of summer crops with machine learning methods. Remote Sens 6(6): 5019–5041

Petrellis N (2021) Measurement of fish morphological features through image processing and deep learning techniques. Appl Sci 11:1–23

Prabhu T (2019) Automated enhancement of aquaculture species growth by observing the water quality using IoT. Int Res J Multidiscip Technovation 1:156–164

Pramunendar RA, Wibirama S, Santosa PI (2019) Fish classification based on underwater image interpolation and back-propagation neural network. In: Proceedings of the 2019 5th international conference on science and technology (ICST), Yogyakarta, Indonesia

Priya CA, Balasaravanan T, Thanamani AS (2012) An efficient leaf recognition algorithm for plant classification using support vector machine. In: International conference on pattern recognition, informatics and medical engineering (PRIME-2012), pp 428–432

Rahman A, Akter S, Khan MM, Rhaman K (2019) Relation between aquaculture with fish disease and health management: a review note. Bangladesh J Fish 31:253–260

Rahman M, Bapery C, Hossain MJ, Hassan Z, Hossain GMJ, Islam M (2020) Internet of Things (IoT) based water quality monitoring system. Int J Multidisc Curr Educ Res 2(4):168–180

Raman V, Perumal S, Navaratnam S, Fazilah S (2016) Computer assisted counter system for larvae and juvenile fish in Malaysian fishing hatcheries by machine learning approach. J Comput 11: 423–431

Rashid M, Nayan AA, Simi SA, Saha J, Rahman O, Kibria MG (2021) IoT based smart water quality prediction for biofloc aquaculture. Int J Adv Comput Sci Applic 12(6):56–62

Rauf HT, Lali MIU, Zahoor S, Shah SZH, Rehman AU, Bukhari SAC (2019) Visual features based automated identification of fish species using deep convolutional neural networks. Comput Electron Agric 167:105075

Reardon, Echeverria T, Berdegué B, Minten J, Liverpool-Tasie S, Tschirley D, Zilberman D (2019) Rapid transformation of food systems in developing regions: highlighting the role of agricultural research & innovations. Agric Syst 172:47–59

Ren Q, Wang X, Li W, Wei Y, An D (2020) Research of dissolved oxygen prediction in recirculating aquaculture systems based on deep belief network. Aquac Eng 90:102085

Rum SNM, Nawawi FAZ (2021) FishDeTec: a fish identification application using image recognition approach. Int J Adv Comput Sci Applic 12(3):102–106

Sanchez-Torres G, Ceballos-Arroyo A, Robles-Serrano S (2018) Automatic measurement of fish weight and size by processing underwater hatchery images. Eng Lett 26:461–472

Sandino J, Gonzalez F, Mengersen K, Gaston KJ (2018) UAVs and machine learning revolutionizing invasive grass and vegetation surveys in remote arid lands. Sensors 18(2):605

Searle JR (1980) Mind, brains, and programs. Behav Brain Sci 3:417

Sharma D, Kumar R (2021) Smart aquaculture: integration of sensors, biosensors, and artificial intelligence. In: Pudake RN, Jain U, Kole C (eds) Biosensors in agriculture: recent trends and future perspectives. Concepts and strategies in plant sciences. Springer, Cham, Switzerland

Shin D (2014) A socio-technical framework for Internet-of-things design: a human-centered design for the Internet of things. Telematics Inform 31(4):519–531

Shubhaker B, Prasad D, Manikanta S, Ba PA (2020) Automated fish farm aquaculture monitoring system using IoT. J Eng Sci 11:1170–1172

Siddiqui SA, Salman A, Malik MI, Shafait F, Mian A, Shortis MR, Harvey ES (2018) Automatic fish species classification in underwater videos: exploiting pre-trained deep neural network models to compensate for limited labelled data. ICES J Mar Sci 75(1):374–389

Sivabalan KN, Anandkumar V, Balakrishnan S (2020) IoT based smart farming for effective utilization of water and energy. Int J Adv Sci Technol 2020(29):2496–2500

Stankovic JA (2008) When sensor and actuator networks cover the world. ETRI J 30(5):627–633

Stokes DE (1997) Pasteur's quadrant – basic science and technological innovation. Brookings Institution Press

Sung HJ, Park MK, Choi JW (2020) Automatic grader for flatfishes using machine vision. Int J Control Autom Syst 18:3073–3082

Szolovits P, Pauker SG (1978) Categorical and probabilistic reasoning in medical diagnosis. Artif Intell 11:115

Ta X, Wei Y (2018) Research on a dissolved oxygen prediction method for recirculating aquaculture systems based on a convolution neural network. Comput Electron Agric 145:302–310

Tang Q, Qiu W, Zhou Y (2020) Classification of complex power quality disturbances using optimized S-transform and kernel SVM. IEEE Trans Industr Electron 67:9715–9723

Tharwat A, Hemedan AA, Hassanien AE, Gabel TA (2018) Biometric-based model for fish species classification. Fish Res 204:324–336

Tsai S-Y, Sou SI, Tsai M-H (2012) Effect of data aggregation in M2M networks. In: The 15th international symposium on wireless personal multimedia communications. IEEE, pp 95–99

Tsouros DC, Bibi S, Sarigiannidis PG (2019) A review on UAV-based applications for precision agriculture. Information 10(11):349

Turing A (1950) Computing machinery and intelligence. Mind 59:433

Vermesan O, Friess P, Guillemin P, Sundmaeker H et al (2013) Internet of things strategic research and innovation agenda. River Publishers

Villon S, Mouillot D, Chaumont M, Darling ES, Subsol G, Claverie T, Villéger SA (2018) Deep learning method for accurate and fast identification of coral reef fishes in underwater images. Ecol Inform 48:238–244

Wan L, Li Y, Cen H, Zhu J, Yin W, Wu W, Zhu H, Sun D, Zhou W, He Y (2018) Combining UAV-based vegetation indices and image classification to estimate flower number in oilseed rape. Remote Sens 10(9):1484

Wang Y, Leblanc D (2016) Integrating SaaS and SaaP with Dew Computing, 2016 IEEE International Conferences on Big Data and Cloud Computing (BDCloud), Social Computing and Networking (SocialCom), Sustainable Computing and Communications (SustainCom) (BDCloud-SocialCom-SustainCom), Atlanta, GA, USA, pp 590–594. https://doi.org/10.1109/BDCloud-SocialCom-SustainCom.2016.92

Wang Q, Nuske S, Bergerman M, Singh S (2013) Automated crop yield estimation for apple orchards. In: Experimental robotics. Springer, pp 745–758

Wang G, Lan Y, Qi H, Chen P, Hewitt A, Han Y (2019) Field evaluation of an unmanned aerial vehicle (UAV) sprayer: effect of spray volume on deposition and the control of pests and disease in wheat. Pest Manage Sci 75(6):1546–1555

Wang C, Li Z, Wang T, Xu X, Zhang X, Li D (2021) Intelligent fish farm - the future of aquaculture. Aquacult Int 29:2681–2711

Webb MAH, Van Eenennaam JP, Crossman JA, Chapman FA (2019) A practical guide for assigning sex and stage of maturity in sturgeons and paddlefish. J Appl Ichthyol 35:169–186

White DJ, Svellingen C, Strachan NJC (2006) Automated measurement of species and length of fish by computer vision. Fish Res 80:203–210

Winograd T (1972) Understanding natural language. Cogn Psychol 3(1):1

Winston P (1973) Progress in vision and robotics. M.I.T. Artificial Intelligence TR-281
Xu W, Matzner S (2018) Underwater fish detection using deep learning for waterpower applications. In: Proceedings of the 2018 international conference on computational science and computational intelligence (CSCI), Las Vegas, NV, USA
Xu F, Pan Z, Xia R (2020) E-commerce product review sentiment classification based on a naïve Bayes continuous learning framework. Inf Process Manag 57:102221
Yang X, Zhang S, Liu J, Gao Q, Dong S, Zhou C (2020a) Deep learning for smart fish farming: applications, opportunities, and challenges. Rev Aquac 13(1):66–90
Yang L, Liu Y, Yu H, Fang X, Song L, Li D, Chen Y (2020b) Computer vision models in intelligent aquaculture with emphasis on fish detection and behavior analysis: a review. Arch Comput Methods Eng 28:2785–2816
Zermas D, Teng D, Stanitsas P, Bazakos M, Kaiser D, Morellas V et al (2015) Automation solutions for the evaluation of plant health in corn fields, 2015 IEEE/RSJ International Conference on Intelligent Robots and Systems (IROS), pp 6521–6527
Zhakov A, Zhu H, Siegel A, Rank S, Schmidt T, Fienhold L, Hummel S (2020) Application of ANN for fault detection in overhead transport systems for semiconductor fab. IEEE Trans Semicond Manuf 33:337–345
Zhang L, Wang J, Duan Q (2020a) Estimation for fish mass using image analysis and neural network. Comput Electron Agric 173:105439
Zhang S, Yang X, Wang Y, Zhao Z, Liu J, Liu Y, Sun C, Zhou C (2020b) Automatic fish population counting by machine vision and a hybrid deep neural network model. Animals 10:1–17
Zhang Y, Fitch P, Thorburn PJ (2020c) Predicting the trend of dissolved oxygen based on the kPCA-RNN model. Water 12:585
Zhao S (2003) Toward a taxonomy of copresence. Presence Teleoperators Virtual Environ 12(5): 445–455
Zhao J, Bao W, Zhang F, Zhu S, Liu Y, Lu H, Shen M, Ye Z (2018) Modified motion influence map and recurrent neural network-based monitoring of the local unusual behaviors for fish school in intensive aquaculture. Aquaculture 493:165–175
Zhou C, Sun C, Lin K, Xu D, Guo Q, Chen L, Yang X (2018) Handling water reflections for computer vision in aquaculture. Trans ASABE 61:469–479
Zhydic O (2021) What is Deep Tech?. https://www.techworks.org.uk/about/what-is-deep-tech, Web. May 10th, 2021

Further Reading

Ali SA (2019) Nutritional feeding of fish and shrimps in India. MJP Publisher, India
De Corte W, Sackett PR, Lievens F (2020) Robustness, sensitivity, and sampling variability of Pareto-optimal selection system solutions to address the quality-diversity trade-off. Organ Res Methods 23(3):535–568
Holt GJ (2011) Larval fish nutrition. Wiley, Hoboken, NJ
The Digital Aquaculture Revolution, Technology. 2021

3 Poseidon-AI, Where Aquatic Intelligence Meets Artificial Intelligence

The impact of climate change on the one hand and global population growth on the other highlight the need for innovative technologies to ensure the world's future. Scarcity of resources, rising protein demand, and depleted ocean stocks all add to the value of the aquaculture sector. As a result, there are numerous companies that address the issues that this valuable sector faces. From disease control to feed optimization, innovative deep tech companies can be found in the countries with the highest aquaculture production in Asia, Europe, and Latin America.

However, as mentioned in the examples in this chapter, most of these technologies are expensive, difficult to install, and/or difficult to scale to other countries around the world. This is primarily due to a lack of direct tech developers and founders, as well as the reality that exists around the world; for example, feed waste is one of the most significant challenges faced in the aquaculture sector. There are numerous types of feed for different species and life stages of species. As a result, developing tailor-made algorithms for feed optimization necessitates multidisciplinary knowledge, and sole expertise in AI/ML cannot provide long-term and cost-effective solutions.

Furthermore, there is a large knowledge gap between what is happening on the ground and academic training, leaving graduates far behind in understanding global needs and a rapidly changing world. Additionally, different cultures, languages, social norms, and conceptions make it more difficult to adapt to new technologies.

Under these conditions, it is critical to present multidisciplinary solutions that are affordable, adaptable, and scalable faster than global changes so that it can deal with the effects of climate change and rapid population growth while contributing to global sustainability.

Poseidon-AI is a Singapore-based company that was founded in 2019 with the goal of using AI/ML and IoT to create affordable, adaptable, and scalable solutions to aid in the sustainability of the aquaculture sector and the world. Poseidon-AI's solutions include an IoT device, image-trained smart algorithms, and IAS.

© The Author(s), under exclusive license to Springer Nature Singapore Pte Ltd. 2023

A. Rahimi-Midani, *Deep Technology for Sustainable Fisheries and Aquaculture*,
https://doi.org/10.1007/978-981-99-4917-5_3

Feeding accounts for 80% of the cost of any aquaculture farm. This is due to the fact that fish feed contains three elements (fish meal, fish oil, and a crop), two of which are primarily sourced from the world's oceans and waters (wild capture). Because the production of these two components in the laboratory raises the cost of the feeds, the industry remains heavily reliant on wild captured fish for aquafeed production. Under these conditions, Poseidon-AI optimizes feeding while saving farmers up to 20% of their feeding costs by combining affordable hardware and intelligent software.

As previously stated, South-East Asia and China produce a large portion of aquaculture production. However, global warming is having a greater impact on this region than others, and because the cultured species are poikilothermic, global warming will have a significant impact on their energy consumption, resulting in higher feed consumption but lower yield for farmers. As a result, Poseidon-AI employs an IoT device to monitor water quality and employs evolutionary algorithms to present to farmers the optimal feeding time, amount of feeding and FCR, growth rate, and potential maturity time of the species. The image processing is carried out using images sent by the IoT device to the Amazon Cloud System (AWS) in order to calibrate the algorithms based on the real-time condition of each pond, tank, or cage, and with this combination, algorithms are trained for different cultured species while farmers save 20% on farm feeding costs; a win-win situation for everyone.

The trained algorithms for various species will then be used in Poseidon-AI® IAS, which will use rainwater, solar energy, and no soil. The Poseidon-AI® IAS is intended for communities and families in landlocked areas, with the goal of generating revenue for vulnerable families while contributing to the sustainability of these areas. The systems are inexpensive, efficient, and simple to use, requiring only a basic understanding of aquaculture.

3.1 Affordable IoT Device and Smart Algorithms

The Poseidon-AI® IoT device (Fig. 3.1) was designed for extreme conditions and outdoor use. It has up to seven sensors and three cameras to monitor the water condition up to 3 m deep and the surface condition for floating contaminants like excess feed and plastics.

The sensors in the Poseidon-AI® IoT device are industrial grade water monitoring sensors that can monitor simple conditions like air and water temperature as well as more complex ones like ORP. These sensors include air and water temperature, ammonia, conductivity, D.O., nitrate, ORP, and pH and can be installed on the IoT device at the request of the farmer.

The device uses a central processing unit for communication between the cameras, sensors, and AWS. The device uses 4G/5G LTE to connect to Internet and holds a SD card for saving the information in case of interruption in Internet connection. The advantages of the device compared to other competitors are easy shipping and deployment to anywhere in the world; no need for installation; carbon

Fig. 3.1 Poseidon-AI IoT device

fiber which makes the device firm to survive the harsh environment but light enough to ship to anywhere in the world; cameras are used for both security (smart alert) and image processing; it uses solar energy to save energy in a small possible lithium battery that can be shipped via DHL; and the Poseidon-AI® smart software can be used both Android and IOS devices.

To clarify the condition of the pond, tank, or cage, the cameras begin taking images an hour before feeding, during feeding, and an hour after feeding. The images are processed an hour before and after to detect dead fish, potential external pollutants such as branches and leaves that fall into the water, and leftover feed from previously scheduled feedings. The overall architecture of the Poseidon-AI system is depicted in Fig. 3.2.

The overall structure of the device is as follows: The IoT device collects the environmental information from the ponds, tanks, and cages using up to seven sensors and demonstrates the information live on a user dashboard. The user dashboard can be installed on IOS, Android, or Windows devices. The farms are medium/large size and located in remote areas, thus, IOS or Android devices are more practical to be used by the farmers. Additionally, there are no water treatment facilities near these farms, hence, farmers use the water from the nearby water bodies such as rivers, deltas, and lagoons. The water in these waterbodies goes through farms, factories, towns, and villages, and with no water treatment facilities, waters are filled with wastes that might have negative impacts on the cultured species. Live water monitoring in farms will allow farmers to be informed about the possible

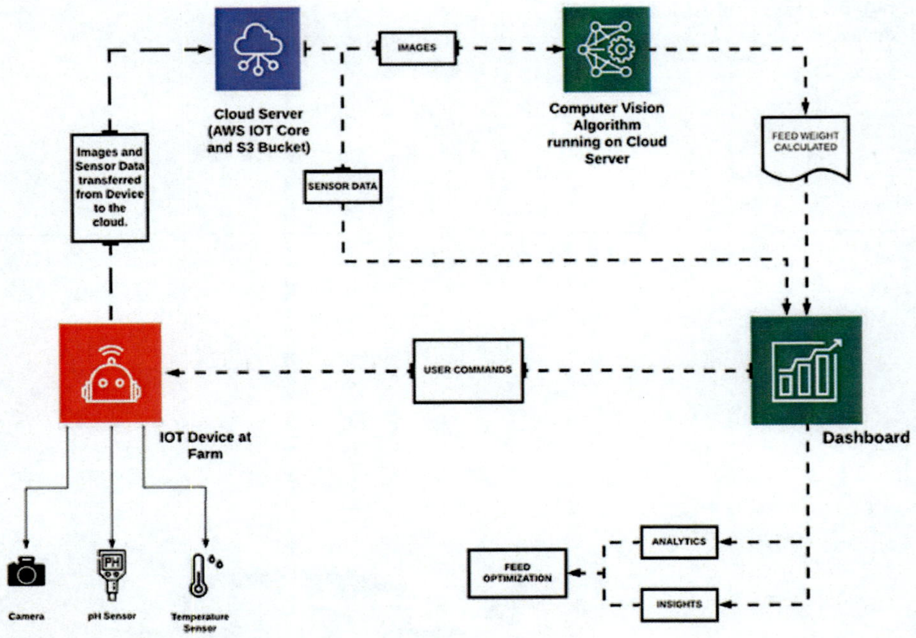

Fig. 3.2 Overall architecture of Poseidon-AI system

pollution before stocking and prevent high mortality in their farms. On the other hand, the sudden change in environmental condition will directly impact the appetite of the species, meaning that cultured species will not consume feed at the same time as always, or even the same amount. For this reason, by looking at the dashboard, farmers can see the best daily schedule for feeding the species on sunny, windy, or rainy days. However, how much feed must be given and how the farmers can trust the result?

As previously stated, the Poseidon-AI® IoT employs three solar panels to charge the smallest lithium battery possible, which can be transported via airplanes or postal services. The device's structure is designed so that the sun can hit the solar panels from all angles and charge the battery at the same time using two or all of the solar panels. The Poseidon-AI® device is built with precise calculations to ensure that small energy consumption is taken into account, and each device can survive in a pond, tank, or cage without the need for human intervention. The cameras start taking pictures of the pond's state an hour before, during, and after the feeding operation. The images are sent to the AWS Cloud for image processing to analyze the condition of each area. The images will be processed so that the amount of leftover feeds, number of death fish, plastic bottles, and floating branches and leaves can be calculated. This aids in calibrating the evolutionary algorithms based on the amount of feed provided as well as training the algorithms based on real-world conditions.

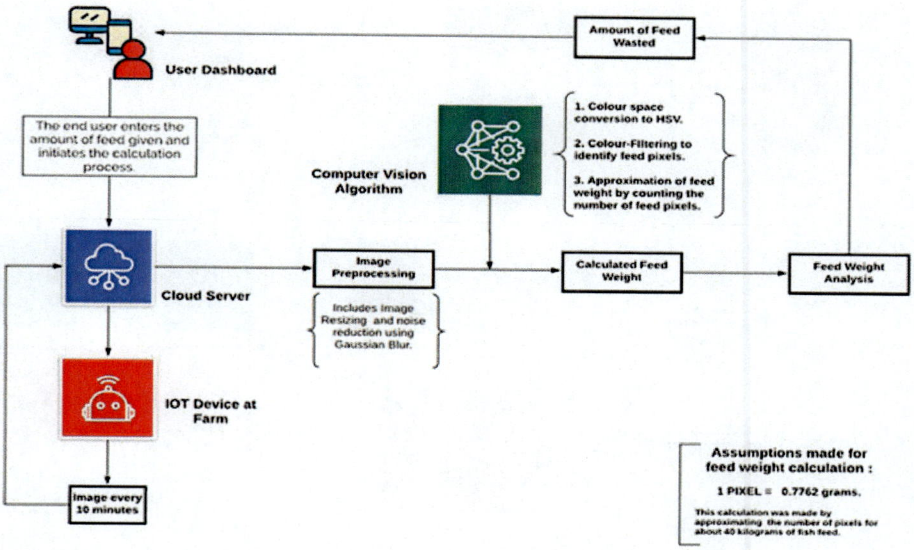

Fig. 3.3 Poseidon-AI's feed wastage calculation procedure

The images are analyzed by segmentation of water and feeds. These images can be segmented with the help of ML, and the amount of feed wastage can be converted in equivalent amount (Kilograms) by the covered area and the specific color gradient after segmentation. Figure 3.3 shows Poseidon-AI's feed wastage calculation procedure.

Poseidon-AI's Weight Calculation Algorithms (WCAs) for feed waste will begin to work after the first images are transferred to the AWS Cloud. The image quality is improved so that the most information can be extracted from each image. The images are then sent to different channels via HSV conversion, such as the HUE channel, saturation channel, and value channel, so that specific color gradients and filtering can be applied to the feed and the surrounding environment. The next step is to count the feed pixels and convert them into kilograms of wasted feed (Fig. 3.4). Because this process will be repeated at least three times per day and up to six times per day (depending on the number of feeding times), one might expect a large amount of data to be used each time. However, because of ML, even low-quality images can be sent to the Clouds and accurately be analyzed by these algorithms. In other words, machine learning will allow for less data usage with greater accuracy, allowing farmers to not only save money on data but also reduce waste.

The CPU in the Poseidon-AI® IoT device manages a variety of tasks. The first task is to allow sensors data from a pond, tank, or cage to be collected in a timely manner; the second task is to manage and optimize energy usage based on the needs of the device; and finally, the CPU holds a 4G/5G connection LTE Sim card to communicate with the AWS Cloud and pump up the information. In the event that the 4G/5G connection is lost, the CPU stores the data on a memory card while attempting to communicate with the Cloud. This will prevent data loss and notify the

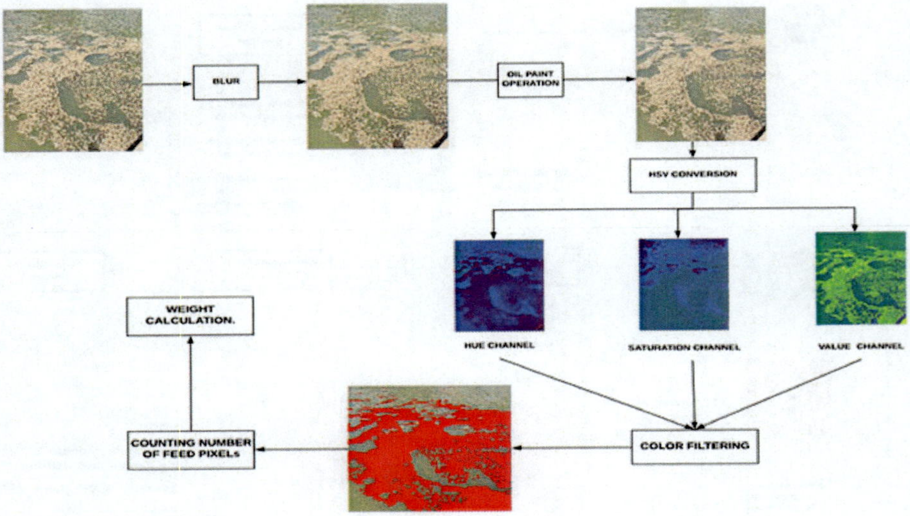

Fig. 3.4 Poseidon-AI weight calculation algorithm

farmer or farm manager of any potential Internet connection issues. The CPU then restarts the system to make sure the issue is not from the sensors, cameras, or the battery, before proceeding to send a smart alert message to the mobile device of the farmer or manager of the farm.

3.2 Evolutionary Algorithms for Feed Optimization

Evolutionary Algorithms (EA) are metaheuristic algorithms which are nature inspired (Fogel 1997; Banzhaf et al. 1998; Talbi 2009; Yu and Gan 2010; Simon 2013; Rahman 2014; Rahman et al. 2016, 2017; Nissen 2018; Nakane et al. 2020; Kumar and Davim 2020). In the late 90s, EA was relatively unidentified (Back et al. 1997; Hertz and Kobler 2000; Bai et al. 2021). The unavailability of platforms for powerful computers was the reason behind the lack of identification of EA in this period (Back et al. 1997; Fogel 1995). Decades after 1950s (Lim 2015; Back et al. 1997), the EA gained growing research interests in Combinatorial optimization problem solving (Hertz and Kobler 2000; Fogel 1992, 1993; Al-Salami 2009; Slowik and Kwasnicka 2020; Zhou et al. 2021a). Complex optimization problems, which are intractable by traditional methods (Back et al. 1997; Zhou et al. 2021a; Deb 1998), have been applied by EA and successfully been solved due to its strong adaptability and easy implementation (Fogel 1997; Back et al. 1997; Slowik and Kwasnicka 2020; Zhou et al. 2021a; Grosan and Abraham 2007).

EA evolved from Genetic Algorithm (GA), Evolution Strategy (ES), Evolutionary Programming (EP), and Genetic Programming (GP) over time (Rahman 2014; Rahman et al. 2016, 2017; Lim 2015; Back et al. 1997; Hertz and Kobler 2000;

Al-Salami 2009; Ojstersek et al. 2020). GA was invented by John Holland in 1975, and ES was invented by Rechenberg and Schwefel in 1981. Additionally, Lawrence J. Fogel discovered EP in 1962, and GP was invented by John Koza in 1994 (Lim 2015; Al-Salami 2009; Grefenstette 1986; Yao et al. 1999; Whitley 2001).

EA is claimed as metaheuristic since heuristics was built in higher-level frameworks (Fogel 1997; Banzhaf et al. 1998; Talbi 2009; Yu and Gan 2010; Simon 2013; Rahman 2014; Rahman et al. 2016, 2017; Nissen 2018; Kumar and Davim 2020; Lim 2015; Ojstersek et al. 2020; Blum and Roli 2003). But what is the difference between heuristics and metaheuristics? The ultimate characteristic of the heuristic's optimization algorithm is to generate satisfactorily good solutions but does not guarantee optimal solutions (Rahman 2014; Rahman et al. 2016, 2017; Grefenstette 1986; Burke and Kendall 2005; Hien and Gillis 2020; Bashab et al. 2020) while metaheuristics can generate and guarantee optimal solutions to applicative optimization problems (Grefenstette 1986; Zhao et al. 2021c; Glover and Kochenberger 2006; Zhang et al. 2016). Heuristic algorithms must be mathematically well explained and designed to solve specific problems (Ojstersek et al. 2020; Siddique 2013; Sundar et al. 2017), whereas metaheuristic algorithms are designed to solve demanding NP-hard problems (Ojstersek et al. 2020; Siddique 2013; Sundar et al. 2017; Nakane et al. 2020; Li et al. 2016; Marinakis and Marinaki 2012; Goodarzian et al. 2020, 2021). Metaheuristic algorithms are typically used to solve complex real-world NP-hard problems (Zhou et al. 2021b; Ojstersek et al. 2020; Zhang et al. 2016; Zhao et al. 2021a; Liu et al. 2020).

The four operators in EA are initialization, selection, crossover, and mutation operators. The selection operator plays the most important role in EA because it defines the direction of search, whereas the other genetic operators search for new search points in a non-directed manner (Zhang and Kim 2000).

The selection operator's role is to differentiate between individuals based on their quality and to allow the better individuals to become parents of the next generation (Deb 2000; Lim 2015; Figueiredo 2020; Xu et al. 2021). The selection operator selects the best candidates for the next generation and keeps the worst candidates out of the next generation (Sivaraj and Ravichandran 2011; Kumar and Jyotishree 2012). It is expected that with the selection process of selection operator, the next generation population will become increasingly fit (Talbi 2009; Simon 2013; Nakane et al. 2020; Lim 2015; Figueiredo 2020; Sivaraj and Ravichandran 2011; Back et al. 1997).

The first studies on selection operators were carried out in the 1990s (Goldberg and Deb 1991; Blickle and Thiele 1995). Several different selection operator applications were proposed during this time period in 1995 and 1996, and various selection operator approaches were implemented (Blickle and Thiele 1995, 1996; Chakravorty et al. 1996; Miller and Goldberg 1996; Veerapen et al. 2012; Rahman and Ramli 2013a, b; Craven and Jimbo 2013; Yadav and Sohal 2017; Ari and Gencer 2020; De Corte et al. 2020; Hussain and Muhammad 2020; Owais et al. 2020; Li et al. 2020). Roulette Wheel Selection (RWS), for example, has been applied to shrimp diet formulation (Rahman et al. 2017; Rahman and Ramli 2013a, b), as well as vehicle routing problems (Sabar et al. 2019; Pasha et al.

2020) and berth scheduling problems (Dulebenets 2018; Kavoosi et al. 2019). Furthermore, tournament selection has been used in the formulation of fish feed (Soong et al. 2016, 2018) and berth scheduling problems (Soong et al. 2018). Binary Tournament (BT) (Dulebenets 2018, 2020) selection has been used in berth scheduling problems (Dulebenets 2018), shrimp diet formulation problems (Rahman 2014; Rahman and Ramli 2013a, b), vehicle routing problems (Dulebenets 2018; Pasha et al. 2020), and the traveling salesman problem (Dulebenets 2018; Gottlieb and Kruse 2000; Shuai et al. 2019; Zhou et al. 2021b).

There are several recognizable aspects that show advantages of EA such as, population-based optimal solutions, robustness, ML, the feasibility of providing fast solutions, multi-objective optimization and constraint handling (Zhou et al. 2021a; Grosan and Abraham 2007; Maenhout and Vanhoucke 2011; Ramli 2004), with the most significant advantages being the flexibility and adaptability (Fogel 1997; Back et al. 1997; Al-Salami 2009; Zhou et al. 2021a; Grosan and Abraham 2007). For these reasons, EA is widely utilized, providing practical advantages in solving complex combinatorial optimization problem (Hertz and Kobler 2000; Fogel 1992, 1993; Al-Salami 2009; Slowik and Kwasnicka 2020; Zhou et al. 2021a).

Many researchers believe that, when compared to other global optimization techniques, EAs are simple to use and provide satisfactory results. Many of their studies have demonstrated that EAs can be used to solve problems where heuristic solutions produce unsatisfactory results because EAs have global search characteristics (Fogel 1997; Burke and Kendall 2005; Grosan and Abraham 2007; Rahman 2014; Rahman et al. 2016, 2017; Hien and Gillis 2020; Bashab et al. 2020; Ojstersek et al. 2020). Furthermore, when compared to other traditional methods, EAs are more efficient at solving combinatorial optimization problems or learning tasks (Fogel 1997; Banzhaf et al. 1998; Grosan and Abraham 2007; Simon 2013; Zhou et al. 2021a). EAs can also handle complex problems like discontinuities, multimodality, disjoin feasible spaces, and noisy function evaluations, which reinforces the potential effectiveness of EAs in search and optimization (Schwefel 1997; Kumar and Singh 2007; Simon 2013). EA is primarily used in many landscapes to solve problems related to management, optimization, and scheduling (Zhao et al. 2021b; Slowik and Kwasnicka 2020).

The advantages of EA are due to the strength of the selection operator which prohibits the worst-fit chromosomes (Nakane et al. 2020; Xu et al. 2021; Sivaraj and Ravichandran 2011; Kumar and Jyotishree 2012). Deliberate parent selection is made by selection pressures. Higher selection intensity can be raised, making an advantage for a large population (Lim 2015; Legg et al. 2004; Saeed et al. 2020). While in low diversity, high selection pressures can result in rapid fall to the base optimum. Furthermore, different pairs of parents have the potential to have more offspring than similar looking parents, and vice versa. According to Numerus research, the preferred operator uses an exploitation strategy (Nakane et al. 2020; Lim 2015; Veerapan et al. 2012; Hussain and Muhammad 2020; Ashlock 2005; Al-Naqi et al. 2010; Ramachandra and Chavan 2010). However, some studies disregard this significance and simply copy a few individuals into a mating pool with no selection strategy (Lim 2015; Hussain and Muhammad 2020; Tsai and Li

2009; Yang and Wu 2012). The selection operator responds by making room in the reproduction operator for permutation potential. It is necessary to control and manage the search for diversity in order to improve the selection operators.

Goldberg (1990) used Boltzmann tournament selection in Pascal applications and demonstrated that this selection method is practical for parallel hardware. The purpose of the Boltzmann tournament selection process was to provide stable distribution across space and time in population structures. The concept of proportionate selection schemes is used in this selection. A distribution mechanism imposes a group of people who share limited resources. The Boltzmann tournament selection process works by forcing individuals to compete for their potential. The recovery mechanism is unknown until the assessment of different individuals leads to more competition among undesirable individuals when selected randomly from the distribution.

Goldberg and Deb (1991) conducted a comparative study on the performance of proportionate reproduction, ranking selection, BT selection, and genitor selection. In 1995 and 1996, a similar study was carried out in terms of fitness distribution performance for ranking selection operator, BT selection operator, truncation selection operator, and exponential ranking selection operator (Blickle and Thiele 1995, 1996). Four types of selection pressure for populations were studied: elitist, linear ranking, BT, fitness proportionate, and genitor selection operator (Chakravorty et al. 1996). Chakravorty et al. (1996) investigated the possible values that can be achieved, the most likely change of utmost importance, the most likely maximum value, and the probabilities that consider the time distribution.

A similar study was carried out to provide practitioners with a logical approach to analyze the impact of various noise levels (Miller and Goldberg 1996). According to the findings of this study, proportionate selection never achieves absolute convergence. In this study, tournament selection was utilized to estimate the convergence time more quickly in small, medium, and large noisy environments. Many studies use tournament selection because larger tournament sizes increase individual competition and improve Pareto-approximation quality (Hussain and Muhammad 2020; Horn et al. 1994; Srinivasan and Rachmawati 2006). Solving uni-model problems can be solved best with BT selection (Goldberg 1990; Hussain and Muhammad 2020; Zitzler and Thiele 1999). In addition, for achieving the best solution quality with low computational time, BT selection with replacement serves better (Goldberg 1990; Zitzler and Thiele 1999; Harik 1995; Razali and Geraghty 2011; Prayudani et al. 2020). The complexity of the tournament selection is lower than complexity of other selection, and the selective pressure is higher, which allows to measure crossover for keeping the population diversity (Goldberg and Deb 1991; Hussain and Muhammad 2020; Back 1996).

Due to complex nature of the food formulation, this problem has been NP-hard problem and nondeterministic polynomial (Florian et al. 1980; Pathumnakul et al. 2011; Sihananto et al. 2019; Altun and Sahman 2013). Studies to related animal feed formulation modeling are very limited, and similar studies on aquaculture are even more limited (Rahman et al. 2016; Soong et al. 2018). For this reason, combining EA and ML for tackling feed optimization problem in aquaculture can play a crucial role in sustainability of our oceans as well as the aquaculture industry around the world.

3.2.1 Poseidon-AI® Feed Optimization Algorithms

Because fish are cold blooded, whatever happens in their environment has a direct impact on their physiological, biological, and welfare aspects. Fish, on the other hand, like any other animal class, can be studied both individually and in groups. Each fish must be separated and considered separately in order to study the individual aspects of fish. This will allow the analysis of growth rate, maturity time, potential diseases, and mortality. The other aspect is analyzing the fish species in groups and taking into account the growth rate, potential maturity time, disease epidemic, and mortality rate of the groups. Individual analysis of species is usually useful in the laboratory for understanding the physiological and biological changes that occur during a species' life cycle. However, group analysis on species is more useful for understanding on a larger scale.

Individual analysis of fish species in the aquaculture industry is time consuming, labor intensive, and expensive, especially since keeping the species in ponds, tanks, and cages adds to the costs. As a result, and because time is valuable to both farmers and markets, species group analysis may be most appropriate for the industry.

Because species live in similar conditions, compete for feeds, and share a limited number of mates during their growth phase, every deep tech company active in the aquaculture sector must use a combination of sensing, image processing, and smart algorithms to analyze the condition of each farm.

Throughout the species' life cycle, farms face a variety of bottlenecks that can affect their revenues, costs, and, ultimately, the sustainable development of farms. Farms' main bottlenecks are feed, disease spread, slow growth, faster maturity time, high mortality rate, lack of skilled labor, and fingerling and larvae quality. Because the species are kept in captivity, the only way to meet the nutritional needs of the fish is to provide scheduled and routine feeds based on the species' life stages. As a result, nearly 80% of the total costs of every farm are allocated to feed. Unsuitable environmental conditions and the spread of diseases (which cause high mortality or reduce the quality of the final product) are other factors that contribute to the farm's total costs. However, overcoming these abrupt changes is difficult, and in many cases, the assistance of a qualified veterinarian is required.

As a result, Poseidon-AI focuses on improving farm feeding conditions in order to achieve two main goals: first, lowering farm feeding costs by applying EA algorithms to the school of fish in every pond, tank, and cage; and second, increasing farm revenue by combining the EA algorithms with physiological and biological aspects of the fish, in order to accurately estimate the exact time for taking the species to market.

Feed ingredients are typically chosen based on economic status, suitability for the fish digestive system, and nutritional value required based on expert recommendations. In addition, nutrients are required for feed formulation and are typically crude protein, crude fat, and crude fiber. However, feeding the school of fish at the same time does not guarantee that all of them will consume the same amount because there are always some levels of competition among the fish.

Fig. 3.5 According to EA models, smaller, weaker, less aggressive, and unfit fish will be eliminated, and only stronger and healthier fish will win the competition

EA models can analyze competition among a school of fish by assuming that some fish are stronger than others. Furthermore, the size of male and female species in the pond, tank, and cage can play an important role in competing for larger amounts of feed. As a result, it is not feasible to analyze the individual growth rates, maturity, and even mortality without taking the competition factor into account (winning of the stronger, healthier, and bigger fingerlings).

Poseidon-AI uses environmental factors such as temperature, D.O., and pH, in conjunction with physiological and biological algorithms, to inform farmers about the growth rate, maturity time, best time of day to feed, and when species have the greatest appetite. Furthermore, the algorithms can estimate and demonstrate the FCR based on the feed ingredients and growth rate. However, estimating the growth rate solely based on environmental factors and fish physiology will not yield an accurate result; thus, EA models can assist in providing the maximum number of mature and healthy stock to be converted into kilograms.

Because the growth rate and FCR have correlations with environmental conditions, the precise timing of feeding can play an important role in the species' efficient consumption of feeds. When the environmental variables are favorable, the species are more active and thus more competitive, and the EA can predict and show the chances of healthier, fine-growing schools of fish. This means that the weaker fish will die because they are unable to compete, and the healthier, more aggressive fish will survive among the millions of fingerlings growing in a pond, tank, or cage (Fig. 3.5). But, with no expert on hand, how can farmers know and understand that these algorithms will help them significantly and save them money?

Once the daily feeding time(s) are displayed on the dashboard of the farmer's or farm manager's device(s), the cameras will begin imaging an hour before the displayed time. Traditional farmers usually mention the feeding time by the time when the fish nibble more aggressively and show signs of hunger. This behavior is influenced not only by hunger, but also by environmental factors. The disadvantage of this behavior is that farmers will believe that feeding more or keeping the same

Fig. 3.6 Excess feeding increases the risk of disease outbreak, higher mortality, and faster maturity

amount will result in larger fish culture. Excessive feeding will alter water quality and increase the likelihood of disease outbreaks, higher mortality rates, and even faster maturation of the species (Fig. 3.6).

Traditional farming employs the strategy of overstocking ponds, tanks, or cages in the hope that with luck, they will achieve the maximum survival rate, but this is not sustainable, and unexpected events will increase the risk for traditional farming practices. As a result, monitoring environmental conditions can provide an overview and reasoning for current and future changes in farm conditions, allowing farmers or farm managers to make precise and efficient decisions.

Images from ponds, tanks, or cages are processed to distinguish between dead fish, feed remnants, leaves, and branches. As shown in Fig. 3.7, the images will be color segmented, and any external objects detected.

Each feed pellet will be assigned a weight and a pixel size to calculate the wasted feed. The number of captured pixels will be used to analyze and process each image. These images will be multiplied by the weights assigned to each pellet to determine the amount of wasted feed. This information is then used to calibrate the feeding algorithms, allowing the amount of given feed to be corrected under similar environmental conditions. The algorithms can distinguish the correct feeding amount

Fig. 3.7 Color segmentation for analysis of external objects in the pond, tank, and cage

under the learned and documented environmental conditions thanks to ML and after one life cycle of the species.

Maturation time will also have a direct impact on the amount of feed consumed by the species. Traditional farmers will keep and grow the species according to the timeline taught to them by their fathers or used by everyone in the same region. Several studies, however, have demonstrated the impact of climate change and global warming on the rapid maturation of species, particularly poikilothermic species. As a result, it is safe to assume that environmental data can demonstrate whether the cultured species is showing signs of rapid maturation or not. This is significant because once the species reaches maturity, its length will stop increasing while its body weight will continue to increase. This does not mean that the species will gain more body weight than they would under normal environmental conditions, but it may help to shorten the time it takes to bring the cultured species to market. This reduces feed consumption, increases farms' capacity to culture more than once a year, and increases revenue. In other words, farmers will reduce their feed consumption while selling their product earlier than expected and by valuing the discount rate of time (The Golden Rule).

Another critical aspect is the FCR algorithm. Essentially, this algorithm shows the amount of feed consumed by the school of fish and how much of that feed is converted into pure stock weight. Higher FCRs are thought to be a function of two variables: first, the quality of the feed, how it is made, and what level of nutrients are used to produce it, and second, environmental variables that can impact the species' physiology. As previously stated, the quality of the feed is determined by the ingredients and nutrients used in its production. There have been studies that use EA models to optimize the production of fish and shrimp feeds for different life stages of the species (Holt 2011; Davis 2015; Ali 2019; Muhammadar et al. 2011), but in the real world, large feed producing companies do not provide all of this information for analysis. As a result, Poseidon-AI® algorithms estimate the FCR based on the physiological model of the fish as well as visual presentation and confirmation from the ponds, tank, and cage. The process involves assessing the impact of environmental variables on the species' growth rate and feed consumption to provide a reasonable and accurate FCR for different types of feed and different life stages of the species. It should be noted that image processing is critical in accurately measuring feed consumption, whereas species (L) and (W) can only be monitored individually and during feeding time when detected by the cameras.

The feed ingredients will have a positive effect on fish growth rate. There are various combinations of feed for increasing production, such as the use of algae, but the impact of different types of feed on different species are unknown. As a result, Poseidon-AI® growth rate algorithms can demonstrate a positive impact in rapid production and growth under a variety of environmental conditions based on specific feed used in various areas and for different species.

References

Ali SA (2019) Nutritional feeding of fish and shrimps in India. MJP Publisher, Chennai

Al-Naqi A, Erdogan AT, Arslan T (2010) Fault tolerance through automatic cell isolation using three-dimensional cellular genetic algorithms. In: Proceedings of the 2010 IEEE congress on evolutionary computation (CEC). IEEE, New York, pp 1–8

Al-Salami NMA (2009) Evolutionary algorithm definition. Am J Eng Appl Sci 2(4):789–795

Altun AA, Sahman MA (2013) Cost optimization of mixed feeds with the particle swarm optimization method. Neural Comput Applic 22(2):383–390

Ari ES, Gencer C (2020) The use and comparison of a deterministic, a stochastic, and a hybrid multiple-criteria decision-making method for site selection of wind power plants: an application in Turkey. Wind Eng 44(1):60–74

Ashlock D (2005) Evolutionary computation for modeling and optimization. Springer, New York

Back JH (1996) Evolutionary algorithms in theory and practice. Oxford University Press, Oxford

Back T, Hammel U, Schwefel HP (1997) Evolutionary computation: comments on the history and current state. IEEE Trans Evol Comput 1(1):3–17

Bai J, Tian M, Li J (2021) An elite adaptive niche evolutionary algorithm for duty clustering problem in SoWSN. J Phys Conf Ser 1769(1):1–7

Banzhaf W, Nordin P, Keller RE, Francone FD (1998) Genetic programming: an introduction on the automatic evolution of computer programs and its applications. Dpunkt Verlag and Morgan Kaufmann Publishers, Inc

Bashab A, Ibrahim AO, AbedElgabar EE et al (2020) A systematic mapping study on solving university timetabling problems using meta-heuristic algorithms. Neural Comput Applic 32(23): 17397–17432

Blickle T, Thiele LA (1995) Comparison of selection schemes used in genetic algorithms. Computer Engineering and Communication Networks Lab (TIK)-report

Blickle T, Thiele L (1996) A comparison of selection schemes used in evolutionary algorithms. Evol Comput 4(4):361–394

Blum C, Roli A (2003) Metaheuristics in combinatorial optimization: overview and conceptual comparison. ACM Comput Surv 35(3):268–308

Burke EK, Kendall G (2005) Introduction. In: Burke EK, Kendall G (eds) Search methodologies: introductory tutorials in optimization and decision support techniques. Springer, New York, pp 5–18

Chakravorty UK, Deb K, Chakraborty M (1996) Analysis of selection algorithms: a Markov chain approach. Evol Comput 4(2):133–167

Craven MJ, Jimbo HC (2013) An EA for portfolio selection over multiple investment periods with exponential transaction costs. In: Proceedings of the 15th annual conference companion on Genetic and evolutionary computation. ACM, New York, pp 115–116

Davis DA (2015) Feed and feeding practices in aquaculture. Elsevier, Waltham, MA

De Corte W, Sackett PR, Lievens F (2020) Robustness, sensitivity, and sampling variability of Pareto-optimal selection system solutions to address the quality-diversity trade-off. Organ Res Methods 23(3):535–568

Deb K (1998) Genetic algorithm in search and optimization: the technique and application. In: Proceedings of the international workshop on soft computing and intelligent systems, pp 58–87

Deb K (2000) Introduction to selection. In: Back T, Fogel DB, Michalewicz T (eds) Evolutionary computation 1: basic algorithms and operators. Institute of Physics Publishing, Bristol, pp 166–171

Dulebenets MA (2018) A comprehensive evaluation of weak and strong mutation mechanisms in evolutionary algorithms for truck scheduling at cross-docking terminals. IEEE Access 6:65635–65650

Dulebenets MA (2020) An adaptive island evolutionary algorithm for the berth scheduling problem. Memetic Comput 12(1):51–72

Figueiredo LM (2020) An overview of the main machine learning models-from theory to algorithms. Unpublished Doctoral Thesis, NOVA University Lisbon, Campolide, Lisbon

Florian M, Lenstra JK, Rinnooy Kan AHG (1980) Deterministic production planning: algorithms and complexity. Manag Sci 26(7):669–679

Fogel DB (1992) Evolving artificial intelligence. Doctoral thesis, University of California, San Diego, La Jolla

Fogel DB (1993) Applying evolutionary programming to selected traveling salesman problems. Cybern Syst 24(1):27–36

Fogel DB (1995) Evolutionary computation: towards a new philosophy of machine intelligence. IEEE Press, New York

Fogel DB (1997) The advantages of evolutionary computation. In: Lundth D, Olsson B, Naraganan A (eds) Bio-computing and emergent computation. World Scientific Press, pp 1–11

Glover FW, Kochenberger GA (2006) Handbook of metaheuristics. Springer Science & Business Media, New York

Goldberg DE (1990) A note on Boltzmann tournament selection for genetic algorithms and population-oriented simulated annealing. Complex Systems 4(4):445–460

Goldberg DE, Deb K (1991) A comparative analysis of selection schemes used in genetic algorithms. Found Genet Algorithms 1:69–93

Goodarzian F, Hosseini-Nasab H, Fakhrzad MB (2020) A multi-objective sustainable medicine supply chain network design using a novel hybrid multi-objective metaheuristic algorithm. Int J Eng 33:1986–1995

Goodarzian F, Shishebori D, Nasseri H, Dadvar F (2021) A bi-objective production-distribution problem in a supply chain network under grey flexible conditions. RAIRO Oper Res 55:1287–1316

Gottlieb J, Kruse T (2000) Selection in evolutionary algorithms for the traveling salesman problem. In: Proceedings of the 2000 ACM symposium on Applied computing, vol 1, pp 415–421

Grefenstette JJ (1986) Optimization of control parameters for genetic algorithms. IEEE Trans Syst Man Cybern 16(1):122–128

Grosan C, Abraham A (2007) Hybrid evolutionary algorithm: methodologies, architectures, and reviews. Stud Computat Int 75:1–17

Harik GR (1995) Finding multimodal solutions using restricted tournament selection. In: Proceedings of the 6th international conference on genetic algorithms, pp 24–31

Hertz A, Kobler D (2000) A framework for the description of evolutionary algorithms. Eur J Oper Res 126:1–12

Hien LTK, Gillis N (2020) Algorithms for nonnegative matrix factorization with the Kullback-Leibler divergence. arXiv 2010:Article ID 01935

Holland J (1975) Adaptation in natural and artificial systems. University of Michigan Press

Holt GJ (2011) Larval fish nutrition. John Wiley & Sons, Inc, Hoboken

Horn J, Nafpliotis N, Goldberg DE (1994) A niched Pareto genetic algorithm for multi-objective optimization. In: Proceedings of the 1st IEEE conference on evolutionary computation. IEEE World Congress on Computational Intelligence, pp 82–87

Hussain A, Muhammad YS (2020) Trade-off between exploration and exploitation with genetic algorithm using a novel selection operator. Complex Intell Syst 6(1):1–14

Kavoosi M, Dulebenets MA, Mikijeljevic M (2019) A universal island-based metaheuristic for effective berth scheduling. In: Proceedings of the XXIII international conference on material handling, constructions and logistics (MHCL 2019), Bar (Montenegro)

Kumar K, Davim JP (2020) Optimization using evolutionary algorithms and metaheuristics: applications in engineering. CRC Press, Boca Raton, FL

Kumar R, Jyotishree (2012) Novel encoding scheme in genetic algorithms for better fitness. Int J Eng Adv Technol 1(6):214–218

Kumar R, Singh PK (2007) Pareto evolutionary algorithm hybridized with local search for bio-objective TSP. Stud Comput Int 75:361–398

Legg S, Hutter M, Kumar A (2004) Tournament versus fitness uniform selection. In: Proceedings of the 2004 congress on evolutionary computation (CEC '04). IEEE, Portland, OR, pp 2144–2151

Li J, Pan Q, Duan P (2016) An improved artificial bee colony algorithm for solving hybrid flexible flow shop with dynamic operation skipping. IEEE Trans Cybern 46(6):1311–1324

Li J, Kang H, Sun G et al (2020) IBDA: improved binary dragonly algorithm with evolutionary population dynamics and adaptive crossover for feature selection. IEEE Access 8:108032–108051

Lim HT (2015) Enhanced Evolutionary algorithm with cuckoo search for nurse scheduling and rescheduling problem. Bus driver scheduling system via genetic algorithm with enhanced crossover mechanism

Liu WL, Gong YJ, Chen WN, Liu Z, Wang H, Zhang J (2020) Coordinated charging scheduling of electric vehicles: a mixed-variable differential evolution approach. IEEE Trans Intell Transp Syst 21(12):5094–5109

Maenhout B, Vanhoucke M (2011) An evolutionary approach for the nurse re-rostering problem. Comput Oper Res 38(10):1400–1411

Marinakis Y, Marinaki M (2012) A hybrid particle swarm optimization algorithm for the open vehicle routing problem. In: Swarm intelligence: 8th international conference, ANTS 2012. Springer, Brussels, pp 180–187

Miller BL, Goldberg DE (1996) Genetic algorithms, selection schemes, and the varying effects of noise. Evol Comput 4(2):113–131

Muhammadar AA, Mazlan AG, Samat A, Muchlin ZA, Simon KD (2011) Crude protein and amino acids content in some common feeds of tiger grouper (Epinephelus fuscoguttatus) juvenile. Aquac Aquar Conserv Legis 4(4):499–504

Nakane T, Bold N, Sun H, Lu X, Akashi T, Zhang C (2020) Application of evolutionary and swarm optimization in computer vision: a literature survey. IPSJ Trans Comput Vis Applic 12(1):3–34

Nissen V (2018) Applications of evolutionary algorithms to management problems. In: Innovative research methodologies in management. Palgrave Macmillan, Cham, pp 211–235

Ojstersek R, Brezocnik M, Buchmeister B (2020) Multi-objective optimization of production scheduling with evolutionary computation: a review. Int J Ind Eng Comput 11(3):359–376

Owais WB, Alkhazendar IW, Saleh D (2020) Evaluating the impact of different types of crossover and selection methods on the convergence of 0/1 Knapsack using Genetic Algorithm. arXiv preprint arXiv 2010:Article ID 03483

Pasha J, Dulebenets MA, Kavoosi M, Abioye OF, Wang H, Guo W (2020) An optimization model and solution algorithms for the vehicle routing problem with a “factory-in-a-box”. IEEE Access 8:134743–134763

Pathumnakul S, Ittiphalin M, Piewthongngam K, Rujikietkumjorn S (2011) Should feed mills go beyond traditional least cost formulation? Comput Electron Agric 75(2):243–249

Prayudani S, Hizriadi A, Nababan EB, Suwilo S (2020) Analysis effect of tournament selection on genetic algorithm performance in Traveling Salesman Problem (TSP). J Phys Conf Ser 1566(1): 012131

Rahman AR (2014) Evolutionary algorithms with average crossover and power heuristics for aquaculture diet formulation. Doctoral thesis, University Utara Malaysia, Malaysia

Rahman AR, Ramli R (2013a) Average concept of crossover operator in real coded genetic algorithm. Int Proc Econ Dev Res 63(15):73–77

Rahman AR, Ramli R (2013b) Roulette-tournament selection for shrimp diet formulation problem. In: Proceedings of the 4th international conference on computing and informatics (ICOCI), Sarawak, Malaysia, pp 28–29

Rahman RA, Ramli R, Jamari Z, Ku-Mahamud KR (2016) Evolutionary algorithm with roulette-tournament selection for solving aquaculture diet formulation. Math Probl Eng 2016:1–10

Rahman AR, Kendall G, Ramli R, Jamari Z, Ku-Mahamud KR (2017) Shrimp feed formulation via evolutionary algorithm with power heuristics for handling constraints. Complexity. In: Eiben AE, Smith JE (eds) Introduction to evolutionary computing. Springer, Berlin, p 1

Ramachandra S, Chavan V (2010) Enhanced-Hybrid-age layered population structure (E-Hybrid-ALPS): a genetic algorithm with adaptive crossover for molecular docking studies of drug discovery process. Int J Curr Res Rev 12(15):07–15

Ramli R (2004) An evolutionary algorithm for the nurse scheduling problem with circadian rhythms. Unpublished doctoral thesis, Universiti Sains Malaysia, Malaysia

Razali NM, Geraghty J (2011) Genetic algorithm performance with different selection strategies in solving TSP. In: Proceedings of the World Congress on Engineering WCE, London, UK

Sabar NR, Bhaskar A, Chung E, Turky A, Song A (2019) A self-adaptive evolutionary algorithm for dynamic vehicle routing problems with traffic congestion. Swarm Evol Comput 44:1018–1027

Saeed F, Mohammed F, Gazem N (2020) Emerging trends in intelligent computing and informatics: data science. In: Intelligent information systems and smart computing, vol 1073

Schwefel HP (1997) Advantages (and disadvantages) of evolutionary computation over other approaches. In: De-Jong K, Fogel L, Schwefel HP (eds) Handbook of evolutionary computation. IOP Publishing Ltd and Oxford University Press, England, pp 1–10

Shuai Y, Yunfeng S, Kai Z (2019) An effective method for solving multiple travelling salesman problem based on NSGA-II. Syst Sci Cont Eng 7(2):108–116

Siddique N (2013) Intelligent Control: a hybrid approach based on fuzzy logic, neural networks and genetic algorithms. Springer, New York

Sihananto AN, Burhan MS, Mahmudy WF (2019) Chicken feed optimization using evolution strategies and firefly algorithm. Int J Electr Comput Eng 9(1):585–592

Simon D (2013) Evolutionary optimization algorithms: biologically inspired and population-based approaches to computer intelligence. Wiley, Hoboken, NJ

Sivaraj R, Ravichandran T (2011) A review of selection methods in Genetic Algorithm. Int J Eng Sci Technol 3(5):3792–3797

Slowik A, Kwasnicka H (2020) Evolutionary algorithms and their applications to engineering problems. Neural Comput Applic 32(16):12363–12379

Soong C-J, Razamin R, Rosshairy AR (2016) A standard deviation selection in evolutionary algorithm for grouper fish feed formulation. In: Proceedings of the 4th international conference on quantitative sciences and its application (ICOQSIA), Putrajaya, Malaysia

Soong C-J, Razamin R, Rosshairy AR (2018) Potential grouper feed formulation based on evolutionary algorithm concept with a unique selection operator. J Eng Sci Technol 13(2): 332–346

Srinivasan D, Rachmawati L (2006) An efficient multi-objective evolutionary algorithm with steady-state replacement model. In: Proceedings of the 8th annual conference on Genetic and evolutionary computation, pp 715–722

Sundar S, Suganthan PN, Jin CT, Xiang CT, Soon CC (2017) A hybrid artificial bee colony algorithm for the job-shop scheduling problem with no-wait constraint. Soft Comput 21(5): 1193–1202

Talbi EG (2009) Metaheuristics: from design to implementation. Wiley, Hoboken, NJ

Tsai CC, Li SH (2009) A two-stage modeling with genetic algorithms for the nurse scheduling problem. Expert Syst Appl 36(5):9506–9512

Veerapen N, Maturana J, Saubion F (2012) An exploration-exploitation compromise-based adaptive operator selection for local search. In: Proceedings of the 2012 genetic and evolutionary computation conference GECCO, New York, USA, pp 1277–1284

Whitley D (2001) An overview of evolutionary algorithms: practical issues and common pitfalls. Inf Softw Technol 43(14):817–831

Xu P, Luo W, Lin X, Qiao Y (2021) Evolutionary continuous constrained optimization using random direction repair. Inf Sci 566(1):80–102

Yadav SL, Sohal A (2017) Comparative study of different selection techniques in genetic algorithm. Int J Energy Sector Manag 6(3):1–7

Yang FC, Wu WT (2012) A genetic algorithm-based method for creating impartial work schedules for nurses. Int J Electron Bus Manag 10(3):182–193

Yao X, Liu Y, Lin GM (1999) Evolutionary programming made faster. IEEE Trans Evol Comput 3(2):82–102

Yu XJ, Gan M (2010) Introduction to evolutionary algorithms. Springer, Berlin

Zhang BT, Kim JJ (2000) Comparison of selection methods for evolutionary optimization. Evol Optimiz 2(1):55–70

Zhang H, Liu S, Moraca S, Ojstersek R (2016) An effective use of hybrid metaheuristics algorithm for job shop scheduling problem. Int J Simul Modell 16(4):644–657

Zhao F, Zhang L, Cao J, Tang J (2021a) A cooperative water wave optimization algorithm with reinforcement learning for the distributed assembly no-idle flowshop scheduling problem. Comput Ind Eng 153:107082

Zhao F, He X, Wang L (2021b) A two-stage cooperative evolutionary algorithm with problem-specific knowledge for energy-efficient scheduling of no-wait flow-shop problem. IEEE Trans Cybern 51(11):5291–5303

Zhao F, Ma R, Wang L (2021c) A self-learning discrete Jaya algorithm for multi-objective energy-efficient distributed no-idle flow-shop scheduling problem in heterogeneous factory system. IEEE Trans Cybernetics 52:1–12

Zhou L, Feng L, Gupta A, Ong YS (2021a) Learnable evolutionary search across heterogeneous problems via kernelized autoencoding. IEEE Trans Evol Comput 25(3):567–581

Zhou S, Xing L, Zheng X, Du N, Wang L, Zhang Q (2021b) A self-adaptive differential evolution algorithm for scheduling a single batch-processing machine with arbitrary job sizes and release times. IEEE Trans Cybern 51(3):1430–1442

Zitzler E, Thiele L (1999) Multi-objective evolutionary algorithms: a comparative case study and the strength Pareto approach. IEEE Trans Evol Comput 3(4):257–271

4 Comparison of Aquaculture Practices with and Without Deep Tech

The COVID-19 pandemic has had a profound impact not only on human lives but also on global food production. Consumer expectations have changed as a result of the pandemic, which has also disrupted markets and logistics and caught everyone off guard. Lockdowns have also decreased demand, which has led to an increase in product costs. The aquaculture sector, which produces a large portion of the world's required protein for a growing population, has also seen a decline. Due to the unprofitability of their production, many aquaculture operations ceased operations or reduced their activities. The pandemic has caused several production inputs to be discontinued at the farm level (both grow-out and seed production), delayed stocking, interfered with normal feeding and management, prolonged the culture time, and impacted the quality of the produced seafoods (Yuan et al. 2022). Other specific issues caused by the COVID-19 pandemic included low market demand for farmed aquatic products, difficulty in selling the seafood products, delayed harvesting (Kakoolaki et al. 2020), disrupted transportation of fresh and live fish products, and capital loss for some farmers and business operators; this last issue is consistent with that reported in a case study incorporating data from Bangladesh, Scotland, and the USA (Hasan et al. 2021; Murray et al. 2021; Van Senten et al. 2021). A shortage of production capital, difficulties employing workers or staffs, rising labor prices, and trouble obtaining loans and investments are further obstacles that farms and related enterprises must overcome before they can resume full production recovery. However, Islam et al. (2021) discovered some positive effects on the ecosystem and fish stocks (e.g., an increase in fish stock) as a result of reduced disturbance from fishing activities.

According to FAO (2022), the aquaculture sector declined by 3% in African countries and 5% in Oceania countries. During the pandemic, China produced 35% of total aquaculture production, while India, Indonesia, and Vietnam produced 20%. Asian countries, particularly those in the South-East, play a significant role in aquaculture production. In 2020, Asia alone produced 91.6% of the world's aquatic animals and algae (FAO 2022). Many developing and low-income countries are

© The Author(s), under exclusive license to Springer Nature Singapore Pte Ltd. 2023

A. Rahimi-Midani, *Deep Technology for Sustainable Fisheries and Aquaculture*,
https://doi.org/10.1007/978-981-99-4917-5_4

struggling to meet their national aquaculture development goals in order to support national food production and feeding their growing populations. Norway, Chile, and China are the leaders in finfish mariculture using sea cages. Globally, middle-income countries dominate inland finfish aquaculture production.

The majority of major aquaculture producing countries are densely populated developing countries. These countries serve as models for aquaculture development in other countries with aquaculture development potential. However, the pandemic created unpredictable conditions that hampered aquaculture development worldwide. Having said that, not everything about the 2022 pandemic was bad. This pandemic resulted in the need for paradigm shifts in food production, involving new and innovative technologies.

Inland aquaculture employs a variety of culture methods and facilities. The operations differ depending on the level of technology and management, as well as the degree of integration with other farm activities. Finfish culture is typically done in constructed earthen ponds, which is by far the most common culture method. To a lesser extent, cage culture and pen culture are used.

Cage and pen culture in rivers, lakes, and reservoirs is gaining popularity in Indonesia and the Philippines. In China, the vast majority of cages and pens were removed due to policy plans. However, in some provinces, a limited number of licenses are issued based on the carrying capacity of the waterbodies.

Poseidon-AI is based in Singapore because more than 90% of global production is produced in Southeast Asia and China. As previously stated, Poseidon-AI's vision is to incorporate deep technology into the aquaculture sector and address the major bottleneck concerns such as feed wastage. Hence, the best possible environment for fish culture in South-East Asia and China is used to use Poseidon-AI. Poseidon-AI® algorithms are used to optimize feeding on freshwater species such as catfish, carp, and tilapia, as well as saltwater species such as grouper, red snapper, and seabass. China, Indonesia, Laos, Malaysia, Thailand, Singapore, and Vietnam are the major countries where these species are cultivated. This chapter describes the process of incorporating Poseidon-AI's innovative approach in some of these countries and saving farmers 20% on feed costs while training algorithms under the real-world conditions. The approach is to confront environmental changes with a business perspective that can benefit farmers, adapt to environmental changes, and ensure the economy's long-term development.

4.1 China

4.1.1 Aquaculture Production in China

China has a total land area of 9.6 million km^2 and a sea area of 3.0 million km^2, which includes the Bohai Sea, Yellow Sea, East China Sea, and South China Sea, as well as approximately 6500 islands and a total coastline of 18,000 km.

Following the open-door policy and economic reform in China, rapid aquaculture production began in the 1970s. The Chinese government placed greater emphasis on increasing the supply of animal protein for the population while ensuring staple food self-sufficiency. The government began to reform seafood production, establishing the fisheries and aquaculture sector as China's first value-added food production sector (Qi 2002). The Chinese central government set a goal of improving fish supply in urban areas within 35 years in the 1980s, and the People's Republic of China passed its first fisheries and aquaculture law in 1986 (FAO 2003).

Aquaculture in China grew rapidly in the 1980s and 1990s, from 640,000 hectares in 1985 to 1,860,000 hectares in 1995 (FAO 2003). At the same time period, increasing efforts were made to improve the survival rates of these stocks and increase fish output from small- and medium-sized lakes and reservoirs through extensive aquaculture, with the major input being large, artificially propagated fish seed. Cage culture was one of the new farming systems introduced in China in the 1970s and 1980s to improve fish production in lakes and reservoirs. Pen culture was introduced in China's inland culture in the 1980s, and it later spread to coastal and sheltered inshore areas. Additionally, RAS systems began in China around the same time.

In the 1980s, the Chinese government made every effort to increase aquaculture unit production. Standardization of fishponds through modification of traditional fishponds for improving pond productivity was one of the major inputs. Additionally, supporting research and development of feed ingredients and production, with a focus on nutritional requirements, energy metabolism, digestive enzymes, and FCR for high-value species were the other implemented strategies. According to Liu (1999), the research outputs effectively supported the development of compound aquafeeds for various cultured species as well as the rapid expansion of the Chinese feed industry. Because of these policies and subsidies, China's feed production reached 19,000,000 tonnes in 2014 (Bo and Lin 2015; Han et al. 2016). From 1978 to 1990, China's aquaculture sector experienced an average annual growth rate of 16% (BoF 2016).

Pond aquaculture unit production was 3270 kg per hectare in 1990, but it has since increased to 4899 kg per hectare (China Fishery Statistical Yearbook 2000). Furthermore, at the end of the 1980s, aquaculture was limited to about 25 aquatic species, but in 10 years, this number had increased to 50 cultured species. In addition, the number of aquatic species has increased to 100, with 88 species included in the FAO report (FAO 2022).

Another important program that increased the income of traditional rice farmers was the promotion of rice-fish farming. In 1990, integrated rice and fish farming produced 1,300,837 tonnes on a total area of 740,000 hectares (BoF 1991). In 2000, production increased to 745,770 tonnes on 1,532,381 hectares (BoF 2001). The overall economic return from fish and other aquatic animals in rice-fish farming far exceeded the income from rice (Miao 2009; Xie et al. 2011; BoF 2012; Hu et al. 2015). Rapid aquaculture development has also fueled the growth of aquaculture product processing in China. In 2000, the total number of seafood processing plants reached 6922, with a total processing capacity of 9,340,000 tonnes (BoF 2001). By 2001, the total number of people employed in primary aquaculture and fisheries had risen to 13,740,000, up from 3,070,000 in 1979, with aquaculture accounting for approximately 85% of the total (Qi 2002).

Development of aquaculture not only increased rural populations' income but also benefited local and national economies. Fisheries and aquaculture contributed 16.8% of Chinese GDP as a subsector of agriculture, up from less than 2% in the 1970s (China Fishery Statistical Yearbook 2015; NSB 2016; Li 2002). Aquaculture contributed nearly half of total Fisheries and Aquaculture GDP, which included aquaculture, capture fisheries, and fishery related manufacturing, construction, logistics, and services.

Due to challenges in both maintaining sustainable aquaculture development and overall social and economic development in China, Chinese aquaculture entered a new stage of development at the start of the twenty-first century. Competition in conservation of national resources, environmental impact control, stringent food safety requirements, and climate change, as with any other development, forced the Chinese government to implement new development strategies for the aquaculture sector during this period. These strategies include changes to the sector's structural framework as well as quality enhancements achieved through the development and application of a range of standards and efficient monitoring, which will improve the environmental protection of fisheries resources and the ecologically sound benefits of the aquaculture sector. Lu (2013) claims that the specified methods for aquaculture development are to move toward ecological sustainability and technology-based aquaculture by following new patterns, which calls for the use of cutting-edge biological and engineering theories and technologies.

Inland farming in lakes and reservoirs has been used as an important approach to environmental improvement and ecological rehabilitation. As a result, farming on land has given way to farming on water. Examples include multitrophic mariculture systems, which mix intense aquaculture of fed aquatic animals with filter-feeding mollusks and autotrophic seaweeds, as well as recirculation pond culture systems that incorporated wastewater treatment into conventional intensive pond culture. The government has introduced several new regulations to address the structural changes and functional transition of aquaculture for sustainable development and greater ecological and social benefits. The Regulation on Certification of Aquaculture Rights in Inland Waters and Tidal Zones, the List of Allowed Drugs in Aquaculture and Their Use, and the Code of Conduct for Safety and Hygiene in Aquaculture are among the important regulations enacted.

Several food safety incidents occurred in the early 2000s, which had a negative impact on public perception of aquaculture products. To address this issue and ensure product safety and public health, the Chinese government focused more on food safety regulations to improve seafood safety and quality, resulting in the adoption of the national food safety law in 2009. Furthermore, various registration and certification schemes have been put in place, and in 2004, 957 brands of aquaculture products from 517 enterprises were certified as hazardless products, with an annual production of 730,000 (China Fishery Statistical Yearbook 2005). Furthermore, 1241 aquaculture sites totaling 1,200,000 hectares of water surface

Fig. 4.1 Fresh and live seafood restaurants are not only limited to coastal areas and can be seen everywhere

were designated as hazardless farming sites (China Fishery Statistical Yearbook 2005).

Social and economic activities were integrated as another tendency in the development of aquaculture. By 2004, over 5000 tourism, sport fishing, and restaurant businesses had been established (Fig. 4.1) (China Fishery Statistical Yearbook 2005). Another recognized overall sustainable development goal in China is high production efficiency with low carbon footprints in the aquaculture sector (Dong 2011). Some Chinese aquaculture practices have been recognized for their carbon sequestration function. Aquaculture has contributed to lower carbon emissions and less eutrophication of inland water bodies (Tang 2012). According to Tang (2012), Chinese mollusk and seaweed farming extracted 1,380,000 tonnes of carbon from coastal seas, while freshwater farming extracted more than 1,300,000 tonnes of carbon from waterbodies, the equivalent of planting a million hectares of forest.

In the 1980s, per capita fish consumption in China was 4.4 kg per person per year, which was less than the world average (11.5 kg per person per year) at the time. Per capita consumption increased significantly, reaching a high of 33.5 kg per person per year, nearly twice the previous high of 19.1 kg per person per year (world average fish consumption).

China continues to support aquaculture development by providing technical assistance to developing countries. Human capacity building, technology transfer, demonstration, and onsite technical guidance are all part of the assistance, as is the sharing of genetic resources (Han and Lu 2010). The Chinese government began providing education and technical training in fisheries and aquaculture sciences in 1995. Shanghai Fisheries College received the first batch of students in 1955, and it is regarded as the first capacity building program. Until now, nearly 3000 people from 127 countries have been trained through the Technical Cooperation among Developing Countries program (TCDC). According to Zhu (2014), joint programs with FAO have significantly improved aquaculture development in many countries.

Despite significant progress in China's aquaculture sector, major constraints to China's aquaculture development still remain in terms of aquaculture health management, genetic improvement in light of germplasm deterioration in cultured species, and environmental sustainability (Zhang 2013).

4.1.1.1 Environmental Impacts

To reduce environmental impacts, Chinese authorities implemented policy guidelines to control overall culture density, feeding regimes, and waste discharge from the aquaculture facilities. It is believed that aquaculture systems that are well planned and managed can provide numerous benefits to society while not exacerbating environmental degradation. To reduce environmental impacts, the farms are undergoing various biological, ecological, and engineering interventions. In Chinese inland and offshore aquaculture, paradigm shifts have been reported (Wang et al. 2017).

4.1.1.2 Genetic Improvement

In China, 200 aquatic species have been cultured over the last two millennia. The production of aquaculture in China today depends heavily on many of these species, including those that have been relocated to other regions of the nation. One of China's remarkable achievements has been the expansion of cultured species from 30 in the 1970s to nearly 200 in 2014. Unfortunately, genetic management and hatchery practices for many cultivated species have largely been insufficient and unsystematic, which appears to have negatively impacted the performance of many farmed species through inbreeding, genetic drift, and uncontrolled hybridization. Some alien species have become very important farmed species, either in volume or in terms of commercial value, and contributed to production and consumption nationwide; among these are the Nile tilapia (*Oreochromis niloticus*), bay scallop (*Argopecten irradians*), vannamei shrimp (*Penaeus vannamei*), flounder or turbot (*Scophthalmus maxima*), largemouth bass (*Micropterus salmoides*), giant freshwater prawn (*Macrobrachium rosenbergii*), and crayfish (*Procambarus clarkii*). These species contribute significantly to Chinese aquaculture industry while fitting best with Chinese ecological condition (Liu and Li 2010). However, in China there is still controversy about alien species production and their consumption (Lin et al. 2015).

4.1.1.3 Aquatic Animal Health Management

Veterinary medicine has been adopted in China, and numerous remarkable advances have been made in restoring disease control and disease curing through the use of drugs and medication. In China, there have been several reports of antibiotic and other chemical overuse in the aquaculture farms. This compelled the authorities to improve and promote good health management practices, as well as strengthen institutional, policy, and regulatory frameworks, and monitoring and control systems. New pathogens have emerged in recent decades as a result of increased production, diversification, and the introduction of new species into the farming systems. In recent years, for example, freshwater species have experienced Epizootic Ulcerative Syndrome (EUS) and shrimp have experienced Early Mortality Syndrome (EMS) or Acute Hepatopancreatic Necrosis (AHPN) (Bureau of Fisheries 2012; Chinese Ministry of Agriculture (n.d.)). China established a nationwide network of aquatic disease prevention and control centers at the provincial and national levels in 2013. Major areas and species groups have been covered by the centers. The system has identified, isolated, and purified several new pathogens in the field of aquatic animal disease epidemic surveys, including the new genetic type of shrimp Yellow Head Virus (YHV), the second Cyprinid Herpesvirus (CyHV2), Oyster Herpesvirus 1 var (OsHV1 var), and the Infectious Hypodermal and Hematopoietic Necrosis Virus (IHHNV) from penaeid shrimp. The World Organization for Animal Health (OIE) officially designated Yellow Sea Fisheries Institute (YSFI) and Shenzhen Quarantine Laboratory (SQL) as global reference laboratories for White Spot Disease (WSD) and Carp Viraemia Disease (CVD), respectively. The establishment of a "Disease Free Zone and Controllable Zone" for the Grass Carp Retrovirus (GCRV) disease in the vast Zhujiang River delta in Guangdong Province and Hainan Province, as well as the widespread use of vaccination, were two additional noteworthy achievements of the Chinese government (Bureau of Fisheries 2012).

4.1.1.4 Feed Management

Ren and Zhou (2001) reported that in China's aquaculture sector, the use of pellet or compound feeds increased from 750,000 tonnes in 1991 to 19,000,000 tonnes in 20 years (China Feed Industry Association 2016; Han et al. 2016). According to FAO (2010), the world's total production of aquaculture feed was around 50,000,000 tonnes, of which around 50% were used in the Chinese aquaculture industry. Bai and Lin (2015) noted that the Chinese aquaculture industry's quick rise was greatly aided by the availability of high-quality feed supplies. Nonetheless, the FAO calculated that 800,000 tonnes of fish oil and 4,000,000 tonnes of fishmeal were used globally in the feed sector, which posed serious concerns. The use of low-value fish in aquaculture is a significant issue as well (approximately 5–7 million tonnes). It is estimated that China has utilized half of this amount, primarily in the marine culture.

Fish meal consumption in China remained constant at 1–1.5 tonnes annually, according to Han et al. (2016), despite increased aquaculture and feed production. There are plans in place to lessen the reliance of Chinese aquaculture on fish meal, but China continues to be the world's largest consumer of fish meal (Han et al. 2016).

4.1.1.5 Low Input Aquaculture Systems

Seaweed and mollusk farming, as well as the cultivation of valuable filter-feeding freshwater species, have enormous potential for improving the socioeconomic standing of rural communities in China. China can impart some valuable lessons on these farming techniques to other nations. Because the communities lack access to alternate, higher return economic pursuits, these species groups are being less actively cultured in many rural areas in other developing countries. The aquaculture industry also faces numerous challenges that call for technical or policy support, which can be summed up as follows: access to or allocation of limited resources, technological advancement, improvement in policy and legal frameworks, service support availability including credit and insurance coverage, strict market and trade requirements, standards of product quality and safety including labeling and traceability issues, potential impacts of climate change.

4.1.2 The Impact of the Pandemic on the Aquaculture of China

As previously indicated, China's aquaculture industry has a highly specialized workforce and strong ties between upstream and downstream stakeholders (Gui et al. 2018). Production of food and logistics were significantly impacted by the COVID-19 epidemic (Minahal et al. 2020). Tilapia and catfish are the most popular aquaculture products in China, where they are also exported (Yuan et al. 2017). The production of these species is mainly in two provinces with highest aquaculture rank in China (BoF 2020). Guangdong and Hubel provinces are ranked first and fourth in terms of all aquaculture productions of China (BoF 2020). In 2019, the overall production of farmed tilapia in Guangdong Province surpassed 740,000 metric tonnes, making it the country's largest producer and exporter of fish. The most significant channel catfish farming regions in China are in Hubei Province, which is also the main source of channel catfish seeds for the entire nation. Hubei is also the largest freshwater aquaculture province in the world. In 2019, Hubei Province produced 40,000 metric tons of farmed channel catfish overall (BoF 2020). Because of this, utilizing these two provinces as examples can help illustrate the effects of the pandemic in China. Aside from that, the outbreak was concentrated in the province of Hubel, and Guangdong was one of the hardest-hit provinces (Yuan et al. 2022).

The catfish and tilapia value chains were specifically impacted by the pandemic, according to a study done by Yuan et al. (2022). The study demonstrated that due to a delay in harvesting, fish stocks were held up, the farming cycle was extended, and the regular management was interrupted, which resulted in poor fish growth, lack of fish traders' supplies, panic in the international markets, higher production costs, and delay in revenues. Also, the pandemic increased the strain on women by putting greater responsibility on them to care for their children, educate them, and maintain the necessities of life for their families (Chanrachkij et al. 2020).

To assist local governments in maintaining normal operations throughout the pandemic, the central and provincial governments of China issued several directives and put relevant policies into place (Yuan et al. 2022). The most significant steps taken by the Chinese government are listed below, according to Yuan et al. (2022):

- Provide emergency relief support to ensure the livelihood of stakeholders.
- Establish unrestricted logistics to ensure normal production and marketing in the agriculture sector.
- Support the normal operation of the sector through strengthened sectoral monitoring and analysis and information dissemination.
- Encourage processing plants and farmers to develop new forms of products that can adapt to new marketing modes and market needs.
- Provide financial assistance to ease the difficulty experienced by the industrial chain.

The COVID-19 pandemic will have a long-term effect on consumers' spending and consumption patterns as well as the cost of aquatic products. There will be a great deal of uncertainty over China's overall aquaculture performance because of the pandemic, but as has been seen, there are technological gaps that can be exploited in the event of sudden occurrences like the pandemic.

4.1.3 Poseidon-AI in China

Since China is the world's largest producer of aquatic animals, Poseidon-AI is attempting to address concerns with aquaculture there. Saving 20% of the cost of feeding in Chinese aquaculture farms can considerably improve global sustainability through resource preservation. Nonetheless, there are high-value species that hold a special position among Chinese and Asian customers. As was noted, China is the biggest producer of freshwater species for both local and international consumption. Grouper is a prime example of a species whose price per kilogram in China can reach 30 USD and which is served on special occasions like the Lunar New Year.

Because of this, Poseidon-AI set out to apply its Poseidon-AI® Algorithms for Grouper which is considered as a high-value species. The primary species raised in Hong Kong, Taiwan, and Hainan is grouper, which is transported to mainland China. Due to the difficulties between Taiwan and Mainland China and the ferocious protests in Hong Kong, Poseidon-AI was forced to install in Hainan.

The smallest and most southern province of China is Hainan, which is situated in the South China Sea (Fig. 4.2). Hainan Island represents 32,900 km^2 of the province's 33,920 km^2 total land area. Wanquan River is in the east, Changhua River is in the west, and Sanya River is in the south of the island. The Nandu River is in the northern section of the island and Songtao Reservoir is the name of the island's biggest reservoir.

Hainan has a predominantly tropical climate, with the coolest months of January and February having temperatures between 16 and 21 °C. During the hottest months of July through August, temperatures range from 25 to 29 °C. In the steep regions of the island's center, Hainan experiences daily average temperatures of about 10 °C all

Fig. 4.2 Map of China and the location of Hainan province

year long. The annual rainfall in Hainan's central and eastern regions ranges from 1500 to 2000 mm, with peaks of 2400 mm. Rainfall in the South West's coastal sections may only amount to 900 mm.

The Islands' predominant economic activity is agriculture followed by tourism thanks to the tropical beaches and forests (Fig. 4.3). Paddy rice is grown in the lowlands of the north and in the valleys of the southern highlands. In addition, Hainan province produces sugarcane, coffee, tea, cashews, coconut, palm oil, sisal, pineapples, black pepper, and tropical fruits. Water buffaloes, ducks, geese, chickens, goats, and cows are among the domesticated agricultural animals.

The economy of Hainan is significantly influenced by aquaculture and fishing (Fig. 4.4). Grouper, Spanish Mackerel, and Tuna are species that can be caught in the wild, while shrimp, scallops, and snails are farmed for both domestic and international markets. Hainan is home to an estimated 100,000 fish farmers, many of whom specialize in growing grouper and tilapia.

4.1.3.1 Grouper Culture

The marine finfish species known as groupers or rockcods are members of the *Epinephelinae* subfamily. The tropical and subtropical coastal waters are home to a large population of groupers. The Indo-Pacific region has 110 species of groupers, and there are roughly 159 species globally (Randall 1987; Kohno et al. 1990;

Fig. 4.3 Old Town of Haikou, Hainan Province, China

Fig. 4.4 Aquaculture Farm located in Hainan Province

Heemstra 1991; Heemstra and Randall 1993). In Asia, groupers are one of the most expensive fish, particularly in Hong Kong, Taiwan, and mainland China. They are a highly coveted species.

The family of *Serranidae* is divided into subfamilies of *Anthiinae*, *Epinephelinae*, *Grammistinae*, *Nephoninae,* and *Serranidae.* All groupers belong to subfamily *Epinephelinae* with 15 genera: *Aethaloperca*, *Alphestes*, *Anyperodon*, *Cephalopholis*, *Cromileptes*, *Dermatolepis*, *Epinephelus*, *Gonioplectrus*, *Gracila*, *Mycteroperca*, *Paranthias*, *Plectropomus*, *Saloptia*, *Triso, and Variola.* According to the reports from Randall (1987), Randall and Heemstra (1991) and Heemstra and Randall (1993), the most cultured groupers in Asia are Orange-spotted grouper (*Epinephelus coioides*), Malabar grouper (*Epinephelus malabaricus*), Giant grouper

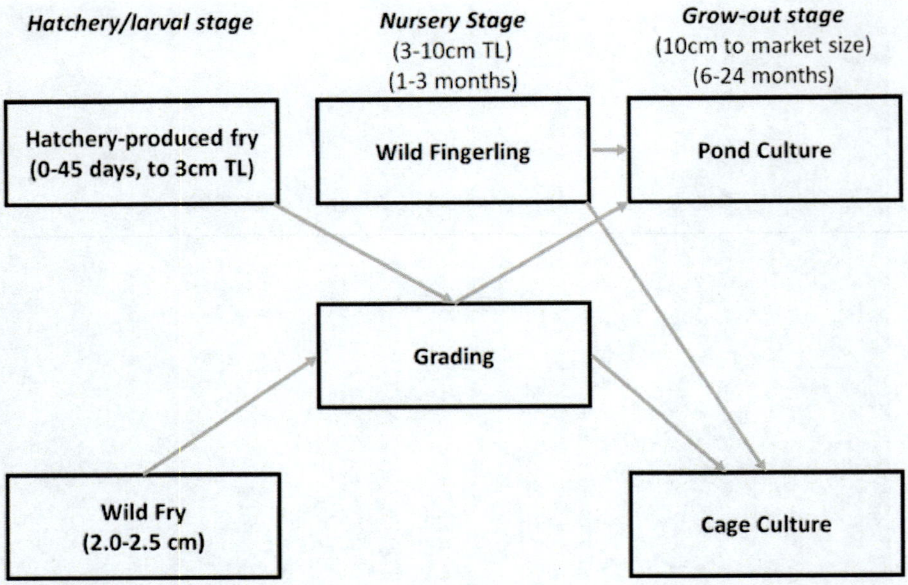

Fig. 4.5 General procedures for grouper culture in Asia (Liao et al. 1995)

(*Epinephelus lanceolatus*) (Sirimontaporn 1993; Main and Rosenfeld 1996; Chao and Chow 1996; Su and Liao 1996).

The majority of cultivated grouper seeds come from the wild; very few are generated in the hatcheries of Hong Kong and Taiwan. The natural seedstocks are collected everywhere throughout the tropical coasts, but they are particularly abundant in areas with robust coral reef ecosystems. Traps can be used to gather fingerlings at any time of the year, but the greatest period for catching grouper fry is from October to March.

In the 80s, groupers were studied to be cultured in several countries in Asia (Anon 1995; Chu 1996; Su and Liao 1996). The species are found in cage-culture sites in South-East Asia, Hong Kong, Japan, South Korea, and Taiwan. There are 22 species of grouper cultured in Southeast Asia and China, namely *Epinephelus akaara*, *Epinephelus amblycephalus*, *Epinephelus areolatus*, *Epinephelus awoara*, *Epinephelus bleekeri*, *Epinephelus caeruleopunctatus*, *Epinephelus coioides*, *Epinephelus fario*, *Epinephelus fasciatus*, *Epinephelus fuscoguttatus*, *Epinephelus lanceolatus*, *Epinephelus macrospilos*, *Epinephelus malabaricus*, *Epinephelus moara*, *Epinephelus rivulatus*, *Epinephelus septemfasciatus*, *Epinephelus summana*, *Chromileptes altivelis*, *Cromileptes argus*, *Cromileptes miniata*, *Plectropomus leopardus*, *Plectropomus maculatus*. The market size for groupers in Asia varies, but it often falls between 600 and 800 g or 1.2 and 1.5 kg. The growth of smaller fish that can be sold will take 8–10 months, whereas the growth of larger fish will take 18–24 months.

The three stages of grouper culture are the hatchery/larval, nursery, and grow-out stages. The general grouper culture process is shown in Fig. 4.5.

At the hatchery stage, the larvae are kept in fiberglass tanks with constant, mild aeration. The tanks are usually stored in secure areas where only people with clearance from the hatchery operators are allowed to access. In addition to nylon net cages, the nursery stage is conducted in ponds, cement, or fiberglass tanks (kept floating with Styrofoam or Plastic Carboys in rivers). The grown groupers are typically kept in cement tanks, earthen ponds, or floating net cages. The floating net-cage system is the most common culture method in Asia. Ponds are popular in Taiwan, Hong Kong, mainland China, Malaysia, and Thailand (Liao et al. 1995; Ruangpanit and Yashiro 1995).

The majority of groupers are protogynous hermaphrodites (male sex is eventually acquired despite being born female) (Shapiro 1987). This indicates that groupers mature in females first, and by the time they are 2–3 years old, they weigh 1.7 kg. They then mature into males older than 5 years old who weigh more than 5 kg (Moe 1969; Kungvankij et al. 1986; Lim et al. 1986; Doi et al. 1991; Tan-Fermin 1992; Tan-Fermin et al. 1993, 1994; Yashiro et al. 1993). However, some grouper species, including *Epinephelus malabaricus* and *Epinephelus microdon*, can be found to be females even in larger fish (Debas et al. 1989; Yashiro et al. 1993).

Males, which are hard to come by in the net-cage environment or in the wild, are necessary for grouper spawning in captivity. Since sex inversion must be induced through hormonal manipulation in order for grouper spawning to be successful, it was demonstrated in the study done by Tan and Tan (1974) that the germinal epithelium of the gonads in *Epinephelus coioides* contains both ovarian and testicular tissues. Yashiro et al. (1993) asserted that because fish mature as females first in their natural environments, there are more female hormones present in the early stages of the grouper life cycle than male hormones.

Chen et al. (1977) reported the first successful induced spawning of groupers in Singapore. However, once the androgen 17α methyltestosterone (MT) treatment stops, a large proportion of these MT-inversed functional male reverted to being females. Hence, prolonged treatment is necessary to obtain functional males for spawning (Chao and Chow 1990; Tan-Fermin 1992; Tan-Fermin et al. 1993, 1994; Quinitio 1996). Chao and Lim (1991) reported the effective implantation of hormones to transform female groupers to functional males. Since 1991, induced spawning has been conducted on many species of groupers such as *Epinephelus malabaricus* (Pakdee and Tantavanit 1985; Rattanachot et al. 1985; Hamamoto et al. 1986; Huang et al. 1986; Kungvankij et al. 1986; Ruangpanit et al. 1988), *Epinephelus coioides* (Tan-Fermin 1992; Tan-Fermin et al. 1993, 1994; Quinitio 1996), and *Epinephelus akaara* (Ukawa et al. 1966; Xu et al. 1985; Fukunaga et al. 1990; Maruyama et al. 1993).

Several species of groupers, such as *E. fuscoguttatus*, *E. summana*, *E. caeruleopunctatus*, *E. macrospilus*, *E. malabaricus,* and *E. coioides*, have been found to be able to spawn naturally in a captive environment (Ruangpanit et al. 1993; Quinitio 1996). For *E. malabaricus*, Ruangpanit et al. (1993) reported that changing 80% of the seawater in a 150 tonnes concrete tank over a period of 5 days prior to the appearance of the full moon stimulated the groupers. The beginning of the lunar period also coincided with the natural spawning of the *E. coioides* (Toledo et al.

1993). It is believed that the hatching rate of eggs from the brood stock that spawns naturally is higher than that from hormone-induced spawners. According to Abu-Hakima et al. (1983), the study of natural spawning of groupers can play a vital role in increasing the productivity of the fry.

Despite the ease with which seeds are now produced, the quality of the produced larvae has a high mortality rate before metamorphosis (Chen et al. 1977; Abdullah et al. 1983; Akatsu et al. 1983; Xu et al. 1985; Ruangpanit et al. 1986, 1993; Chao and Chow 1996; Hussin and Ahmad 1996; Quinitio 1996). Poor hatchery practices, nutritional deficiencies in the brood stock, and a lack of the proper live larval food organisms are all factors that contribute to high mortality.

Feeding the grouper larvae will start 6 h after the mouths open, normally on the second day (Kitajima et al. 1991; Kungvankij et al. 1986; Ruangpanit et al. 1993; Duray 1994; Doi et al. 1996). The diet at this stage consists of Chlorella, rotifers and artemia. Forty-five days after the groupers are hatched, minced fish is given to the larvae. According to Doi et al. (1996), the chances of survival of grouper larvae increase by providing them with a mixture of nauplii of *Pseudodiaptomus annandalei* and rotifers, compared to the groupers larvae fed only with rotifers.

The environment in the culture tank is important for grouper larviculture. According to Duray et al. (1997), grouper larvae cultured in 3 tonnes tanks have a better survival rate of 19.8% at day 24 compared with only 7.4% for those placed in half tonnes tanks at day 21. Chao and Lim (1991) noticed that larvae swim erratically if there is strong and intense sunlight. According to Chao and Lim (1991), 801 green algal water is needed to achieve a transparency of about 50 cm, and 401 is needed to achieve a transparency of around 70 cm. Newly hatched groupers are very sensitive to stress and handling (Predalumpaburt and Tanvilai 1988). Mortality due to handling is avoided by stocking larvae into culture tanks 2 h before hatching and at the high density of 40 L^{-1} (Lim et al. 1986).

Thyroid hormone levels have been shown to be higher in buoyant estuarine grouper eggs than in non-buoyant eggs (Lam 1994; Lam et al. 1994). Numerous larval marine fish have shorter metamorphosis times when thyroid hormones are present (Inui and Miwa 1985; Miwa and Inui 1987; Miwa et al. 1988; Hirata et al. 1989; Inui et al. 1989; Reddy and Lam 1992; Lam et al. 1994). Fertilized eggs were submerged in 0.5 ppm triiodothyronine (T3) for 6 h after fertilization. This increased larval survival, the rate of hatching, and accelerated the dorsal and anal fin resorption (Tay et al. 1994; de Jesus 1996).

The rearing of grouper larvae comes to an end after 45–60 days, and they are then moved to nursery ponds. The primary purpose of this grouper nursery culture period is size grading. If the cannibalistic behavior is not size graded, there is a high mortality rate. Throughout the nursery stage, grouper fry undergoes continuous size grading every week. They are moved to grow-out ponds and cages when they reach a height of 6–7.5 cm, which takes 4–12 weeks on average.

According to Liao et al. (1995), the stocking density of groupers in intensive pond culture is 2–7 fish/m^2 (Liao et al. 1995); the density is 20–30 fish/m^2 if net cages are used (Ruangpanit and Yashiro 1995). It takes approximately 7–8 months for groupers cultured in either ponds or net cages to reach the marketable size of 600–800 g and 12–16 months for them to reach 1200–1400 g.

The environmental conditions play an important role in grouper culture. The D.O. content needs to be stable and high enough, approximately 5.54 mg/L. According to Nabhitabhata et al. (1988), theoretically it is estimated that 75–457 grouper fry/m^2 can be stocked in a net cage to produce 500 g groupers and 40–244 fry/m^2 to obtain 1200 g groupers. Artificial hides can be placed in the cages to increase the stocking density (Teng and Chua 1979). Under the normal favorable optimum conditions, the production of groupers per cage will vary with different stocking density. The stocking density of 75 m^{-2} provides the highest production at 33.9 kg per 1 × 1 × 1.5 m^2 as compared to 30 m^{-2}, 45 m^{-2}, 60 m^{-2}, and 90 m^{-2}, respectively, 16.82 kg, 23.85 kg, 29.68 kg, and 32.9 kg (Sakaras and Sukbantaung 1985; Sakaras and Kumpang 1987). The stocking density will also affect the FCR. When high stocking density is used, the FCR of groupers is lower and vice versa. According to Tookwinas (1989), the better FCR in a high stocking density situation is because bigger fish groups require less energy for the fish to swim against strong currents, also their appetites are better stimulated (Sakaras and Kumpang 1988).

Carnivorous fish such as groupers need a high protein diet to grow faster and healthier. The level of dietary protein requirement varies with the species of groupers, their size and culture conditions (Chua and Teng 1978, 1982; Hu and Lim 1984; Kohno et al. 1989). Grouper fingerlings require higher protein content in their diet than adult groupers (Wongsomnuk et al. 1978).

Fish juveniles need between 47% and 60% of their body weight in protein to gain the most weight (Sukhanongs et al. 1978; Teng et al. 1978; Wongsomnuk et al. 1978; El-Dakour and George 1982). Grouper diets should contain 14% lipids on average (New 1987). Groupers suffer from deficiencies in vitamins and minerals. For instance, juvenile groupers fed a diet deficient in vitamin C exhibited signs of loss of appetite, short snouts, erosion of the operculate fins and lower jaw, hemorrhagic eyes and fins, loss of scales, exophthalmia, swollen abdomens, abnormal skulls, lethargy, and paralysis (Boonyaratpalin et al. 1993). Table 4.1 displays the results of studies on the effects of different vitamins on grouper (*Epinephelus akaara*) diets in China (Chen 1996).

Traditionally, groupers are fed trash fish. However, with rapid expansion of net cage and decrease in the availability of trash fish, the feed prices have increased significantly. Groupers that have been fed trash fish are also more tolerant to handling (Liao et al. 1995). In Asian countries, the commercial feeds for grouper contain 73% crude protein, 6% fat, 16% ash, 3% fiber, and 12% moisture. The feed components typically used in Taiwan and Thailand are listed in Table 4.2.

According to ADB/NACA (1991), aquaculture diseases cost the industry millions of dollars each year in lost revenue. Along with the cultured fish's actual health, groupers are handled and transported under stressful circumstances, which results in a high mortality rate in the first week following the placement of the fish in the net cage (Chong and Chao 1986; Leong 1994a, b). Groupers are extremely susceptible to illnesses brought on by various pathogens associated with the new culture site because they do not receive prophylaxis at the culture site. The disease problems

Table 4.1 Symptoms of groupers (*E. akaara*) suffering from vitamin deficiencies

Vitamins	Deficiency signs
Choline	Enlarged liver, hemorrhagic kidney
Folic acid	Reduced appetite, reduced growth
Nicotinic acid	Tetany, lethargy, reduced growth
Pantothenic acid	Mucus on gills, clubbed gills
Pyridoxine	Erratic swimming, tetany, reduced growth
Riboflavin	Opaqueness of the eyes, reduced growth
Thiamine	Reduced growth, lethargy, poor equilibrium
Vitamin A	Protruding and opaque eyes, reduced weight gain
Vitamin K	Hemorrhaging of the epithelium
Vitamin E	Lethargy, reduced pigmentation, reduced growth, increased mortality
Vitamin C	Atrophy of the spinal cord, depigmented areas, reduced growth body deformity, increased mortality

Table 4.2 Dietary composition of some commonly used grouper feeds

Ingredient	Taiwan (Liao et al. 1995)	Thailand (Ruangpanit 1993)	Thailand (Anon 1995)
Fishmeal	70	75	45–50
Soybean meal	–	–	30
Fish oil	–	5	3
Soybean oil	–	–	3–5
Binder/bio flour	2	11	10
Tapioca starch	23	–	–
Rice flour	–	5	–
Rice bran	–	–	7–20
Premixed vitamins	1	0.04	2
Premixed minerals	1	4	2
Others	–	–	3–5

occurring during the culture of groupers can be broadly categorized into two phases according to the culture cycle (nursery or grow-out phase). At the nursery stage, grouper diseases are caused predominantly by protozoans, particularly the ciliates, *Cryptocaryon irritans* and *Trichodina* sp. and the dinoflagellate *Amyloodinium* sp. (Chong and Chao 1986; Chua et al. 1993; Chang 1994; Chao and Chung 1994; Leong 1994a, b).

Grouper fry infected by *C. irritans* show typical signs of white spots on the epidermis (Chong and Chao 1986). Infected groupers normally lose their appetite, and their body color becomes dark. There are reported cases of "paralytic syndrome" disease in grouper fingerlings (Danayadol and Kanchanapungka 1989; Chua et al. 1995). The characteristics of affected fish are loss of appetite, dark body color, and floating on the water surface in a lateral curvature position. Histopathological studies reveal vacuolation of the eyes and brain. Electron microscopy shows the presence of numerous cytoplasmic, non-enveloped virus-like particles icosahedral-shaped, about 20 nm in diameter in the eye and the brain. This disease is diagnosed as viral nervous necrosis (Danayadol et al. 1993, 1995).

In Taiwan, swim bladder inflation syndrome disease is recorded as a major limiting factor during metamorphosis in fry production (Liao et al. 1995). When transferring fingerlings, very often imported fry are found infected with a large variety and high density of parasites (Leong and Wong 1987, 1990, 1993; Ruangpan and Tabkaew 1993; Chao and Chung 1994; Leong 1995, 1997). The most common parasites detected on groupers are the pathogenic ciliates, *cryptocaryon irritans* and *Trichodina* sp., the diplectanid monogeneans, *Pseudorhabdosynochus* spp., as well as capsalid monogeneans, *Benedenia* sp., and *Neobenedenia* sp.

Groupers are susceptible to *hemorrhagic septicemia* caused by vibrios during the grow-out phase. As normal flora in cultured groupers and surrounding environment, various species of vibrio including *V. parahaemolyticus* and *V. alginolyticus* are found to exist (Leong and Wong 1987, 1990, 1993; Ong 1988). Vibrio (*hemorrhagic septicemia*) is often associated with another disease referred to as the "red boil disease." The bacteria, *Streptococcus* sp. is the main cause of this disease.

Another illness that has been found in groupers kept in cages for 3–4 weeks is called "sleepy grouper." Ninety percent of the affected fish die, refuse to eat, have darkened skin, and exhibit a lack of energy (Chua et al. 1993, 1994; Arthur and Ogawa 1996). Red grouper spleens were found to contain a reovirus in *Plectropomus maculatus* imported from Indonesia to Singapore, according to May et al. (1992). The affected fish have anorexia, darkened skin, and a complete lack of appetite.

Groupers also have a condition in which the fish cannot regulate the amount of air in their swim bladder and float upside down at the water surface. The fish eventually die of starvation and exposure to sunlight. Leong (1994a, b) mentioned that swim bladder disease coincides with the monsoon season when there is upwelling of bottom sediments under the net cages.

Excess feeding, especially in net cage culture has led to serious concerns about the environmental impacts. Studies showed grouper culture exerted a localized environmental impact in the Asian-Pacific regions (Muller and Varadi 1980; Bergheim et al. 1982; Beveridge and Muir 1982; Enell 1982; Penczak et al. 1982; Wienbeck 1983; Beveridge 1985; Bohl 1985; Phillips and Beveridge 1986; Anon 1987; Gowen and Bradbury 1987; Molver et al. 1988). The impact of excess feed is greater on the sea-bottom sediments than at the water column (Songsangjinda et al. 1993; Wu et al. 1994).

It is shown that when a culture site is used for a long time, sediment quality deteriorates. Studies showed that the values of organic nitrogen, total nitrogen, total phosphorus, ignition loss, and total sulfide are significantly higher in cage areas where groupers and seabass have been cultured for more than 3 years than in areas where they have been cultured less than 3 years. The deteriorating condition of sediment quality in culture sites has a direct relation with the age of the fish farm. There is a higher frequency of disease outbreak at the aged fish farms (Leong and

Wong 1990). The major impact is therefore on the sea bottom, where anoxic sediments creating high oxygen demand, production of toxic gases and a decrease in benthic diversity within 1 km of the farm may be found (Wu et al. 1994; Wu 1995).

4.1.3.2 Sustainable Development for Grouper Farms Using Deep Tech

As already mentioned, grouper culture involves several elements that, despite its profitability, make it extremely risky for both farmers and the environment. Environmental factors further increase the complexity of grouper farming as the species is subject to a variety of unpredictable conditions during stocking, transportation, and growth.

The government offers certified fingerlings to the farmers due to the high market value and risk in China. These fingerlings are transported in the best conditions to minimize stress and mortality. They are free of pesticides, bacteria, and potential genetic diseases. The farms in Hainan use the water from the coastal waters because they have the best water quality for the region's inland grouper culture. In covered cement tanks on land and in open ponds, the nursery and growth are carried out while water is constantly moving back and forth from the ocean to the tanks and/or ponds (Fig. 4.6).

However, feed wastage causes changes in water conditions and increases the possibility of spread of diseases, contamination of coastal waters and increase in the farms fish mortality. Depending on the age of the fish, most of the farms use high protein pellets as feed. Figure 4.7 shows the feeds given to grouper species in Hainan, China.

The feeds are given twice per day separately per tank and the excess feed will go through the center of tank through a basic filtration unit, gathering the solid wastes. The environmental conditions such as temperature, D.O., pH, and salinity are monitored manually by the farmers with manual sensing tools. According to the regulations set by the local government, farmers can't overuse medications and antibiotics and thus, monitoring the water quality can give early signs of possible disease outbreak in the tanks or ponds. The best possible depth for grouper culture is 1.5–2.5 m (Kangkan 2006). According to Kangkan (2006), the perfect temperature, salinity, pH, D.O., and nitrate level for grouper culture is 28–30 °C, 30–35 ppt, 6.5–8.5, 5 ppm and 0.9–3.2, respectively.

Poseidon-AI® IoT device was deployed in farms in the Hainan province, for collecting environmental information and providing farmers with live environmental data. The dashboard shows the condition of every tank or pond every hour for the period of 18–24 months. The average maximum temperature collected in the farms was 28.5 °C, and the average minimum recorded temperature is 26 °C. The maximum recorded pH is 8 and minimum is 6, also, the D.O. maximum level was 8 ppm.

The ORP was measured to keep tabs on the tanks' or ponds' capacity to self-cleanse or degrade waste. This indicates that the bacteria are capable of breaking down waste materials and dead tissues with ease. Because ORP is based on the concentration of dissolved oxygen in the water, it can also tell us something about it. The water's minimal voltage, which is caused by the electric charge of reducing

Fig. 4.6 Groupers culture tanks in Hainan, China

and oxidizing agents, is indicated by the ORP level. It is possible for contaminants to become more toxic when ORP and dissolved oxygen levels are low. The majority of the bacteria in the water will perish if there are antibiotics or other medications present, which will result in a low ORP. Due to this, most farmers forbid ORP sensors from being used in their tanks or ponds. The ORP data, however, showed a maximum ORP of 480 mV and a minimum ORP of 300 mV at some farms in Hainan. For various types of water, the ORP level is displayed in millivolts in Table 4.3.

For grouper raised in Hainan, the Poseidon-AI® algorithms were used to calculate the recommended feeding rate. The size of the tanks and/or ponds was taken into consideration when calculating the feeding amounts. A tank of groupers typically holds 75–100 fish, and feeding takes place twice daily. Since the farmers have at least 100–150 tanks or ponds, they can produce between 10,000 and 150,000 grouper fish per cycle. The FCRs on the basis of the pellets fed in the farms were roughly 2.1 $\pm$ 0.7, displaying an approximate period of harvest stated on 22 months, hoping for each fish to achieve 1.6 kg.

The cameras can keep an eye on the pellets as they sink because of the water's high transparency in the tanks. The feeding hours play an important role in

Fig. 4.7 High protein pellets used in grouper farming in Hainan, China

Table 4.3 Approximate ORP level in different water types

Water type	ORP level (mV)
Well water	0
Pure water	200
Mineral water	250
Tap water	220–380
Surface seawater	400
Deep seawater	450
Rainwater	600

monitoring feed wastage since the excess feed will be taken out of the tanks. Sinking pellets are more difficult since the amount of feed recommended by the Poseidon-AI® algorithms need to calculate the exact amount before the feeding time starts. Once feeds are thrown into the water, the image processing cannot send images with high accuracy to the Cloud. For this reason, it was necessary to increase the imaging pixels sent by the cameras. This increased the data transferring cost per device but helped with easier calibration of the algorithms.

Another issue faced was the filtering of AWS Cloud system in China. The data pumped up to the Cloud using 4G/5G Chinese Sim card was blocked or failed due to filtering system set in China on Amazon and Google. Under these circumstances, there were two ways to overcome this obstacle.

First, using 4G/5G LTE Sim card from Singapore or Hong Kong under roaming. This increased the data fee significantly but there was no interruption in data transfer to the AWS Cloud. The second way to overcome this problem was to shift the Cloud to Alibaba Cloud system. The Alibaba Cloud system can receive the data from the farms in Hainan with no interruption but in order to conduct the image processing and/or input the environmental data gathered from the IoT devices into the Poseidon-AI® algorithms, it is necessary to transfer data from Alibaba Cloud to AWS. There is a fixed rate per byte of transferred data from Alibaba Cloud to AWS, which with high resolution images necessary to monitor the sinking pellets, imposes significant costs both for farmers and the company.

An average of 20% in feeding costs were saved per tank or pond over the course of using Poseidon-AI® algorithms. The training process for the algorithms, however, was unable to be successfully finished due to high data transmission costs and severe pandemic restrictions. In addition, some farms used medication and antibiotics, which led to farmers removing the devices without first giving notice. As a result, the number of active devices in Hainan Province decreased, and the objectives set for the 24-month period were not met. In conclusion, despite the Poseidon-AI® systems and algorithms' success in reducing the cost of feeding by 20% in the grouper farms in Hainan, the Chinese government has imposed severe pandemic restrictions that prevent farmers from continuing their operations. According to reports, COVID-19 disrupted the supply chain, preventing farmers from receiving feed, fingerlings, and technical support to continue their regular farming operations. There have also been instances where farms' water circulation systems have failed, significantly lowering the quality of the water in the tanks and ponds. Normally, a system failure could be fixed in 2–3 days, but due to the restrictions, farmers were unable to fix their systems, leading to a high rate of farm mortality. With failure in water circulation systems, the chances of spread of diseases increased, and with no support, many of the farmers started self-medicating their stocks, with the hope of increasing their survival rates.

Recharging the Sim cards have also faced problems due to the pandemic. Since each device carried one Sim card and every farm has 100–150 tanks/ponds, it was nearly impossible to recharge all the Sim cards via the online platform. After the Sim cards were not charged, the devices saved the information on the memory cards until the memory cards were filled with environmental data and images.

4.2 Indonesia

4.2.1 Aquaculture Production in Indonesia

Among producers and consumers of seafood, Indonesia ranks third (FAO 2022). Consequently, it is crucial for Indonesia to produce seafood from a financial and nutritional standpoint. With an annual growth rate of 8.5% until 2030, Indonesia wants to increase its aquaculture species (IDH 2018) and outpace its population growth rate (Directorate General of Aquaculture 2017). Indonesia is the fourth-largest shrimp exporter after Argentina, Ecuador, and India, according to Guillen et al. (2019), and it ranks third among countries that consume seafood after China and Japan. Domestic consumption accounts for 80% of Indonesia's capture production and almost all of its aquaculture production (Belton et al. 2017). However, Usfar and Fahmida (2011) reported that macro- and micronutrient deficiencies are still widespread in Indonesia. Therefore, increasing mean and overall fish consumption is recommended as part of the national diet (Behrens et al. 2017).

In Indonesia, aquaculture production is already well-established. However, a nation like Indonesia, which has many biodiversity hotspots (Maynard et al. 2010; Abood et al. 2015), faces numerous resource conservation challenges (Gaither and Rocha 2013; Murray et al. 2015).

Freshwater species like tilapia, Clarias catfish, carp, and Pangasius catfish are the most common species raised in Indonesia (BPS 2018), followed by backwater species like shrimp and milkfish. Shrimp and tilapia are the two most exported farmed seafood species, both in terms of volume and value, according to FAO (2019). Over the past 10 years, whiteleg shrimp species (*Litopenaeus vannamei*) have been cultivated in Indonesia, enhancing the country's aquaculture output (FAO 2019).

Aquaculture products of superior quality, nutritious, and environmentally friendly produced are becoming more and more in demand among consumers in the Asia-Pacific region. The Indonesia Act of 2004 requires aquaculture to follow quality and product safety standards. The AsiaFish model was used to project aquaculture growth and its environmental implications in order to follow the development of the Indonesian aquaculture industry (Dey et al. 2016; Tran et al. 2017; Henriksson et al. 2017). The most crucial elements in attaining sustainable aquaculture in Indonesia are the accessibility of high-quality seed, sound farming practices, aquaculture environments, fish health management, product quality, and marketing. These models concluded that none of the six alternative scenarios could achieve the growth targets, and the majority would have a significant negative impact on the environment both locally and globally. Based on the findings, it can be highlighted that new and innovative technologies are required to participate in Indonesian aquaculture in order to support sustainable development of this sector while minimizing the environmental effects caused by the sector's rapid development.

4.2.1.1 Fish Health Management

According to the farmers' experiences, the use of medications for disease treatment in aquaculture farms had a negative impact on the environment. Disease prevention-focused first begins with good management practices, particularly high-quality water, certified seeds, and premium feeds. Adding probiotics to the pond water on a daily basis, avoiding changing the water during the first 2 months of culture, and utilizing fish and shrimp seeds Free of Specific Pathogens (SPF) are other ways to control sickness. Farmers in Indonesia can feel secure in the health of their cultured stocks even though SPF seeds are twice as expensive. Additionally, the local government asks farmers to set aside 10–30% of their land for reservoir construction and to use probiotics to maintain good water quality and prevent deadly fish diseases. To stop disease carriers from entering the pond, biosecurity zones are also applied to the farms.

Before flushing the water into the environment, chlorine is used to disinfect the potentially disease-carrying water. The government now gives farmers guidelines on water quality requirements, and the depth and length of aeration are restricted based on the stock density (Sutanto 2005).

Farmers have been made aware of the ecological role that the mangrove ecosystem plays. Local communities have started mangrove management and reforestation projects in a number of shrimp farms. The action will be advantageous for shrimp farms, particularly those that cultivate whiteleg shrimp. Although the rapid expansion of whiteleg shrimp has raised questions about sustainability over the long term. Farmers in East Java received SPF seeds from hatcheries in Hawaii and Florida. However, Taura Syndrome Virus (TSV) has been detected in many shrimp farms. Whiteleg shrimps are not resistant to other viruses such as White Spot Syndrome Virus (WSSV) and Infectious Hypodermal and Hematopoietic Necrosis Virus (IHHNV) (Tauhid et al. 2006). To find hazardous infections, it is necessary to perform quick and early diagnostics on brood stocks and on seeds up until the developing stage in ponds.

Indonesian freshwater aquaculture is hampered by Koi Herpes Virus (KHV). KHV is a contagious disease which infect common carp and Koi carp in Indonesia since 2002. To overcome this disease, government of Indonesia developed strategies such as integrated fish health management; using SPF-KHV fish quarantine system; applying immunoprophylactic to halt status of fish; inducing specific immunity vaccination by cohabitation technique; stress factor avoidance; disinfectant; treatment against secondary infections; biosecurity application; polyculture system; avoidance of macroclimate affect; and altering commodities (Tauhid et al. 2005).

4.2.1.2 Trash Fish to Feed the Farmed Fish

Trash fish is frequently used in Indonesia to feed marine finfish species, including small-scale grouper culture. Trash fish is only accessible at specific periods of the year, and there is evidence that the Malacca Straits are producing less small pelagic and demersal fish (Sumiono 2006). In Indonesia, there are commercial diets for milkfish (*Chanos chanos*), barramundi/seabass (*Lates calcarifer*), and groupers, with 80% of the fish meals coming from Chile and Peru. Due to higher feed wastage in open sea cage culture (Wu 1999), feeding loss and environmental pollution near sea cages are rising. For this reason, the development of feeding by commercial pellets is prioritized. According to research on nutrient needs, juvenile groupers require diets that are high in digestible CP, fairly low in lipid, and provide at least 1% and ideally 1.5% Omega 3 HUFA (Rimmer and Sugama 2005). Seaweed farming allows for the recycling of phosphorus and nitrogen wastes from any source, not just aquaculture farms, which may be highly advantageous (Sorgeloos 2001).

4.2.1.3 Excess Use of Drugs and Contamination in Open Waters

According to Indonesian Act No 31/2004 (2004): "*owner of aquaculture farm, representative of the owner, and/or responsible person who run aquaculture business is not allowed to use chemicals, biological agents, explosive materials, equipment and/or constructions which may be harmful and/or threatens the sustainability of fish resources and/or its environments of fisheries management area in the Republic of Indonesia.*" Manual of Good Aquaculture Practice is developed to prevent farmers from using prohibited and restricted drugs.

Water quality monitoring is conducted in the area where mollusk culture operates. The monitoring program is conducted by Aquaculture Development Centers under Directorate of Aquaculture. Mollusk culture is not permitted in areas where excessive levels of heavy metal or potential poisoning plankton occurred.

4.2.1.4 Carrying Capacity Assessment for Aquaculture Planning

In order to lessen the effects of mariculture, coastal shores must have land-based support infrastructure. The supporting infrastructure is based on estimates of carrying capacity in a variety of habitat types. Assessments of land capability and estimates of carrying capacity are given top priority in Indonesia. According to Wu (2001), improving feed formulation and implementing integrated culture are two ways to lessen the negative effects of fish farming on the environment. In South Sulawesi, Awarange Bay has a carrying capacity of 3 tonnes of fish biomass and has a potential marine culture area of 28 hectares (Rachmansyah 2004).

4.2.2 Indonesia's Sustainability Measures for the Aquaculture Sector

Indonesia focused on two main bottlenecks for the sustainable development of the aquaculture industry. The first bottleneck is associated with stock breeding, and the second is associated with the dietary requirements of these cultured animals.

4.2.2.1 Breeding and Genetics

A crucial component of the aquaculture industry is high-quality seeds. There have been many brood stocking facilities established over the past ten years. Shrimp brood stock centers can be found in Ujung Batee, Takalar, Situbondo, and Jepara (all in Mid Java). White shrimp (*L. vannamei*) breeding programs are currently being carried out by the Gondol Research Institute for Marine Culture. In Sukabumi (West Java), Jambi, Mandiangin (South Kalimantan), and Tatelu (North Sulawesi), tilapia brood stock centers have been established. Common carps, tilapia, and freshwater giant prawn are the priority species for national breeding programs. The seaweed center is developed in Lombok and Grouper Center in Lampung. Sugama (2005) mentioned milkfish (*Chanos chanos*), grouper (*Epinephelus* sp., *Plectropomus* sp., *Cromileptes altivelis*), freshwater prawn (*Macrobrachium rosenbergii*), mud crab (*Scylla* sp.), catfish (*Clarias batrachus*, *Pangasius* sp.), sea cucumber (*Holothuria scabra*), seaweed (*Gracilaria* sp. and *Eucheuma* sp.) as important cultured species in Indonesia.

Small-scale hatcheries in many coastal areas are used to produce milkfish fry, grouper, and Post Larvae (PL) of tiger prawn (*Penaeus monodon*). Milkfish and grouper seed production is the main activity of backyard hatcheries located in Gondol (Bali), while tiger prawn PL is produced by those in South Lampung, Northern Part of Mid Java, and Barru (South Sulawesi).

Artemia is an important natural feed use in the process of shrimp PL production where Indonesia imported this artemia around 100 tonnes every year. Artemia culture is now developing in salt production ponds in Mid Java, Madura, East Java, and East Nusa Tenggara to provide domestic artemia.

In freshwater aquaculture, there are 13 species of Pangasius that have been found in several rivers in Sumatera Java and Kalimantan. Cross breeding between *P. hypophthalmus* and *P. jambal* has been observed, and its offspring has better growth than that of their parents. This *P. jambal* and its hybrid have been released as a good candidate species for national pangasius culture in Indonesia. Additionally, species of freshwater lobster like *Cherax albertisii*, indigenous species of Papua and red claw (*C. quadricarinatus*) has been studied.

In 2004, a certification program on seed production has been initiated to assure good quality seed which are produced by hatcheries. Integrated quality management control on seed production system is applied to create more feasible aquaculture and sustainable production.

4.2.2.2 Feeding and Feed Production

Depleted fish stocks caused higher fish meal prices as well as increasing costs of fish feeds. In order to overcome feeds high prices, several steps are taken:

- Increase the utilization of local fish meals
- Substitution of fish with agricultural wastes
- Synchronizing the input levels with the culture technology
- Special zoning area for aquaculture to provide better marketing facilities
- Improvement of feed composition with lower fish meal usage

IAS and polyculture are other techniques used in rainfed ponds applied by farmers in Lamongan, East Java. Intensive fertilization is applied to stimulate natural feed and hence minimize the cost of feeds. Culture-based fisheries are developing in Mid Java with the aim to empower community-based management on small reservoirs. There is no feed required for these activities, but the availability of seeds is demanded. Organic culture of shrimp is usually conducted by farmers at low intensity feeding in same places in East Java and East Kalimantan. Silvo aquaculture where mangrove area is utilized for stocking crabs or fish is also one good example as there is no feeding aquaculture in Indonesia.

4.2.3 Poseidon-AI in Indonesia

In Indonesia, farms rely heavily on labor to carry out daily tasks like feeding, monitoring the quality of the water, and cleaning. This is mostly a result of the affordable labor and simple aquaculture-related duties. Under these conditions, the use of technology cannot be for the purpose of automating tasks, but rather to carry out related tasks more effectively.

Deep-tech underwater cameras won't be a useful tool to monitor stock health, feed waste, and disease control in waters with low water transparency. However, water monitoring for cultivated species in Indonesia can aid in enhancing both their living and growth conditions.

The feed issue affecting Indonesian farms is addressed by Poseidon-AI. Poseidon-AI® algorithms assist carp farmers in Indonesia in reducing their feeding costs by 20%, which is a major step. In order to assist farmers, the strategy exports environmental data. The carp culture began to become more and more popular over the past 20 years. This is because of the effective rice-fish strategy used in China and adjusted to the situation in Indonesia.

4.2.3.1 Carp Culture

Common Carp generally inhabits freshwater environments such as ponds, lakes, and rivers, but rarely inhabits brackish water environments (Barus et al. 2001). It is very popular in Asia and European countries (Weber and Brown 2011; Kloskowski 2011a, b; Parkos and Wahl 2014) and has high adaptive capability to both environment and food web (Soltani et al. 2010; Manjappa et al. 2011; Rahman 2015). In central and east European countries, carp contributes to more than 80% of total aquaculture production (Woynarovich et al. 2010; Anton-Pardo et al. 2014). Common carp is also known as an ecological engineer because it has the ability to change the ecological characteristics of aquatic systems (Matsuzaki et al. 2009; Bajer and Sorensen 2015; Rahman 2015). However, in nations like the USA, the carp population is thought to be the biggest threat to the biodiversity of wetlands and shallow ecosystems. As a result, numerous studies are being done to reduce the carp population (Baldry 2000; Weber and Brown 2009, 2011).

The third most widely cultivated species of freshwater fish worldwide is the common carp, a well-known benthivore with significant bottom-up effects. It is believed that productivity of freshwater systems is limited by a lack of soluble phosphorus (PO_4-P) which limit the nutrient for phytoplankton growth. Common carp releases nutrients including PO_4-P from the sediments, and this accelerates the nutrient fluxes to the next trophic levels (Rahman et al. 2008a, b, c, d, e, f; Rahman 2015). This will allow better production of planktivorous fish which grow better with common carp rather than in a monoculture condition (Rahman et al. 2006). This allows carp to be reared in polyculture ponds but in colder regions, monoculture is more common (Szucs et al. 2007). Aquatic ecosystem will also be affected by the density of the animals (Soundarapandian and Kannan 2008; Rahman et al. 2012; Khatune-Jannat et al. 2012; Amira et al. 2016). Overstocking of common carp has many negative effects on aquatic ecosystems (Rahman et al. 2006; Kloskowski 2011a, b; Rahman 2015).

Common carp can browse for benthic macroinvertebrates in sediment and by doing this, it affects water visibility by resuspending clay particles in the water, cycling of nutrients, and affecting the abundance of phytoplankton, zooplankton, and benthic macroinvertebrates (Rahman et al. 2008a, b, c, d, e, f, 2009). Common carp consumes chironomid larvae which, depending on the species, size, and type of sediment, lives several centimeters deep in the sediments. According to Rahman and Verdegem (2007), carp uses special techniques to pick up the larvae from the sediments. Common carp digs actively up to 3 cm to reach benthic macroinvertebrates (Ritvo et al. 2004). By increasing the amount of oxygen available in the bottom soil, the carp's action accelerates aerobic breakdown in the sediment. In comparison to anaerobic settings, this process significantly speeds up the mineralization of organic materials (Beristain 2005; Rahman et al. 2008a, b, c, d, e, f). Therefore, in the ponds, this action has a large impact on the abiotic and biotic properties of the overlying water column (Rahman et al. 2015; Rahman 2015).

Through diffusion, oxygen is incorporated into the pond's bottom soils. This happens as a result of wind and water currents mixing the water column and introducing some oxygen to the lower levels. The carp helps this situation by agitating the zone where the subsurface soil meets the surface water (Rahman et al. 2015; Yathavamoorthi et al. 2010). The acceleration of the aerobic conditions leads to increased decomposition and mineralization of organic matter in the bottom soil. Additionally, resuspension of bottom soil by carp induces decomposition by exporting the organic matter to the water column, where oxygen concentration is generally high.

Rahman et al. (2008a, b, c, d, e, f) mentioned that during aerobic decomposition, D.O. is decreased and CO_2 is emitted, lowering the pH through reducing alkalinity. Furthermore, nitrate nitrogen ($NO_3{-}N$), total ammonia nitrogen (TAN), total nitrogen (TN), soluble phosphorus (orthophosphate as phosphorus) ($PO_4{-}P$), and total phosphorus (TP) in the water will increase. Common carp accelerates nitrogen and phosphorus transport via excretion from the bottom sediments (Morgan and Hicks 2013).

The biomass of phytoplankton in the ocean is enhanced by the increased stimulation of photosynthesis brought on by an increase in nutrients. Phosphate phosphorus limits the phytoplankton production in ponds and lakes. According to Rahman et al. (2015), resuspension by carp decreases this limitation by increasing phosphorus flux for the sediments. Enhancement of phytoplankton biomass will lead to an increase in zooplankton production rates.

Carp density is a very critical factor for the aquatic environment. In shallow unfertile lakes and when the common carp density reaches to 200 kg per hectare, submerged macrophytes will significantly be affected (Kloskowski 2011a, b) while in a fertile lake that can support up to 1125 kg per hectare carp is five times more (Baldry 2000). In polyculture ponds when the carp density reaches more than 1000 kg per hectare, water quality, abundance of plankton, and benthic macroinvertebrates and fish growth can be significantly affected (Rahman et al. 2006).

Another important factor impacting fish density is artificial feeding of the stocks (Rahman et al. 2008d; Rajkumar et al. 2013; Wu et al. 2015). In the presence of artificial feed, common carp shifts its preference from zooplankton and benthic macroinvertebrates to artificial feed (Rahman et al. 2006; Rahman and Meyer 2009). It is believed that carp will grow faster and better with pelleted feed than with extruded feed and cereals. The pellets act as sources of nutrients but also indirectly maintain ecological stability, controlling cyanobacterial blooms in ponds (Ciric et al. 2013). Thus, carp can be stocked at high density in polyculture ponds supplied with artificial feed without any negative effects on water quality and food resources. There were observations by Rahman et al. (2006) that supplying feed to 1628 kg per hectare of carp biomass in ponds gave zero negative effects on water quality and food resources. However, excess stocking in common carp has many negative effects such as decrease soluble phosphorus and system production while increased turbidity of water.

Excessive resuspension of sediments will increase the O.R.P. An increase in O.R.P. will have a positive effect on the precipitation of soluble phosphorus through the formation of phosphate rich inorganic particles. Excess carp decreases the concentration of soluble phosphorus (Zambrano et al. 1999; Rahman et al. 2008a, b, c, d, e, f). On the other hand, high turbidity and low phosphorus concentration reduce photosynthesis. As a result of this, primary and secondary and tertiary production in the aquatic ecosystem is reduced (Rahman et al. 2008a, b, c, d, e, f, 2009). Stocking carp in high density levels will increase the pressure on natural food (zooplankton and benthic macroinvertebrates), to the extent that recovery is impossible. There is a threshold in the density of carp (depending on the environment) in which above that zooplankton and benthic macroinvertebrate populations collapse (Rahman et al. 2008a, b, c, d, e, f; Zambrano et al. 2001).

High resuspension of the bottom soil will result in an increase in turbidity. The amount of plankton biomass will decline as the turbidity level rises (Fig. 4.8). High turbidity also has an impact on the development of submerged vegetation in shallow ponds. If macrophytes are reduced by carp, many invertebrate groups may have less of them, and their community composition may change due to the strong linear relationship between carp high density and vegetation level. High density of common carp can have a significant negative impact on the aquatic ecosystem's biodiversity (Zambrano et al. 2001; Kloskowski 2011a, b). The condition of the environment in presence of carp is also size dependent. Different ontogenetic stages of carp will affect the aquatic environment differently (Driver et al. 2005; Kloskowski 2011a, b). However, this effect almost disappears once the fish are more than a year old (Kloskowski 2011a, b).

As mentioned in Fig. 4.8, resuspension of bottom soil also affects the level of fish production. Therefore, increasing fish production by the addition of carp is a common practice around the world, especially in Asia. This technique is most efficient in semi-intensive polyculture systems where fish production depends on natural food (Mohapatra et al. 2007; Rahman 2015). The interactions among fish

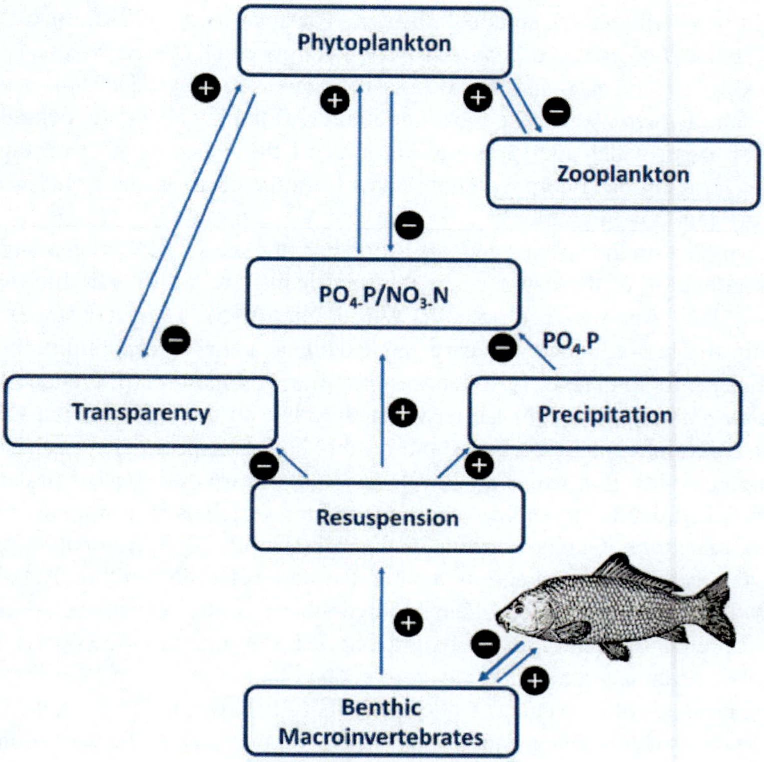

Fig. 4.8 Schematic representation of the effects of common carp on nutrients and natural food availability

species are important in the sustenance of polyculture system (EL-Shebly et al. 2007; El-Sherif and Mervat 2009). If proper combination of two or more fish species be utilized, the polyculture systems increase the natural food consumption in the ponds, especially if the combination is between planktivorous and benthivorous fish (Rahman et al. 2008a, b, c, d, e, f). There are many examples of benthivorous carp species used in polyculture. These include common carp (Rahman et al. 2006, 2008a, b, c, d, e, f), mrigal (*Cirrhinus cirrhosus*) (Rahman and Verdegem 2007), calbasu (*Labeo calbasu*) (Rahman and Verdegem 2007), bream (*Abramis brama*) (Driver et al. 2005), and roach (*Rutilus rutilus*) (Driver et al. 2005). Common carp has a greater impact than any other benthivorous species on aquatic ecology and production (Wahab et al. 2002).

Studies showed that in presence of carp rohu had 1.6 times higher yield in semi-intensive polyculture ponds, while 1.5 times higher growth compared to presence of calbasu (Wahab et al. 2002; Rahman et al. 2008a, b, c, d, e, f). Phytoplankton and total phosphorus availability increase in the presence of carp compared to Channel catfish (*Ictalurus punctatus*). In large lakes, common carp biomass is often negatively related to the abundance of other fish, especially bluegill (*Lepomis macrochirus*), black crappie (*Pomoxis nigromaculatus*), largemouth bass

(*Micropterus salmoides*), smallmouth bass (*Micropterus dolomieu*), black bullhead (*Ameiurus melas*), walleye (*Sander vitreus*), yellow perch (*Perca flavescens*), northern pike (*Esox lucius*), and white bass (*Morone chrysops*) (Jackson et al. 2010; Weber and Brown 2011). The most important reason may be habitat degradation by common carp, which increases the turbidity of the water in the lake through its benthic foraging behavior, switching lakes from the clear- to the turbid-water state (Stewart and Downing 2008).

Depending on the species and age of the fish, the preferred food availability, and the combination of the fish species, the food habits are highly variable (Rajkumar et al. 2013; Antony et al. 2014; Wu et al. 2015). Carp consumes benthic macroinvertebrates. However, carp has excellent adaptive capabilities in case of insufficient food. There is evidence that the common carp predates crayfish (*Cambarellus montezumae*) larvae when they live in the same habitat (Hinojosa-Garro and Zambrano 2004). In the presence of superior species, carp can modify its feeding behavior (Rahman et al. 2009). The bulk of carp diet consists of detritus but as mentioned, feeds on benthic macroinvertebrates such as chironomids, tubificids, and zooplanktons (Garcia-Berthou 2001; Parkos et al. 2003; Rahman et al. 2009). Thus, the feeding niche of carp in a natural system is largely benthic (Rahman et al. 2010). The preference of benthic macroinvertebrates influences the behavior of carp since it prefers to spend most of its grazing near the bottom in the ponds, lakes, or reservoirs (Rahman et al. 2008a, b, c, d, e, f).

The presence of carp can also alter other fish's behaviors. For example, Rohu is a planktivorous fish which spends most of its swimming time in the water column and grazes more than three times longer in the water column than near the bottom. However, in the presence of carp, Rohu spends more time grazing near the pond bottom. Carp also affects the behavior of the crustaceans. For example, in the presence of carp, crayfish displaces its habitat but also displacement speed (Hinojosa-Garro and Zambrano 2004).

There are many external and internal factors responsible for ontogenetic shifts in the diet of fish. In the case of carp, small ones are known to feed preferentially on zooplankton, where larger ones avoid zooplankton and concentrate on benthic macroinvertebrates (Adamek et al. 2003; Rahman et al. 2009).

Fish activities are mainly synchronized with alterations in day and night. Some fish are diurnal feeds which rely on vision or nocturnal feeders which rely more on tactile, chemical, or electrical senses. However, diel activities are species specific, which some fishes search for food during dark and light periods, and others only during daytime. Carp is an active fish which grazes during day and night but prefers daytime (Rahman et al. 2008a, b, c, d, e, f; Rahman and Meyer 2009). Carp also affects the diel feeding rhythms of other fish (Rahman et al. 2008a, b, c, d, e, f). Rohu is a diurnal feeder, spending the majority of its swimming time in the water column during daytime, however, in the presence of carp, the difference between day and night grazing and swimming in the water column decreases, while near the pond bottom increases.

4.2.3.2 Rice-Fish and Poseidon-AI

Rice is a staple food in Indonesia, just like it is in the majority of South-East Asian nations. Smallholder farmers produce most of the rice, which is either kept for personal use or sold in the neighborhood market. The production of rice in Indonesia has increased by 2.2% annually on average over the past 60 years. However, domestic consumption recently surpassed domestic production (USDA 2019). The Java region, which is well known for its rice exports and surpluses, is where the majority of rice is produced.

Rice farmers in Indonesia keep their products for personal use. Farmers continue to increase their output for three main reasons. First, they continue to pay hired labor to harvest rice (Ellis et al. 1992; Nusril and Sukiyono 2007; Ilham et al. 2010). They pay for services by allocating proportions from their production. For example, in Java, farmers use two payment systems by production, open and close system (Ellis et al. 1992). The open system is when the harvest is conducted by only a few hired laborers and whoever can participate in harvesting. Close system on the other hand depends on the availability of labor in the village. According to Ellis et al. (1992), the close system has a proportion of 1:4 to 1:6 and system between 1:9 and 1:10. The more abundant the labor, the less hired the labor received (Ilham et al. 2010). Finally, and in some areas, farmers pay land rent using rice as well as some inputs such as fertilizers (Nusril and Sukiyono 2007). Second, rice will be used as a seed for agriculture in the upcoming season. A little more than 50.6% of farmers use their own seeds (Statistics Indonesia 2014). Third, farmers continue to grow rice for domestic consumption (Ellis et al. 1992; Nusril and Sukiyono 2007; Ilham et al. 2010).

The level of rice commercialization in Indonesia is very low (Noviar et al. 2020). In terms of quantity, 46% of the rice production was consumed by households versus nearly 50% being sold in the market (Statistics Indonesia 2014). Nearly 60% of the rice produced in the Yogyakarta region is sold, and only about 49% is consumed by households (Rini et al. 2021). India has a 78% market surplus compared to other Asian nations that produce rice, while Bangladesh's surpluses range from 38% to 60%. (Alam and Afruz 2002). Ghana sells 70% of the rice it produces in the African region (Amfo et al. 2022).

As already mentioned, farmers, particularly those in the Java region, depend heavily on the production of rice. The Java region has the potential to develop into a major contributor to the nutritional advancement and food security of the region as well as the country by producing rice and fish under irrigation or rainfed conditions.

The development of aquaculture was given top priority by the Indonesian government, and because rice production is crucial for families, rice-fish culture strategies similar to those used in China were created for areas like Java. In order to establish an Integrated Agriculture-Aquaculture System (IAAS) using rice and carp culture, the government gives carp fingerlings to rice farmers. When carp are grown alongside other fish, such as catfish, tilapia, or koi, the result is called polyculture.

In Indonesia, rice is grown in close to 7000 different genuine varieties. Minapadi rice-fish systems are found in West Java, and sawah tambak is a system found in the

East Java coastal regions (FAO 2001). It has been demonstrated that rice varieties like IR 64 produce high yields when grown with fish and in the dry season. These types of rice are grown in 20 × 20 cm plots with urea, triple superphosphate, potassium chloride, and ammonium sulphate fertilizer. A few days after rice is transplanted, small carp are stocked in rice fields at a rate of 2000–3000 per hectare. After 2 months, the field is drained when the fish are about four times larger (about 100 g). In west Java, carp are stocked without feeding, however, when the stocking density increases, supplemental feeds such as rice bran, chopped cassava, corn kernel, poultry feeds, and remaining food are given as fish feed.

The excess fertilizer use, feed wastage and impact of sea level rise, and increased soil salinity prevented the rice-fish systems to reach to the desirable level. Under these circumstances, continuous monitoring of the rice fields can improve the welfare of stocked fish, as well as reducing the number of fertilizers used for rice production. Additionally, increasing soil salinity in the coastal areas as a result of sea level rise puts more stress on the fish as well as reduces amount of produced rice. Poseidon-AI uses industrial grade sensors with the ability to be inserted into soil for monitoring salinity level, nitrate, and ammonia.

The average water temperature in the rice fields was 28 °C with the average air temperature 30 °C. The water salinity level was approximately 4% with 8 pH level. D.O. level was as low as 2 ppm. Adding these to the level of fertilizers caused excess levels of nitrate, ammonia, and phosphate.

Carp on average can grow 3–4% body weight per day, but under saline water, the growth of these species is limited. Although carp can survive up to 5% salinity level, salinity will prevent the species from efficiently growing. Under these circumstances, the information provided to farmers helps them to grow carp in their rice fields more efficiently while effectively using fertilizers.

Aiming to have a kilogram body weight carp within 120 days needs a well-balanced diet, effective density size, and good water quality. On the other hand, healthy growth of rice also needs enough nutrients provided by fertilizers. As mentioned, carp can increase the amount of oxygen in the water by resuspension of bottom soils but can also move the excess fertilizers used in the field and cause environmental issues.

Poseidon-AI® algorithms provide the farmers with the most efficient amount of feed under the rice field water condition. The amount is calculated based on the possible fish excrements that reduces the need of fertilizer usage in the rice fields. Depending on the age of the fish, farmers can remove and grow them into inland tanks or early consumption.

The image processing is used for detecting the growth of the carp and distinguishing between carp and rice paddies. There are cases of detection of other species such as golden snails (*Pomacea* spp.) in the rice fields that traditionally farmers use ducks to control the infestation. Poseidon-AI® algorithms are being considered for the detection of common rice diseases, but with the aid of IoT devices and existing algorithms, the farms have been able to lower their fertilizer costs and add an additional revenue stream by selling kilograms of carps within 120 days.

The local government and the farm owners were both informed of the salinity level that was discovered. The analysis revealed reported mortalities and that the fish did not reach a kilo weight within the predicted time. The rice growth rate was slower than expected, and there were requests for use of saline resistance rice paddies.

Artificial feeds are being used for the stage that the fish reaches 100 g. The stocking tank uses water from the nearby waterbodies. Poseidon-AI® IoT device monitors the condition on the stocking tank to maintain the pH between 7 and 9 and temperature always above 25 °C. Due to sea level rise, it is very difficult to find waterbodies with zero salinity level, but rain collection systems were advised to reduce the salinity level below 2%. The stocks were provided by the local government and so, the stocks are SPF-KHV. Therefore, no disease monitoring was conducted during this period. Additionally, no medications and/or chemicals were added to the water to prevent further environmental complications since these farms have no water treatment facilities.

4.3 Vietnam

4.3.1 Aquaculture Production in Vietnam

Vietnam has a long history of aquaculture, particularly in the lowlands and on the coast. Vietnam aquaculture has enormous development potential because of its extensive coastal regions and abundance of rivers, lakes, and estuaries (Phuong et al. 2006). The aquaculture industry expanded quickly, which boosted the nation's economy and enhanced the standard of living for farmers in rural areas (Duc 2009; Lan 2009, 2013).

Under the direction of the Vietnamese Ministerial Council, the General Fisheries Department was established in 1960. The first phase of sector development lasted from 1960 to 1980, and the second phase began in 1981. Many farmers were drawn to rice-fish farming during the first period as well as marine and brackish aquaculture. Kien An commune was the birthplace of marine culture. In Hai Phong city, the first successful fish seed production facility was constructed in 1962 (Chu et al. 2003). During this time, earthen ponds, lakes, rivers, and rice-fish ponds were also developed. With the introduction of rice, fish farming expanded to 100,000 hectares. Because of its significance in supplying food for people, the aquaculture sector was promoted during the Vietnam War, which lasted from 1963 to 1975. In 1965, there were almost 15,000 aquaculture cooperatives and government-run businesses. Significant progress was made in Hai Phong and Thanh Hoa, particularly in shrimp farming. At the district level, aquaculture is regarded as a major career option in these places (Chu et al. 2003). The total aquaculture production increased from 59,000 tonnes in 1976 to 160,000 tonnes in 1980, according to MoFI (2003).

In 1981, shrimp farming for export has dominated the aquaculture of Vietnam. Aquaculture farmers have started to diversify their farming practices by adapting exportable species such as black tiger shrimp (*Penaeus monodon*), catfishes

(*Pangasius hypophthalmus* and *Pangasius bocourti*), lobster (*Panulirus* spp.), groupers (*Epinephelus* spp.), and bivalves (*Meretrix lyrata* and *Anadara granosa*).

Distributions of aquaculture systems are different from the North, Central, and South of Vietnam. In the Northern part, the culture system includes freshwater fishponds, rice-fish, and marine cages. The most common aquaculture practices in the Central region are black tiger shrimp farming and marine cage culture of finfish or lobster. Most of the existing literature focuses on the Mekong River Delta and the impact of aquaculture on the livelihood of the people in the South (Ha and van Dijk 2013; Joffre and Schmitt 2010; Loc et al. 2010). In the Southern part, culture systems are more diversified. These include pond, fence, and cage culture of catfish and some local species such as snakehead and perch. Shrimp farming is also operated in the Southern part at extensive, semi-intensive, and intensive levels.

Vietnam's aquaculture has a wide range of species that provide a great potential for aquaculture development. In freshwater areas, catfishes (*Pangasius hypophthalmus* and *Pangasius bocourti*) which are the representative species of the Mekong River Delta have the greatest production. There are other five most popular cultured fish species that have high significant contribution to the total freshwater fish production. These consist of species belong to the *Cyprinidae* family such as silver carp (*Hypophthalmichthys molitrix*), grass carp (*Ctenopharyngodon idella*), common carp (*Cyprinus carpio*), big head carp (*Aristichthys nobilis*), and major Indian carps including catla (*Catla catla*), rohu (*Labeo rohita*), and mrigal (*Cirrhinus mrigala*). Nowadays, mono-sex tilapia (*Oreochromis niloticus*) has also been introduced into inland and brackish water aquaculture. Besides, giant freshwater prawn (*Macrobrachium rosenbergii*), climbing perch (*Anabas testudineus*), and snakehead fish (*Channa micropeltes*) are the most popular cultured species in the Southern part of Vietnam.

In marine water areas, the most popular cultured species consist of lobster (*Panulirus* spp.), grouper (*Epinephelus* spp.), and seaweed (*Gracilaria verrucosa*), and these species dominate in the Central part of Viet Nam. Shrimp (*Penaeus monodon*), mud crab (*Scylla* spp.) and bivalves (*Meretrix* spp. and *Anadara* spp.) are the most popular cultured species that have the highest production in brackish water areas, particularly in the South of Vietnam. A number of high potential cultured species have also been focused to add more species to the cultured species population. These new species are Cobia (*Rachycentron canadum*), abalone (*Haliotis* spp.), sweet snail (*Babylonia areolate*), Pearl oyster (*Pinctada maxima* spp.), whiteleg shrimp (*Litopenaeus vannamei*), and seabass (*Lates calcarifer*).

There are various aquaculture practices in Vietnam. These include IAS such as rice-fish, rice-shrimp, mangrove-shrimp, monoculture and polyculture in fresh and marine waters (Kluts et al. 2012; Phong et al. 2007). Most common culture practices are marine shrimp farming. Marine finfish such as groupers are cultured in small cages in Quang Ninh and Hai Phong provinces in the North and in Nghe An, Khanh Hoa, and other coastal provinces in the Central regions. Large cage culture of Cobia has been introduced from Norway. Lobsters (*Panulirus* spp.) is farmed mainly in the Central coastal provinces with small cage using wild collected seeds.

Fig. 4.9 Transport of catfish for processing in Mekong Delta

Catfish culture is an intensive practice in diverse habitats of the Mekong Delta (Fig. 4.9). Catfish are cultured in cages, ponds, and fences with high intensity. The cage culture in the area started to decrease while the pond culture increased significantly. The fence culture in the delta is increasing gradually. The productivity of pond culture varies from 183 to 582 metric tonnes per hectare depending on stocking density (Nguyen et al. 2004; Le 2004) while fence culture's productivity is 345 metric tonnes per hectare. The fish yield from rice-fish farming varied from 482 to 808 kg per hectare whereas yield from livestock-fish polyculture ranges from 467 to 1456 kg per hectare which also depends on stocking densities (Nguyen et al. 2005).

Giant freshwater prawn is another species cultured in Mekong River Delta. The practice is done in ponds fences or rice paddies. The productivity of prawn varied by culture practices ranges from 100 to 887 kg per hectare for integrated rice-prawn culture system, 384–1681 kg per hectare for alternative rice-prawn culture (Nguyen et al. 2005), and from 140 to 160 kg per cubic meters for fence culture (Vu et al. 2005). However, the IAS is very climate dependent with high risk of failure (Adger 1999; Allison et al. 2009). Evidence in Vietnam shows that disasters like floods and droughts have increased in intensity and frequency over the past two decades (Adger 1999; Dasgupta et al. 2009; Schad et al. 2012).

In Vietnam, the ministry of fisheries is a governmental body which is managed under administration of the Vietnamese government. There are three administrative levels in the fisheries sector of Vietnam which includes national, provincial, and district levels.

There are supportive divisions to fulfill state management functions. There are division of aquaculture, division of collective and individual economic sector, division of plan and finance, division of science and technology, division of international relations, division of legislation, division of personnel organization, bureau of capture fishery and aquatic resources management, bureau of quality management, sanity safety and fisheries veterinary and ministerial inspectors and ministerial offices.

The specialized institutions support the ministry in terms of R & D. There are Research Institute for Marine Fisheries, Institute for Fisheries Economic and Planning, Research Institute for Aquaculture based in the North, Research Institute for Aquaculture based in The South, Research Institute for Aquaculture based in the Central, and National Fisheries Information Center. There are also unions and associations which are Labor Union of Vietnam's Fisheries sector, Vietnam Fisheries Association, and Association of Seafood Exporters and producers.

In 2004, the president of Vietnam issued the fisheries governing law. The fisheries law consists of ten chapters about general regulations; protection and development of aquatic resources; capture fishery; aquaculture regulations; regulations of fishing boat and fisheries services; regulations on processing, trading, export, and import of aquatic products; regulations on international cooperation for fisheries operations; regulations on governmental administration for fishery; regulations on reward and punishments and regulations on clause of implementation.

In Vietnam, the Ministry of Fisheries conduct research on national aquaculture development. The research focused on aquatic seed production, improvement on aquaculture technology, feeds for aquaculture, and technological improvement on reservation of aquatic products, aquaculture environment, and other urgent issues in aquaculture practices.

4.3.1.1 Artificial Seed Production

Vietnam produces artificial seeds for major aquatic species that are important for export. The species include marine shrimp (*Penaeus monodon*), climbing perch (*Anabas testudineus*), snakehead fish (*Channa micropeltes*), spotted gourami, mud crab, swimming crab (*Charybdis affinis*), sweet snail (*Babylonia areolata*), cobia (*Rachycentron canadum*), and spotted grouper (*Epinephelus coioides*). In freshwater aquaculture, production of artificial seed, rearing and grow-out of some species in the Mekong River Delta are conducted and supported by the government of Vietnam. The support has led to technological breakthrough in catfish seed production, as well as artificial seed production and commercial grow-out of giant freshwater prawn (*Macrobrachium rosenbergii*), leading to yield improvement in the integrated rice-fish and rice-prawn farming systems.

In marine and brackish water aquaculture, Vietnam can artificially produce seeds of the species such as black tiger shrimp (*Penaeus monodon*) and mud crab (*Scylla* spp). Other species that Vietnam started to seed production and grow-out procedure over the last two decades are spotted grouper (*Epinephelus coioides*), cobia (*Rachycentron canadum*), Red drum (*Sciaenops ocellatus*), seabass (*Lates calcarifer* and *Psammoperca waigiensis*), swimming crab (*Charybdis affinis*), and oyster (*Crassostrea* sp.). Additionally, the seed production for blood cockle (*Anadara granosa*) and brown tiger shrimp (*Penaeus semisulcatus*) is gaining some break throughs.

4.3.1.2 Environmental Impact on Vietnamese Aquaculture

Typically, the pond cultures are constructed in series or in parallel, where water flows through the ponds by means of gravity. The average size of the pond is around 900 m^2, and almost all the farms have trees around their ponds with branches hanging to prevent angling or theft (Fig. 4.10).

The waters in the ponds are shared from the same waterbody or irrigation system. Hence, the application of pesticides in the farms or human activities such as washing clothes have negative influences on the ponds water quality and eventually, cultured species (Lamers et al. 2011; Alcaraz et al. 1993).

Since the species are cold blooded animals, water temperature will have a strong influence in their growth and well-being. The ideal temperature for fish culture for most Asian fish is generally above 25 °C (Cagauan 2001), and the species will stop eating or eat less at the temperature below 20 °C (Ling 1977). The water temperature in most of the areas in Vietnam is close to optimal during the summertime but it reduces during the winter season (Diaz et al. 1998), especially in shallow ponds that do not have a source of fresh water (Dan and Little 2000). In low water temperatures,

Fig. 4.10 Example of an aquaculture pond in Vietnam

not only the growth rate of the species reduces but also immune suppression characteristics are revealed among the species cultured in Vietnam (Yang and Zuo 1997).

D.O. of water decreases when the water temperature increases, this is due to the low oxygen solubility in higher temperatures. With deduction in D.O., the food intake of fish will be affected, impacting the growth of the species (Ross 2000). For this reason, in the ponds and with high temperatures, species are regularly observed gulping at the pond's surface, reflecting low D.O. levels. Most of the water in the ponds are considered turbid due to flow of water with sediments from the nearby waterbodies. Turbidity limits the photosynthesis due to diminished sunlight penetration, reducing the D.O. level.

4.3.1.3 Feed Production in Vietnam

Vietnam can produce both artificial feed pellets and feeds from natural ingredients remained from farms leftovers. In farms, the leftovers from one production, for example, livestock production activities, are used as an input for fish. Farmers use aquatic plants, by-products from the paddy fields, as fish feed. The most important plants used as feed inputs from fields are Banana leaves, cassava, and maize.

In IAS, the system is plant based mainly using aquatic and terrestrial macrophytes as feed (Prein 2002). Some species such as grass carp can process raw plant materials, however, not all the cultured species can process and grow with these materials. Some of the leaves and by-products provided as feeds in the farms, limited or had negative impacts on fish growth (Dongmeza et al. 2009; Dongmeza et al. 2010). This style of feeding is very similar to leaf-based aquaculture systems such as Chinese mulberry dike-carp pond farming (Ruddle and Christensen 1993) and the Napier grass-based tilapia culture (Van Dam et al. 1993).

From banana plant, the leaves and soft parts of stems are used. Other inputs include the by-products of rice cultivation such as rice bran, smaller rice pieces as well as weeds and aquatic plants. In Vietnam, barnyard grass (*Echinochloa crusgalli*) is the weed used from the paddy fields while manures from pigs and buffalos are used as animal derived pond inputs. Table 4.4 shows the plant products used in fish feed in Vietnam.

Most of the feed types mentioned above cannot provide enough nutrients for the culture species and thus, a large amount of feed needs to be consumed by the species. For example, for grass carp nearly 80% plus of these materials need to be consumed to obtain nutrient requirement (Tan 1970; Cui et al. 1992, 1994). According to Dongmeza et al. (2009), most of the feeds applied with similar ingredients have low nutrients. These feeds have high moisture and fiber content but relatively low protein. Certain feeds made from cassava and mulberry leaves contain 20–30% crude protein contents (Dongmeza et al. 2009). However, better protein content does not necessarily indicate a better-quality diet (Shireman et al. 1978; Tan 1970). According to Hajra et al. (1988), the digestibility coefficient for crude protein and gross energy decline significantly with increased fiber content of the plant in the feed. Additionally, gross energy and chemical composition with anti-nutrients may also influence the fish growth (Francis et al. 2001a, b, 2002).

Table 4.4 Plant products used as fish feed in Vietnamese farm (Source: Steinbronn 2009)

Plant type	Specification
Aquatic plants	*Azolla imbricata, Lemna paucicostata, Pistia stratiotes, and Salvinia natans*
Banana (*Musa* sp.)	Leaves, chopped stems
Bamboo (*Bambusa* sp.)	Leaves
Cassava (*Manihot* sp.)	Leaves and chopped stems
Fruits	Fruits and leaves from figs (*Ficus glomerata*) and tamarind (*Tamarindus indica*)
Fodder grass	Napier grass (*Pennisetum purpureum*)
Maize (*Zea* sp.)	Leaves, meal
Mulberry (*Morus* sp.)	Leaves
Rice (*Oryza* sp.)	Bran, blades/straw, sorted grains, husks, fermentation residues
Terrestrial plants	*Alternanthera sessilis, Chromolaena odorata, Commelina nudiflora, Cyperus imbricatus, Cyperus rotundus, Digitaria timorensis, Echinochloa crus-galli, Eclipta prostrata, Kyllinga monocephala, Sagittaria sagittifolia, Sporobolus indicus, Urochloa reptans, and Wedelia calendulacea*
Vegetables	Leaves and stems from, e.g., water dropwort (*Oenanthe javanica*), sweet potato (*Ipomea batatas*), and water morning glory (*Ipomea reptans*)

As shown, these feeds cannot answer to the nutritional needs of the cultured species unless large quantities are given or be used as supplementary feed types. According to FAO (2006), some species of carp and tilapia can use these agricultural byproducts, however, it is highly unlikely that these feeds contribute significantly to fish growth.

4.3.2 Poseidon-AI in Vietnam

In Vietnam, Poseidon-AI works with large- and medium-sized farms. Most of these farms are found in the Mekong River Delta, and each has at least 100 ponds. As previously mentioned, these farms use artificial feeds to culture these species of fish because natural feeds made from agricultural by-products are neither sufficient nor effective for fish culture. The main industry in the Mekong River Delta is the industrial culture of catfish, which has a big impact on the micro and macro economies of the area. The excessive feeding and unknown water quality conditions in these farms, however, result in high costs and a decline in the farmers' income.

Poseidon-AI uses its approach to help farmers sustainably grow their businesses as well as reduce 20% of their feeding costs. The live water monitoring, without the capacity to improve the quality of the main source of water coming from Mekong Delta, created real-world challenges that only involvement of deep tech can face and make meaningful contributions.

4.3.2.1 Catfish Culture

In the Mekong Delta of Vietnam, catfish farming quickly became one of the major aquaculture industries (Phuong and Oanh 2010). Striped catfish (*Pangasianodon hypophthalmus*) is Vietnam's second most important product (MoFi 2005). The processed catfish have been exported to more than 150 countries in the world (MARD 2014). In Vietnam, catfish culture is an intensive culture, raised in 2- to 6-m-deep ponds (Fig. 4.11), with 70–800 metric ton per hectare annually (Phan et al. 2009).

Almost all the ponds are located along the Mekong River which is convenient to discharge nutrients to the river and to transport feed, fingerling, and market size fish to or from the farm (Fig. 4.12).

The amount of water used per kilogram of fish product is between 2.5 and 9.1 cubic meters (Anh et al. 2010; Bosma et al. 2009; Phan et al. 2009). The commercial floating pellets provided to the species contain 22–30% of protein with estimated feed conversion ratios ranging between 1.5 and 1.8 (Bosma et al. 2009; Phan et al. 2009). According to De Silva et al. (2010), a kilogram of catfish can produce waste containing 46 g nitrogen and 14 g phosphorus. About 70% of striped catfish farmers discharge untreated effluents to the river. The remaining 30% of farmers dispose of their waste into gardens or rice fields, where some of the nutrients are reused (Phan et al. 2009). Discharging effluents pollute surface waters and increase the risk of horizontal disease transmission (Nguyen et al. 2007).

In Vietnam, there is a standard of 5000 L of water per kilogram of catfish (The Aquaculture Stewardship Council (ASC), 2012). The studies showed the range of water usage 20–50% lower than standard, around 2500–4050 L per kilogram of

Fig. 4.11 Sample of catfish ponds in along Mekong River

Fig. 4.12 Transport of market size fish from the farm for processing via Mekong River

catfish (Bosma et al. 2009; Phan et al. 2009), or 82% higher than the standard, 9167 L per kilogram (Anh et al. 2010). Nitrogen and phosphorus discharge per fish is calculated 27.5 g and 7.2 g per fish (ASC 2012) which are much lower than the values estimated by Anh et al. (2010) and De Silva et al. (2010). In literature, the estimate values for nitrogen and phosphorus discharge are 38–46 g of nitrogen per kilogram and 9.9 and 14.4 g of phosphorus per kilogram of catfish.

Farmers utilizing both approaches engage in tidal water exchange in the Mekong River, where catfish farming initially began upstream and then spread to the downstream. However, farmers working downstream have two advantages over the upstream ones. First, a larger tidal amplitude with longer part of the day for tidal water exchange, and second, less potential of conflict with other types of the land use (Bosma et al. 2009).

Due to diseases, the survival rate of catfish in ponds is less than 70% and more than 15 diseases and syndromes occur commonly in catfish farming (Phan et al. 2009). These diseases are treated with different antibiotics and chemicals (Bosma et al. 2009; Rico et al. 2013). Chemicals and antibiotics are advised to be banned in the last 2 months before harvesting (BMP 2009; MARD 2011). Sludge removed from catfish ponds has a high mineral content, but they are heavily diluted which limits the option to be reused as fertilizers (Phu and Tinh 2012). With an FCR of 1.6, on average 320 metric tonnes of feed are applied per hectare annually. Use of 26% protein feeds, containing 1.4% phosphorus by weight used, results annually in 13,300 kg nitrogen and 4480 kg of phosphorus input per hectare.

There is always a risk that Mekong River carries high loads of nutrients, residual chemicals, and antibiotics (Toan et al. 2013; Rico et al. 2013; Rico and Van den Brink 2014). Additionally, the Mekong River carries heavy metals (Minh et al. 1997), persistent organic pollutants (Minh et al. 2007), pesticides, and agricultural wastes (Toan et al. 2013). As a result of rising sea levels, inland farms are now more exposed to saltwater intrusion during the dry season than they were in the past (Anh et al. 2014; Nguyen and Savenije 2006), which could eventually result in less catfish being produced (Renaud and Kuenzer 2012). Reliance on high water exchange increases the risk for horizontal disease transmission (Anh et al. 2010; Madsena et al. 2015; Phan et al. 2009). Although trematode treatment of influent water to minimize infection in the striped catfish fingerling ponds has been suggested (Madsena et al. 2015), treatment of intake water is expensive and labor-intensive due to the huge exchangeable volumes in the grow-out ponds. Although it is frequently advocated (Anh et al. 2010; ASC 2012; BMP 2009; Phan et al. 2009), treatment of influent and effluent water to and from striped catfish grow-out ponds is rarely implemented.

It was suggested as a more affordable and practical solution over the past 10 years to treat influent water in a sizable reservoir that serves as a stabilization pond and to send effluents through a sedimentation pond (Anh et al. 2010). However, this solution was not implemented due to the high land costs surrounding the Mekong River.

Feed accounted for 87–91% of the total variable cost in striped catfish pond culture (Ngoc et al. 2016). During grow out, diets containing 22–30% protein are used, which is lower than for most other fish species produced in the aquaculture sector. The average FCR of the industry pellets are 1.86 (Bosma et al. 2011; Phan et al. 2009) which is high compared to carnivorous fish species, but in the same range as for omnivorous species and better than the crustacean culture (Pahlow et al. 2015).

4.3.2.2 Sustainable Catfish Farming with Poseidon-AI

Throughout the Mekong River Delta, there are tens of thousands of catfish farms, all with nearby processing facilities for quick export of the fish. The farms use the same water from the river and powerful pumps to bring water into the ponds used for the cultivated species, primarily catfish (Fig. 4.13).

With nearly a million fingerlings waiting to reach the market size, the average pond is 150 × 60 m in size. The farms are medium to large and have hundreds of these ponds where tons of catfish are raised each year (Fig. 4.14).

Artificial pellets (Fig. 4.15) with a 20–30% protein level are manually fed in the morning and the afternoon for almost an hour. The feed is carried across the ponds using a floating carrier, and pellets are distributed throughout the ponds. Figure 4.16 shows the feeding procedure in a pond in Mekong River Delta. The average amount of feed per day per pond was calculated to be 1.6 tonnes of industrial pellets.

Fig. 4.13 Water brought in from the Mekong River with strong water pumps to fill the ponds use for culturing catfish

Fig. 4.14 A pond from a medium/large size farm located in Mekong River Delta

It is believed that 20% of the feed is being wasted, and each farm has approximately 80% mortality rate. There are no water treatment facilities, and farmers use manual water testing kits to analyze the quality of the water.

Fig. 4.15 Artificial pellets used for catfish culture in Vietnam

Poseidon-AI® IoT device continuously monitors the water quality every 15 min and shows the information on the user dashboard of farmers cellphones. The images are sent to AWS an hour before, during and an hour after the feeding. According to literature and laboratory findings, the optimal temperature for catfish is between 27 and 30 °C. The ideal pH for these species is between 6.5 and 7.5 but sudden pH changes due to rain falls, humans pollute and excess feeding which causes high mortality in catfish species. The optimal level of D.O. is 4 mg per liter to saturation levels. When D.O. level is consistently between 1 mg per liter to 5 mg per liter, fish feed intake will reduce significantly. Table 4.5 shows the summary of environmental conditions for catfish.

The pond's surface temperature will rise with rising air temperature. However, catfish will spend the majority of their life cycle submerged in the pond's water. The life cycle of the fish in the pond can therefore be better understood by keeping track of the water temperature at a depth of 1 m. The average air temperature is 30.54 °C, while the water temperature is 31.87 °C at a depth of 1 m. However, during feeding time, the species must come to the surface, exposing them to the nearly 35 °C surface

Fig. 4.16 Manual feeding procedure for catfish farming in Mekong River Delta

Table 4.5 Summary of environmental condition for catfish

Parameters	Recommended range	References
Temperature (°C)	20–33	Choudhury (2000)
Dissolved oxygen (DO) (mg/L)	5–8	Sarker (2000)
pH	6.7–8.6	Sarker (2000)
Depth (m)	1.2–1.5	Sarker (2000)
Transparency (cm)	25–35	Andrew et al. (1972)

air temperature. The species will go into shock because of the surface heat, which will lead to problems like mortality, increased energy consumption, and increased appetite.

Live monitoring of ponds helps farmers to start the feeding procedure at the right time and based on live condition for each pond. By changing the feeding time to cooler time of the day, the farmers managed to reduce fish mortality.

On the other hand, the suitable pH for catfish would be between 6.7 and 8.6 (Sarker 2000), and pH level has direct relation with photosynthesis and the amount of sunlight. In normal conditions, with increasing photosynthesis, the pH increases while the waters in ponds and lakes have many other factors which can influence the pH level. However, due to environmental events such as acidic rain, discharges of species, high amount of remaining feed and other factors; the pH varies significantly. This fluctuation will impact the movement of the fish and according to farmers, the fish will become lazier. Hence, during the period that device shows extreme changes, it was recommended to stop the feeding or significantly reduce the feeding amount.

Regarding the socioeconomic aspect of the feed optimization in the catfish farming of Vietnam, Poseidon-AI® algorithms combine the environmental data

Fig. 4.17 Overfeeding in catfish farms cause change in water characteristics

from the ponds, comparing them with the laboratory gained data and present the best possible time for feeding, feeding amount, FCR, maturity time, and growth rate. The algorithms also show the mortality rate in the pond, trying to present the best possible condition for reducing the death rate in the farm.

There are areas in each pond where the oxygen level is zero or low. Fish rarely visit these areas, where there is a high likelihood of an accumulation of extra feed. Additionally, the water quality in these areas is far below what is considered to be acceptable and standard. In order to increase the likelihood of an effective pond monitoring system, the Poseidon-AI® IoT devices are placed far from these locations. Without the need to increase the number of pixels in the captured images, each device can monitor an area of 20 × 20 m. Therefore, there are two Poseidon-AI® IoT devices in each pond.

Since there are no clues as to the precise location of the school of fish, pellets are dispersed throughout the farm, as was previously mentioned. The feed moved toward the dead zones (low oxygen zones) due to the movement of the feed carrier, the wind, and species movement. As a result, a significant portion of the feed is collected in these locations without being noticed by the cameras. This makes it crucial that the algorithms are as precise as possible. The farm's costs are increased by the wasted feed, which also has a negative impact on the water's quality. Figure 4.17 depicts how the buildup of extra feed around the farm changes the water's environmental characteristics.

The catfish with the size of 100 g will consume approximately 1.7 g per day under water temperature of 30–32 °C. Currently, for 800,000 catfish, irrespective of the environment and based on farmers aquaculture knowledge, 1.6 tonnes of feed is given. With the high mortality rate, the farmer continues to provide the same amount until the species reach the market size. Feed amount moderation based on the number of living catfish in the pond is done using Poseidon-AI® algorithms and with the help of images sent to the AWS Cloud. Image processing distinguishes between remained feed, dead fish, leaves, and branches, giving the algorithms the chance to calculate the feed amount not only based on the environmental characteristics but also based on the number of live fish species.

Starting with 800 fingerlings and feeding amount of 1.6 tonnes per day, under 32 °C and pH of 8, a fish will consume 1.7 g per day, thus, 800 fish will consume 1.36 tonnes per day which already is 0.24 tonnes less than current feeding amount. Assuming a kilo of feed costs 0.5 USD, on day one the cost of feeding without Poseidon-AI approach will be 800 USD, while with Poseidon-AI approach it will be 680 USD. This cost will fluctuate according to the environmental condition and mortality rate calculated by the image processing. At the end of the life cycle of catfish and when the fish is ready to enter the market, the saved feeding amount is equal to 20% of total feeding without Poseidon-AI approach. In terms of saved cost in USD, this amount according to the data is 1.2 million USD.

Results from the applying Poseidon-AI approach showed that the adaptation techniques can help catfish farmers reduce their losses and/or increase their revenues. The Poseidon-AI extracts the power of IoT and AI/ML to monitor the condition to help farmers take proper management decisions, reduce their costs, and increase their profits.

The Poseidon-AI approach was built based on a strong perception that the only possible solution for environmental challenges faced in many aquaculture farms around the world is by facing these challenges as a business.

References

Abdullah MAS, Akatsu S, Al-Abdul-Elah KM, Teng SK (1983) Refinement of spawning and larval rearing techniques in hamoor (*Epinephelus tauvina*). Annual research report. Kuwait Institute for Scientific Research, Kuwait, pp 55–57

Abood SAA, Lee JSHSH, Burivalova Z, Garcia-Ulloa J, Koh LPP (2015) Relative contributions of the logging, fiber, oil palm, and mining industries to forest loss in Indonesia. Conserv Lett 8:58–67

Abu-Hakima R, Al-Abdul-Elah KM, Teng SK (1983) Reproductive biology of *Epinephelus tauvina* (Forskal) (Family: Serranidae) in Kuwait waters. Kuwait Institute for Scientific Research, Kuwait

Adamek Z, Sukop I, Rendon PM, Kouril J (2003) Food competition between 2+ tench (*Tinca tinca* L.), common carp (*Cyprinus carpio* L.) and bigmouth buffalo (Ictiobus cyprinellus Val.) in pond polyculture. J Appl Ichthyol 6:165–169

ADB/NACA (1991) Fish health management in Asia-Pacific. Report on a Regional Study and Workshop on Fish diseases and fish health management, ADB Agriculture Dep. Rep. Ser. No. 1. Network of Aquaculture Centers in Asia-Pacific, Bangkok, Thailand

Adger WN (1999) Social vulnerability to climate change and extremes in coastal Vietnam. World Dev 27(2):249–269

Akatsu S, Al-Abdul-Elah KM, Teng SK (1983) Effects of salinity and water temperature on the survival and growth of brown spotted grouper larvae (*Epinephelus tauvina*: Serranidae). J World Maricult Soc 14:624–635

Alam S, Afruz S (2002) Marketable and marketed surpluses of some leading crops in Bangladesh: recent trends and policy implications. Bangladesh J Agric Econ 25(2):115–132

Alcaraz G, Rosas C, Espina S (1993) Effect of detergent on the response to temperature and growth of grass carp, Ctenopharyngodon idella. Bull Environ Contam Toxicol 50:659–664

Allison EH, Perry AL, Badjeck M-C et al (2009) Vulnerability of national economies to the impacts of climate change on fisheries. Fish Fish 10(2):173–196

Amfo B, Aidoo R, Mensah JO, Adzawla W, Appiah-Twumasi M, Akey EA, Bannor RK (2022) Rice marketing outlets, commercialization, and welfare: insights from rural Ghana. J Int Food Agribus Market 35. https://doi.org/10.1080/08974438.2021.2022556

Amira FS, Rahman MM, Kamaruzzaman Y, Jalal KCA, Hossain MY, Khan NS (2016) Relative abundance and growth of male and female Nemipterus furcosus population. Sains Malays 45(1): 79–86

Andrew AW, Moore KD, LeRoy AC (1972) A guide to the study of environmental pollution. Prentice-Hall, Inc., Englewood Cliffs, NJ, pp 23–32

Anh PT, Kroeze C, Bush SR, Mol APJ (2010) Water pollution by pangasius production in the Mekong delta, Vietnam: causes and options for control. Aquac Res 42:108–128

Anh NL, Vinh DH, Bosma R, Verret J, Leemans R, De Silva S (2014) Simulated impacts of climate change on current farming locations of striped catfish (Pangasianodon hypophthalmus Sauvage) in the Mekong Delta, Vietnam. Ambio 43:1059–1068

Anon (1987) Fish-farm impact study. Mar Pollut Bull 18:573

Anon (1995) Fishery statistical bulletin for the South China sea area in 1993. Southeast Asia Fisheries Development Center, Bangkok, Thailand

Anton-Pardo M, Hlaváč D, Másílko J, Hartman P, Adámek Z (2014) Natural diet of mirror and scaly carp (Cyprinus carpio) phenotypes in earth ponds. Folia Zool 63(4):229–237

Antony PJ, Rahman MM, Rajkumar M, Kamaruzzaman BY, Khan SA (2014) Relative growth of *Harpiosquilla raphidea* (Fabricius, 1798) (Crustacea: Stomatopoda) male and female populations. Sains Malays 43:1305–1310

Arthur JR, Ogawa K (1996) A brief overview of disease problems in the culture of marine finfishes in East and Southeast Asia. In: Main KL, Rosenfeld C (eds) Aquaculture health management strategies for marine fishes. Proceedings of a workshop in Honolulu, Hawaii, October 9–13, 1995. The Oceanic Institute, Hawaii, pp 9–32

ASC (2012) ASC pangasius standard. http://www.asc-aqua.org, 70 pp

Bai YQ, Lin GF (2015) Up-to-date analysis on feed, culture, and demand for feed ingredients in China 2015. Guangdong Feed 6:22–24

Bajer PG, Sorensen PW (2015) Effects of common carp on phosphorus concentrations, water clarity, and vegetation density: a whole system experiment in a thermally stratified lake. Hydrobiologia 746:303–311

Baldry I (2000) Effect of common carp (*Cyprinus carpio*) on aquatic restorations. University of Minnesota, Department of Horticultural Science

Barus V, Peaz M, Kohlmann K (2001) *Cyprinus carpio* (Linnaeus, 1758). In: Banarescu PM, Paepke HJ (eds) The freshwater fishes of Europe, v. 5/III; Cyprinidae 2/III, and Gasterosteidae. AULA-G GmbH Wiebelsheim, Germany, pp 85–179

Behrens P, Kiefte-de Jong JC, Bosker T, Rodrigues JFD, de Koning A, Tukker A (2017) Evaluating the environmental impacts of dietary recommendations. Proc Natl Acad Sci U S A 114:13412–13417

Belton B, Bush SR, Little DC (2017) Not just for the wealthy: rethinking farmed fish consumption in the Global South. Glob Food Sec 16:85–92

Bergheim A, Hustveit H, Kittelsen A, Selmer-Olsen A (1982) Estimated pollution loadings from Norwegian fish farms. I. Investigation 1978–79. Aquaculture 28:347–361

Beristain BT (2005) Organic matter decomposition in simulated aquaculture ponds. PhD thesis, Fish Culture and Fisheries Group, Department of Animal Science, Wageningen University, The Netherlands

Beveridge MCM (1985) Cage and pen fish farming. Carrying capacity models and environmental impact, FAO Fisheries Technical Paper No. 255. FAO, Rome

Beveridge MCM, Muir JM (1982) An evaluation on proposed cage fish culture on Loch Lomond, an important reservoir in Central Scotland. Canadian Water Resour J 7:181–196

BMP (2009) Better management practices (BMPs) for striped catfish (tra catfish) farming practices in the Mekong Delta, Vietnam. Project No 001/07VIE. In Version 2

Bo Y, Lin G (2015) Better profitability and increased marketing pressure of feed industry in 2015. Guangdong Feed 24(6):22–24

Bohl M (1985) Impact of fish production on the receiving water. Muench Beitr Abwasser-fisch Flussbiologie 39:297–323

Boonyaratpalin M, Wannagowat J, Burisut C (1993) Nutritional requirements of grouper Epinephelus. In: The proceedings of grouper culture, November 30–December 1. National Institute of Coastal Aquaculture, and Japan International Cooperation Agency, Department of Fisheries, Songkhla, Thailand and Bangkok, Thailand, pp 50–55

Bosma RH, Chau TTH, Potting J (2009) Environmental impact assessment of the pangasius sector in the Mekong delta. Wageningen University report to the Dutch Ministry of Agriculture, Nature and Food Quality and the Vietnamese Ministry of Agriculture and Rural Development

Bosma RH, Anh PT, Potting J (2011) Life cycle assessment of intensive striped catfish farming in the Mekong Delta for screening hotspots as input to environmental policy and research agenda. Int J Life Cycle Assess 16:903–915

BPS (2018) Badan Pusat Statistik (Satu Data Kelautan dan Perikanan). https://bps.go.id/subject/56/perikanan.html#subjekViewTab4

Bureau of Fisheries (BoF) (1991) Chinese fisheries in the past fifty years. In: China fisheries yearbook 1990

Bureau of Fisheries (BoF) (2001) Chinese fisheries in the past fifty years. In: China fisheries yearbook 2000

Bureau of Fisheries (BoF) (2012) National annual report on status of aquatic animal health. Chinese Ministry of Agriculture, Beijing, China

Bureau of Fisheries (BoF) (2016) China fishery statistical yearbook 2016. Bureau of Fisheries, Chinese Ministry of Agriculture, Beijing, China

Bureau of Fisheries, China (BOF) (2020) China fishery statistical yearbook. China Agriculture Press, Beijing, China

Cagauan AG (2001) Water quality management for freshwater fish culture. In: Utilizing different aquatic resources for livelihoods in Asia, a resource book. International Institute of Rural Reconstruction, International Development Research Centre, Food and Agriculture Organization of the United Nations, Network for Aquaculture Centers in Asia-Pacific and International Center for Living Aquatic Resources Management, Cavite, Philippines, pp 294–299

Chang CF (1994) Disease control of grouper. National Sun Yat-Sen University Ext. Rep. No. 12, pp 1–18

Chanrachkij I, Laongmanee P, Lanmeen J, Suasi T, Sornkliang J, Tiaye R, Chumchuen SV (2020) Severity of the impacts of COVID-19 pandemic on smallscale fisheries of Thailand: a preliminary assessment. Fish for the People 18(2):43–47

Chao TM, Chow M (1990) Effect of methyltestosterone on gonadal development of *Epinephelus tauvina* (Forskal). Singap J Prim Ind 18:1–14

Chao TM, Chow M (1996) Grouper culture and a review of the grouper breeding programme in Singapore. Paper presented at NACA Workshop on Aquaculture of coral reef fishes and sustainable reef fisheries, 4–8 December 1996, Kota Kinabalu, Sabah, Malaysia

Chao CB, Chung HY (1994) Study on *Cryptocaryon irritans* infection on captive grouper, *Epinephelus* spp. life cycle and pathogenicity. Council Agric Fish Ser 46:3140

Chao TM, Lim LC (1991) Recent developments in the breeding of grouper (*Epinephelus spp.*) in Singapore. Singap J Prim Ind 19:78–93

Chen BS (1996) An overview of the disease situation, diagnostic techniques, preventives, and treatments for cage-cultured, high value marine fishes in China. In: Main KL, Rosenfeld C (eds)

Aquaculture health management strategies for marine fishes. Proceedings of a Workshop in Honolulu, Hawaii, Oct 9–13 1995. The Oceanic Institute, Hawaii, pp 109–116

Chen FY, Chow M, Chao TM, Lim R (1977) Artificial spawning and larval rearing of the grouper, *Epinephelus tauvina* (Forksal) in Singapore. Singap J Prim Ind 5:1–21

China Feed Industry Association (2016) China feed industry yearbook 2001–2015. China Commercial Publishing House, Beijing, China

China Fishery Statistical Yearbook (2000) Ministry of Agriculture. China Agriculture Press, Beijing, p 2001

China Fishery Statistical Yearbook (2005) Ministry of Agriculture. China Agriculture Press, Beijing

China Fishery Statistical Yearbook (2015) Ministry of Agriculture. China Agriculture Press, Beijing, p 2016

China National Statistical Bureau (NSB) (2016) 2015 China national economy and social development statistical communique. China National Statistical Bureau, Beijing

Chinese Ministry of Agriculture (n.d.) China Agriculture Press, Beijing, China, pp 1–4

Chong YC, Chao TM (1986) Common diseases of marine food fish, Fishery handbook No. 2. Primary Production Department, Singapore

Choudhury HA (2000) Polyculture system of Thai pangus at Alpha Fisheries, Trishal, Mymensingh. Unpublished MS Thesis, Bangladesh Agricultural University, Mymensingh, Bangladesh

Chu TW (1996) Grouper culture in Taiwan. Paper presented at the workshop on Aquaculture of coral reef fishes and sustainable reef fisheries, 4–8 December 1996, Kota Kinabalu, Sabah, Malaysia

Chu VL, Luong QH, Pham MD, Ha T, Truong DH, Trieu LQ, Nguyen TH (2003) Viet Nam's fisheries development and integration. National Politics Publishing House

Chua TE, Teng SK (1978) Effects of feeding frequency on the growth of young estuary grouper, Epinephelus tauvina (Forskal) cultured in floating cages. Aquaculture 14:31–47

Chua TE, Teng SK (1982) Effects of food ration on growth, conditions factor, food conversion efficiency and net yield in floating net cages. Aquaculture 27:273–283

Chua F, Loo JJ, Wee JY, Ng M (1993) Findings from a fish disease survey: an overview of the marine fish disease situation in Singapore. Singap J Prim Ind 2:26–37

Chua FHC, Ng ML, Loo JJ, Wee JY (1994) Investigation of outbreaks of a viral disease 'sleepy grouper disease' affecting the brown-spotted grouper (*Epinephelus tauvina* Forskal). J Fish Dis 17:417–427

Chua FHC, Loo JJ, Wee JY (1995) Mass mortality in juvenile greasy grouper, *Epinephelus tauvina* associated with vacuolating encephalopathy and retinopathy. In: Shariff M, Arthur JR, Subasinghe RP (eds) Diseases in Asian aquaculture 11. Fish Health Section, Asian Fisheries Society, Manila, Philippines, p 235141

Ciric M, Subakov-Simic G, Dulic Z, Bjelanovi K, Cicovacki S, Markovic Z (2013) Effect of supplemental feed type on water quality, plankton and benthos availability and carp (*Cyprinus carpio L.*) growth in semi-intensive monoculture ponds. Aquac Res 46:777–788

Cui Y, Liu X, Wang S, Chen S (1992) Growth and energy budget in young grass carp, *Ctenopharyngodon idella* Val., fed plant and animal diets. J Fish Biol 41:231–238

Cui Y, Chen S, Wang S (1994) Effect of ration size on the growth and energy budget of the grass carp, *Ctenopharyngodon idella* Val. Aquaculture 123:95–107

Dan NC, Little DC (2000) Overwintering performance of Nile tilapia Oreochromis niloticus (L.) broodfish and seed at ambient temperatures in Northern Vietnam. Aquac Res 31:485–493

Danayadol Y, Kanchanapungka S (1989) Swim bladder syndrome disease in grouper (*Epinephelus malabaricus*). I. Occurrence and proposed primary cause. (Abstract). The seminar of fisheries 1989. Department of Fisheries

Danayadol Y, Direkbusarakom S, Supamattaya K (1993) Viral nervous necrosis in brown-spotted grouper (*Epinephelus malabaricus*) cultured in Thailand. In: The proceedings of grouper culture, November 30–December 1. National Institute of Coastal Aquaculture, and Japan

International Cooperation Agency, Department of Fisheries, Songkhla, Thailand and Bangkok, Thailand, pp 81–88

Danayadol Y, Direkbusarakom S, Supamattaya K (1995) Viral nervous necrosis in brown-spotted grouper (*Epinephelus malabaricus*) cultured in Thailand. In: Shariff M, Arthur JR, Subasinghe RP (eds) Diseases in Asian aquaculture 11. Fish Health Section, Asian Fisheries Society, Manila, Philippines, pp 227–233

Dasgupta S, Laplante B, Meisner C et al (2009) The impact of sea level rise on developing countries: a comparative analysis. Clim Chang 93(3–4):379–388

de Jesus EG (1996) Do thyroid hormones play a role in the metamorphosis of the grouper, *Epinephelus coicoides*? In: Joss J (ed) The Third Congress of the Asia and Oceania Society for Comparative Endocrinology, 22–26 Jan 1991. Macquarie University, Sydney, Australia, pp 259–260

De Silva SS, Brett AI, Phuong TN, Tam MB, Geoff JG, Giovanni MT (2010) Estimation of nitrogen and phosphorus in effluent from the striped catfish farming sector in the Mekong delta, Vietnam. Ambio:504–514

Debas L, Fostier A, Fuchs J, Weppe M, Nedelec G, Benett A (1989) Advances in tropical aquaculture, Tahiti, Feb 20–Mar 4, 1989. Aquacop Tfsemer Actes de Callogue 9:543–557

Dey MM, Rosegrant MW, Gosh K, Chen OL, Valmonte-Santos R (2016) Analysis of the economic impact of climate change and climate change adaptation strategies for fisheries sector in Pacific coral triangle countries: model, estimation strategy, and baseline results. Mar Policy 67:156–163

Diaz F, Espina S, Rodriguez C, Soto F (1998) Preferred temperature of grass carp, *Ctenopharyngodon idella* (Valenciennes), and brema carp, *Megalobrama amblycephala* (Yih), (Pisces, Cyprinidae) in horizontal and vertical gradients. Aquac Res 29:643–648

Directorate General of Aquaculture (2017) Roadmap, aquaculture development 2017–2021. Directorate General of Aquaculture, Jakarta

Doi M, Nawi MHM, Nik Lab NR, Talib I (1991) Artificial propagation of the grouper *Epinephelus suillus* at the marine finfish hatchery in Tanjung Demong, Terengganu, Malaysia. Dept of Fisheries, Ministry of Agriculture, Malaysia

Doi M, Toledo JD, Golez SN, de los Santos M, Ohno A (1996) Feeding performance of early red-spotted grouper, *Epinephelus coioides* larvae reared with mixed zooplankton. In: (Abstract only.) A paper presented at the International symposium on Live food organisms and environmental control for larviculture of marine animals, Nagasaki, Japan

Dong S (2011) History, principle and classification of integrated aquaculture in China. China Fish Sci 18(5):1202–1209

Dongmeza E, Steinbronn S, Francis G, Focken U, Becker K (2009) Investigations on the nutrient and antinutrient content of typical plants used as fish feed in small scale aquaculture in the mountainous regions of Northern Vietnam. Anim Feed Sci Technol 149:162–178

Dongmeza E, Francis G, Steinbronn S, Focken U, Becker K (2010) Investigation on the digestibility and metabolizability of the major nutrients and energy of maize leaves and barnyard grass in grass carp (Ctenopharyngodon idella). Aquac Nutr 16(3):313–326

Driver PD, Closs GP, Koen T (2005) The effects of size and density of carp (*Cyprinus carpio L.*) on water quality in an experimental pond. Arch Hydrobiol 163:117–131

Duc NM (2009) Economic contribution of fish culture to farm income in Southeast Vietnam. Aquac Int 17(1):15–29

Duray MM (1994) Daily rates of ingestion of rotifers and artemia nauplii by laboratory-reared grouper larvae *Epinephelus suillus*. Philippines Sci 31:3241

Duray MM, Estudillo CB, Alpasan LG (1997) Larval rearing of the grouper Epinephelus suillus under laboratory conditions. Aquaculture 150:63–76

El-Dakour S, George KA (1982) Growth of hamoor (*Epinephelus tauvina*) fed on different protein: energy ratios. Animal Research Report, 1981. Kuwait Institute of Scientific Research, pp 75–77

Ellis F, Trotter B, Magrath P (1992) Rice marketing in Indonesia: methodology, result and implications of a research study. Natural Resources Institute

El-Shebly AA, El-Kady MAH, Hussin AB, Hossain MY (2007) Preliminary observations on the pond culture of meagre, Argyrosomus regius (Asso, 1801) (Sciaenidae) in Egypt. J Fish Aquat Sci 2:345–352

El-Sherif MS, Mervat AMA (2009) Effect of rearing systems (mono-and polyculture) on the performance of freshwater prawn (*M. rosenbergii*) juveniles. J Fish Aquat Sci 4:117–128

Enell M (1982) Changes in sediment dynamics caused by cage culture activities. In: Proceedings of the 10th Nordic Symposium on Sediments, pp 72–78

Food and Agriculture Organization (FAO) (2001) Fisheries Technical Paper No. 407, Integrated agriculture-aquaculture. FAO, Rome, Italy

Food and Agriculture Organization (FAO) (2003) Fisheries Technical Paper No. 427, Aquaculture development in China – the role of public sector policies. FAO, Rome, Italy

Food and Agriculture Organization (FAO) (2006) Search aquaculture fact sheets: cultured aquatic species. FAO, Rome, Italy

Food and Agriculture Organization (FAO) (2010) State of world Fisheries and Aquaculture. FAO, Rome, Italy

Food and Agriculture Organization (FAO) (2019) State of world Fisheries and Aquaculture. FAO, Rome, Italy

Food and Agriculture Organization (FAO) (2022) State of world Fisheries and Aquaculture. FAO, Rome, Italy

Francis G, Makkar HPS, Becker K (2001a) Antinutritional factors present in plant-derived alternate fish feed ingredients and their effects in fish. Aquaculture 199:197–227

Francis G, Makkar HPS, Becker K (2001b) Dietary supplementation with a Quillaja saponin mixture improves growth performance and metabolic efficiency in common carp (*Cyprinus carpio* L.). Aquaculture 203:311–320

Francis G, Kerem Z, Makkar HPS, Becker K (2002) The biological action of saponins in animal systems – a review. Br J Nutr 88(6):587–605

Fukunaga K, Nogami Y, Yoshida K, Hamazaki S, Maruyama K (1990) Recent increase of *E. akaara* fry production and its problems at the Tamano Branch Station of the Japan Sea-Farming Association. Saibai Giken 19:33–40

Gaither MR, Rocha LA (2013) Origins of species richness in the Indo-Malay-Philippine biodiversity hotspot: evidence for the center of overlap hypothesis. J Biogeogr 40:1638–1648

Garcia-Berthou E (2001) Size- and depth-dependent variation in habitat and diet of the common carp (*Cyprinus carpio*). Aquat Sci 63:466–476

Gowen RJ, Bradbury NB (1987) The ecological impact of salmonid farming in coastal waters: a review. Oceanogr Mar Biol Annu Rev 25:563–575

Gui J-F, Tang Q, Li Z, Liu J, De Silva SS (eds) (2018) Aquaculture in China: success stories and modern trends. Wiley, Hoboken

Guillen J, Natale F, Carvalho N, Casey J, Hofherr J, Druon JN, Fiore G, Gibin M, Zanzi A, Martinsohn JT (2019) Global seafood consumption footprint. Ambio 48:111–122

Ha TTP, van Dijk H (2013) Fishery livelihoods and (non-)compliance with fishery regulations—a case study in Ca Mau Province, Mekong Delta. Viet Nam Mar Policy 38:417–427

Hainan Province: Economic News and Statistics for Hainan's Economy. (n.d.). Thechinaperspective.com

Hajra A, Ghosh A, Mandal SK (1988) Biochemical studies on the determination of optimum dietary protein to energy ratio for tiger prawn, Penaeus monodon (Fab.), juveniles. Aquaculture 71(1–2):71–79

Hamamoto S, Manade S, Kasuga A, Nosaka K (1986) Natural spawning and early life history of *Epinephelus salmonoides*. Technical report. Farm Fish 15:143–155

Han J, Lu H (2010) Effectiveness of China-Aid international short term technical training program. J China Univ Min Technol 1:83–88

Han D, Shan X, Zhang W, Chen Y, Wang Q, Li Z, Zhang G, Xu P et al (2016) A revisit to fishmeal usage and associated consequences in Chinese aquaculture. Rev Aquacult 2016:1–16

Hasan NA, Heal RD, Bashar A, Bablee AL, Haque MM (2021) Impacts of COVID-19 on the finfish aquaculture industry of Bangladesh: a case study. Mar Policy 130:104577

Heemstra PC (1991) A taxonomic revision of the Eastern Atlantic groupers (Pisces: Serranidae). Biol Museum Mun Funchal 43:5–71

Heemstra PC, Randall JE (1993) FAO Species Catalogue. Groupers of the world (family: Serranidae, subfamily: Epinephelinae), FAO Fisheries Synopsis No. 125. FAO, Rome, Italy

Henriksson PJG, Tran N, Mohan CV, Chan CY, Rodriguez U-P, Suri S, Mateos LD, Utomo NBP, Hall S, Phillips MJ (2017) Indonesian aquaculture futures—evaluating environmental and socioeconomic potentials and limitations. J Clean Prod 162:1482–1490

Hinojosa-Garro D, Zambrano L (2004) Interactions of common carp (*Cyprinus carpio*) with benthic crayfish decapods in shallow ponds. Hydrobiologia 515:115–122

Hirata Y, Kurokura H, Kasahara S (1989) Effects of thyroxine and thiourea on the development of larval red sea bream Pagrus major. Nippon Suisan Gakkaishi 55:1189–1195

Hu SH, Lim KJ (1984) Food and feeding frequency in rearing grouper Epinephelus sp. fry, Penhu Fisheries Research Institute Report, 4, pp 40–52. Penhu, Taiwan

Hu L, Tang J, Zhang J, Ren W, Guo L, Halwart M, Li K, Zhu Z et al (2015) Development of rice-fish systems: today and tomorrow. Chin J Eco-Agric 23(3):268–275

Huang TS, Lim KJ, Yen CL, Lim CY, Chen CL (1986) Experiments on the artificial propagation of black spotted grouper, *Epinephelus salmonoides* (Lacepede). I. Hormone treatment, ovulation of spawners and embryonic development. Bull Taiwan Fish Res Inst 40:241–258

Hussin MA, Ahmad A (1996) Status report on aquaculture of coral reef fishes in Peninsular Malaysia. Paper presented at the Workshop on Aquaculture of coral reef fishes and sustainable reef fisheries, 4–8 December 1966, Kota Kinabalu, Sabah, Malaysia

IDH (2018) Investment guideline for sustainable aquaculture in Indonesia. The Sustainable Trade Initiative, Utrecht, The Netherlands

Ilham N, Kusnadi N, Friyatno S, Suryani S (2010) Factors that determine the surplus marketed Gabah. Informatika Pertanian 19(2):45–75

Indonesian Act (2004) Undang-Undang Republik Indonesia No 31 Tahun 2004. DKP, Jakarta, p 68

Inui Y, Miwa S (1985) Thyroid hormone induces metamorphosis of flounder larvae. Gen Comp Endocrinol 60:450–454

Inui Y, Tagawa M, Miwa S, Hirano T (1989) Effects of bovine TSH on the tissue thyroxine level and metamorphosis in pre-metamorphic flounder larvae. Gen Comp Endocrinol 74:406–410

Islam MM, Khan MI, Barman A (2021) Impact of novel coronavirus pandemic on aquaculture and fisheries in developing countries and sustainable recovery plans: case of Bangladesh. Mar Policy 131:104611

Jackson ZJ, Quist MC, Downing JA, Larscheid JG (2010) Common carp (Cyprinus carpio), sport fishes, and water quality: ecological thresholds in agriculturally eutrophic lakes. Lake Reserv Manage 26:14–22

Joffre OM, Schmitt K (2010) Community livelihood and patterns of natural resources uses in the shrimp-farm impacted Mekong Delta. Aquac Res 41(12):1855–1866

Kakoolaki S, Ebne al-Torab SA, Ghajari A, Anvar AA, Sepahdari A, Ahari H, Hoseinzadeh H (2020) Socioeconomic impacts of coronavirus (COVID-19) outbreak on world shrimp aquaculture sector. Iran J Aquat Anim Health 6(1):1–18

Kangkan AL (2006) Studi Penentuan Lokasi Untuk Pengembangan Budidaya Laut Berdasarkan Parameter Fisika, Kimia dan Biologi Di Teluk Kupang, Nusa Tenggara Timur. Tesis, Universitas Diponegoro

Khatune-Jannat M, Rahman MM, Bashar MA, Hasan MD, Ahamed F, Hossain MY (2012) Effects of stocking density on survival, growth and production of Thai climbing perch (*Anabas testudineus*) under fed ponds. Sains Malays 41:1205–1210

Kitajima C, Takaya M, Tsukashima Y, Arakawa T (1991) Development of eggs, larvae, and juveniles of the grouper, *Epinephelus septemfasciatus* reared in the laboratory. Japanese J lchthyol 3S:47–53

Kloskowski J (2011a) Impact of common carp Cyprinus carpio on aquatic communities: direct trophic effects versus habitat deterioration. Fundam Appl Limnol 178:245–255

Kloskowski J (2011b) Differential effects of age-structured common carp (*Cyprinus carpio*) stocks on pond invertebrate communities: implications for recreational and wildlife use of farm ponds. Aquac Int 19:1151–1164

Kluts IN, Potting J, Bosma RH et al (2012) Environmental comparison of intensive and integrated agriculture-aquaculture systems for striped catfish production in the Mekong Delta, Vietnam, based on two existing case studies using life cycle assessment. Rev Aquac 4(4):195–208

Kohno H, Tririo A, Girochi D, Duray M (1989) Effects of feeding frequency and amount of feeding on the growth of grouper, Epinephelus malabaricus. Philippine J Sci 118:89–100

Kohno H, Duray M, Sunyoto P (1990) A field guide to groupers of Southeast Asia. Central Research Institute for Fisheries and Japan International Cooperative Agency, Jakarta, Indonesia

Kungvankij P, Tiro LB, Pudadera BP, Potestas IO (1986) Induced spawning and larval rearing of grouper (*Epinephelus salmoides*, Maxwell). In: Maclean JL, Dizon LB, Hosillos LV (eds) The First Asian Fisheries Forum. Asian Fisheries Society, Manila, Philippines, pp 663–666

Lam TJ (1994) Hormones and egg/larval quality in fish. J World Aquacult Soc 25:2–12

Lam TJ, Chao TM, Lim LC, Nugegoda D, Yong AM (1994) Thyroid hormone levels are higher in buoyant than in non-buoyant eggs in greasy grouper, *Epinephelus tauvina*. Singap J Prim Ind 22:29–33

Lamers M, Anyusheva M, La N, Nguyen VV, Streck T (2011) Pesticide pollution in surface- and groundwater by paddy rice cultivation: a case study from Northern Vietnam (short communication). Clean Soil Air Water 39(4):356–361

Lan TD (2009) Coastal aquaculture and shrimp farming in North Vietnam and environmental cost estimation. Aquat Ecosyst Health Manage 12(3):235–242

Lan NTP (2013) Social and ecological challenges of market-oriented shrimp farming in Vietnam. Springerplus 2(1):675

Le BN (2004) Study on the nutrient dynamics in intensive catfish (ca tra) culture ponds. Master thesis, Can Tho University, Viet Nam

Leong TS (1994a) A history of diseases in marine finfishes cultured in floating cages in Malaysia. In: Vidyadavan MK, Aziz MT, Shariff M (eds) International Congress in Quality Veterinary Services for the 21st Century. Dept of Veterinary Services, Kuala Lumpur, Malaysia, pp 65–66

Leong TS (1994b) Parasites and diseases of cultured marine finishes in Southeast Asia. University Sains Malaysia, Penang, Malaysia

Leong TS (1995) Bacteria and parasite population size as potential indicators of stress in seabass cage culture. In: Main KL, Rosenfeld C (eds) Aquaculture health management strategies for marine fishes. Proceedings of a Workshop in Honolulu, Hawaii, Oct. 9–13. The Oceanic Institute, Hawaii, pp 121–130

Leong TS (1997) Control of parasites in cultured marine finfishes in Southeast Asia - an overview. Int J Parasitol 27:1177–1184

Leong TS, Wong SY (1987) Bacterial pathogens and parasites in Greasy Grouper, *Epinephelus salmoides* Maxwell, silver seabass, Lates calcarifer Bloch and golden snapper Lutjanusjohni Bloch cultured in Penang, Perak and Kedah, Peninsular Malaysia, Final report. International Development Research Centre, Canada

Leong TS, Wong SY (1990) Parasites of healthy and diseased juvenile grouper (*Epinephelus malabaricus* Bloch and Schneider) and seabass (*Lates calcarifer* Bloch) in floating cages in Penang, Malaysia. Asian Fish Sci 3:319–327

Leong TS, Wong SY (1993) Environmental induced mortality of culture grouper *Epinephelus malabaricus* infected with high density of monogeneans and Vibrio. In: Second Symposium on Disease in Asian aquaculture, 25–29 Oct 1993, Phuket, Thailand

Li J (2002) Status and outlook of aquaculture industry in China. Aquaculture 20:7–9

Liao IC, Mao-Sen S, Chang SL (1995) A review of the nursery and grow out technique of high-value marine finfishes in Taiwan. In: Main KL, Rosenfeld C (eds) Culture of high-value marine fishes in Asia and the United States. The Oceanic Institute, Hawaii, pp 121–138

Lim KJ, Yen CL, Huang TS, Lim CY, Chen CL (1986) Experiment of fry nursing of *E. salmonoides* and its morphological study. Bull Taiwan Fish Res Inst 40:219140

Lin Y, Gao Z, Zhan A (2015) Introduction and use of non-native species for aquaculture in China: status, risks and management solutions. Rev Aquac 7:28–58

Ling SW (1977) Aquaculture in Southeast Asia, a historical overview. University of Washington Press, Seattle/London

Liu H (1999) Development and scientific and technological achievements of China Fisheries and Aquaculture industry. J Dalian Fish College 14(3):51–63

Liu J, Li Z (2010) The role of exotics in Chinese aquaculture. In: De Silva SS, Davy BF (eds) Success stories in Asian aquaculture. Springer, Dordrecht, The Netherlands, pp 173–185

Loc V, Bush S, Sinh L et al (2010) High and low value fish chains in the Mekong Delta: challenges for livelihoods and governance. Environ Dev Sustain 12(6):889–908

Lu Z (2013) Status and recommendations – Chinese aqua-feed industry. Fishery Guide to be Rich 10:18–21

Madsena H, Thien PC, Nga HTN, Clausen JH, Dalsgaarda A, Murrell KD (2015) Two-year intervention trial to control of fish-borne zoonotic trematodes in giant gourami (Osphronemus goramy) and striped catfish (Pangasianodon hypophthalmus) in nursery ponds in the Mekong Delta, Vietnam. Acta Tropica 152:201–207

Main KL, Rosenfeld C (eds) (1996) Aquaculture health management strategies for marine fishes. The Oceanic Institute, Hawaii

Manjappa K, Keshavanath P, Gangadhara B (2011) Influence of sardine oil supplemented fish meal free diets on common carp (*Cyprinus carpio*) growth, carcass composition and digestive enzyme activity. J Fish Aquat Sci 6:604–613

MARD (2011) VietGAP guideline for striped catfish (*Pangasianodon hypophthalmus*), tiger shrimp (*Penaeus monodon*) and white leg shrimp (*Penaeus vannamei*)

MARD (2014) Annual report of striped catfish production in Mekong delta, Vietnam. (in Vietnamese), 10 pp

Maruyama K, Nogamo K, Yoshida Y, Fukunaga K (1993) Hatchery technology on the breeding and seed production of red spotted grouper *Epinephelus akaara* in Japan. Japan Sea-Farming Association, Goto Station, Tamanuora, Nagasaki

Matsuzaki SS, Usio N, Takamura N, Washitani I (2009) Contrasting impacts of invasive engineers on freshwater ecosystems: an experiment and meta-analysis. Oecologia 158:673–686

May CL, Ngoh GH, Chong SY, Chua HCF, Chan YC, Howe LCJ et al (1992) Description of a virus isolated from the grouper *Plectropomus maculatus*. J Aquat Anim Health 4:222–226

Maynard JA, Anthony KRN, Afatta S, Dahl-Tacconi N, Hoegh-Guldberg O (2010) Making a model meaningful to coral reef managers in a developing nation: a case study of overfishing and rock anchoring in Indonesia. Conserv Biol 24:1316–1326

Miao W (2009) Recent developments in rice-fish culture in China: a holistic approach for livelihood improvement in rural areas. In: De Silva SS, Davy FB (eds) Success stories in Asian aquaculture. Springer, Dordrecht, The Netherlands, pp 15–40

Minahal Q, Munir S, Kumal W, Fatima S, Liaqat R, Shehdzadi I (2020) Global impact of COVID-19 on aquaculture and fisheries: a review. Int J Fish Aquat Stud 8(6):1–7

Minh LQ, Tuong TP, van Mensvoort MEF, Bouma J (1997) Contamination of surface water as affected by land use in acid sulfate soils in the Mekong River Delta, Vietnam. Agric Ecosyst Environ 61(1):19–27

Minh NH, Minh TB, Kajiwara N, Kunisue T, Iwata H, Viet PH, Tu NPC, Tuyen BC, Tanabe S (2007) Pollution sources and occurrences of selected persistent organic pollutants (POPs) in sediments of the Mekong River delta, South Vietnam. Chemosphere 67(9):1794–1801

Miwa S, Inui Y (1987) Effects of various doses of thyroxine and triiodothyronine on the metamorphosis of flounder (Paralichthys olivaceus). Gen Comp Endocrinol 67:356–363

Miwa S, Tagawa W, Inui Y, Hirano T (1988) Thyroxine surge in metamorphosing flounder larvae. Gen Comp Endocrinol 70:158–163

Moe MA (1969) Biology of red snapper *Epinephelus mario* (Val.) from the eastern Gulf of Mexico, Professional paper series, 10. Florida Department of Natural Resources, Marine Research Laboratory, St Petersburg, pp 1–95

MoFI (2003) Annual report on Aquaculture achievements in 2002 and implementation plan for 2003, 12 p

MoFI (2005) Annual report on the state achievements of 2004 and action plan of 2005 of fisheries sector

Mohapatra BC, Singh SK, Sarkar B, Majhi D, Sarangi N (2007) Observation of carp polyculture with giant freshwater prawn in solar heated fishpond. J Fish Aquat Sci 2:149–155

Molver J, Stigobrandt A, Bjerkenes V (1988) On the excretion of nitrogen and phosphorus from salmon. In: Proceedings Aquaculture International Congress, Vancouver, p 80

Morgan DK, Hicks BJ (2013) A metabolic theory of ecology applied to temperature and mass dependence of N and P excretion by common carp. Hydrobiologia 705:135–145

Muller F, Varadi L (1980) The results of cage fish culture in Hungary. Aquac Hungary 2:154–167

Murray JP, Grenyer R, Wunder S, Raes N, Jones JPG (2015) Spatial patterns of carbon, biodiversity, deforestation threat, and REDD + projects in Indonesia. Conserv Biol 29:1434–1445

Murray AG, Ives SC, Smith RJ, Moriarty M (2021) A preliminary assessment of indirect impacts on aquaculture species health and welfare in Scotland during COVID-19 lockdown. Vet Anim Sci 11:100167

Nabhitabhata J, Prempijawat R, Klaskliang K, Kaberirum S (1988) Estimation on optimum stocking density of grouper (Epinephelus tauvina Forskal) in cages on the basis of dissolved oxygen budget: Pang-rat River. Rayong Brackish Water Fisheries Station, Brackish Water Fisheries Division, Department of Fisheries, Thailand

New MB (1987) Feed and feeding of fish and shrimp. A manual on the preparation and presentation of compound feeds for shrimp and fish in aquaculture. United Nations Development Program. FAO, Rome

Ngoc PTA, Miranda PMM, Tru LT, Bosma RH, Verreth J, Lansin AO (2016) Economic feasibility of recirculating aquaculture systems in pangasius farming. Aquac Econ Manag 20:185–200

Nguyen AD, Savenije HH (2006) Salt intrusion in multi-channel estuaries: a case study in the Mekong Delta, Vietnam. Hydrol Earth Syst Sci 10:743–754

Nguyen, T.P., Pham, M. D., Vu, N. S., Tran, V. B., Au, T. A. N. (2004) Application of high technologies for quality improvement of catfish, tilapia and giant freshwater prawn products and reduction of their production costs. Report submitted to An Giang Department of Science and Technology

Nguyen TP, Duong NL, Ly VK (2005) Integrated aquaculture system in the Mekong River delta. In: Proceeding of the National workshop on research and technologies applied in aquaculture. Agricultural Publishing House, Ha Noi, pp 299–313

Nguyen, T.P., Tran, T.T.H., Vu, N.U., Huynh, T.G., Cao, T.A., Nguyen, T.T.H., (2007). Study on environment and disease pathogens of catfish farming Tra (Pangasianodon hypophthalmus) and basa (Pangasius bocourti) and giant freshwater prawn (Macrobrachium rosenbergii) in An Giang province. Report submitted to An Giang Science and Technology Department

Noviar H, Masbar R, Aliasuddin SS, Saputra J, Saputra J, Saputra J (2020) The agricultural commercialisation and its impact on economy management: an application of duality-neoclassic and Stochastic Frontier Approach. Ind Eng Manag Syst 19(3):510–519

Nusril HHS, Sukiyono K (2007) Marketable surplus of rice analysis (Case Study in Dusun Muara Aman, North Lebong Sub-district, Lebong District). Jurnal Akta Agrosia 10(1):32–39

Ong B (1988) Characteristics of bacteria isolated from diseased groupers, *Epinephelus salmoides*. Aquaculture 73:7–17

Pahlow M, van Oel PR, Mekonnen MM, Hoekstra AY (2015) Increasing pressure on freshwater resources due to terrestrial feed ingredients for aquaculture production. Sci Total Environ 536: 847–857

Pakdee K, Tantavanit S (1985) Brown spotted grouper (Epinephelus tauvina) seed production at Phuket Brackwater Fisheries Station. In: Proceedings of the Second research seminar (fisheries). Kasetsart University, Bangkok, pp 129–137

Parkos JJ, Wahl D (2014) Effects of common carp (*Cyprinus carpio*), an exotic fish, on Aquatic Ecosystems. Illinois Natural History Survey report of January/February 2000. University of Illinois Board of Trustees, Center for Aquatic Ecology; Victor Santucci, Jr., Max McGraw Wildlife Foundation

Parkos JJ, Santucci VJ Jr, Wahl DH (2003) Effects of adult common carp (Cyprinus carpio) on multiple trophic levels in shallow mesocosms. Can J Fish Aquat Sci 60:182–192

Penczak T, Galieka W, Molinski M, Kusto E, Zalewski M (1982) The enrichment of a mesotrophic lake by carbon, phosphorus, and nitrogen from the cage aquaculture of rainbow trout *Salmo gairdneri*. J Appl Ecol 19:37–393

Phan LT, Bui TM, Nguyen TTT, Gooley GJ, Ingram BA, Nguyen HV, Nguyen PT, De Silva SS (2009) Current status of farming practices of striped catfish, *Pangasianodon hypophthalmus* in the Mekong Delta, Vietnam. Aquaculture 296:227–236

Phillips M, Beveridge M (1986) Cages and the effect on water condition. Fish Farmer 9(3):17–19

Phong LT, Udo HMJ, Mensvoort MEF et al (2007) Integrated agriculture-aquaculture systems in the Mekong Delta, Vietnam: an analysis of recent trends. AJAD 4(2):51–66

Phu TQ, Tinh TK (2012) Chemical compositions of sludge from intensive striped catfish (*Pangasianodon hypophthalmus*) culture pond. Sci J Cantho Univ 22:290–299

Phuong NT, Oanh D (2010) Striped catfish aquaculture in Vietnam: a decade of unprecedented development. In: De Silva SS, Davy FB (eds) Success stories in Asian aquaculture. Springer, Netherlands, pp 131–147

Phuong NT, Hai TN, Hien TTT (2006) Current status of freshwater prawn culture in Vietnam and the development and transfer of seed production technology. Fish Sci 72(1):1–12

Predalumpaburt Y, Tanvilai D (1988) Morphological development and the early life history of grouper, *Epinephelus malabaricus*, Bloch and Schneider (Pisces: Serranidae). Report of Thailand and Japan Joint Coastal Aquaculture Research Project, 3. National Institute of Coastal Aquaculture, Songkhla, Thailand, pp 49–72

Prein M (2002) Integration of aquaculture into crop-animal systems in Asia. Agric Syst 71:127–146

Qi J (2002) Development and prospect of aquaculture industry in China. Sci Fish Farm 6:4–5

Quinitio GF (1996) The status of grouper and other coral reef fishes seed production in the Philippines. Paper presented at Workshop on Aquaculture of Coral Fishes and Sustainable Reef Fisheries, 4–8 December, Kota Kinabalu, Sabah, Malaysia

Rachmansyah (2004) Carrying capacity analysis of Awarange Bay for milkfish culture development in floating net cages. PhD Dissertation, Bogor Agricultural University

Rahman MM (2015) Effects of co-cultured common carp on nutrients and food web dynamics in rohu aquaculture ponds. Aquac Environ Interact 6:223–232

Rahman MM, Meyer CG (2009) Effects of food type on diel behaviors of common carp *Cyprinus carpio* L. in simulated aquaculture pond conditions. J Fish Biol 74:2269–2278

Rahman MM, Verdegem MCJ (2007) Multi-species fishpond and nutrients balance. In: ven der Zijpp AJ, Verreth AJA, Tri LQ, ven Mensvoort MEF, Bosma RH, Beveridge MCM (eds) Fishponds in farming systems. Wageningen Academic Publishers, Wageningen, pp 79–88

Rahman MM, Verdegem MCJ, Nagelkerke LAJ, Wahab MA, Milstein A, Verreth JAJ (2006) Growth, production and food preference of rohu *Labeo rohita* (H.) in monoculture and in polyculture with common carp *Cyprinus carpio* (L.) under fed and non-fed ponds. Aquaculture 257:359–372

Rahman MM, Verdegem MCJ, Nagelkerke LAJ, Wahab MA, Milstein A, Verreth JAJ (2008a) Effects of common carp *Cyprinus carpio* (L.) and feed addition in rohu *Labeo rohita* (Hamilton) ponds on nutrient partitioning among fish, plankton and benthos. Aquac Res 39:85–95

Rahman MM, Verdegem MCJ, Nagelkerke LAJ, Wahab MA, Verreth JAJ (2008b) Relationships among water quality, food resources, fish diet and fish growth in polyculture ponds: a multivariate approach. Aquaculture 275:108–115

Rahman MM, Jo Q, Gong YG, Miller SA, Hossain MY (2008c) A comparative study of common carp (*Cyprinus carpio* L.) and calbasu (*Labeo calbasu* Hamilton) on bottom soil resuspension, water quality, nutrient accumulations, food intake and growth of fish in simulated rohu (*Labeo rohita* Hamilton) ponds. Aquaculture 285:78–83

Rahman MM, Verdegem MCJ, Wahab MA (2008d) Effects of tilapia (*Oreochromis nilotica L.*) addition and artificial feeding on water quality, and fish growth and production in rohu-common carp bi-culture ponds. Aquac Res 39:1579–1587

Rahman MM, Verdegem MCJ, Nagelkerke LAJ, Wahab MA, Verreth JAJ (2008e) Swimming, grazing and social behaviour of rohu Labeo rohita (Hamilton) and common carp *Cyprinus carpio* (L.) in tanks under fed and non-fed conditions. Appl Anim Behav Sci 113:255–264

Rahman MM, Verdegem MCJ, Wahab MA, Hossain MY, Jo Q (2008f) Effects of day and night on swimming, grazing and social behaviours of rohu *Labeo rohita* (Hamilton) and common carp *Cyprinus carpio* (L.) in simulated ponds. Aquac Res 39:1383–1392

Rahman MM, Hossain MY, Jo Q, Kim SK, Ohtomi J, Meyer C (2009) Ontogenetic shift in dietary preference and low dietary overlap in rohu (*Labeo rohita*) and common carp (*Cyprinus carpio*) in semi-intensive polyculture ponds. Ichthyol Res 56:28–36

Rahman MM, Kadowaki S, Balcombe SR, Wahab MA (2010) Common carp (Cyprinus carpio L.) alter their feeding niche in response to changing food resources: direct observatsions in simulated ponds. Ecol Res 25:303–309

Rahman MM, Kadowaki S, Linn SM, Yohei Y (2012) Effects of protein skimming on water quality, bacterial abundance and abalone growth in land based recirculating aquaculture systems. J Fish Aquat Sci 7:150–161

Rahman MM, Normawaty MN, Shahbudin S, Kamaruzzaman Y (2015) Coastal water quality of Tioman Island: effects of human activity and the distance from shoreline. Desalin Water Treat 57(1):83–87

Rajkumar M, Rahman MM, Reni Prabha A, Phukan B (2013) Effect of cholymbi on growth, proximate composition, and digestive enzyme activity of fingerlings of long whiskered catfish, Mystus gulio (Actinopterygii: Siluriformes: Bagridae). Acta Ichthyol Piscat 43:15–20

Randall JE (1987) A preliminary synopsis of the groupers (Perciformes: Serranidae: Epinephelinae) of the Indo-Pacific region. In: Polovina JJ, Ralston S (eds) Tropical snappers and groupers. Westview Press, Boulder, CO, pp 89–186

Randall JE, Heemstra PC (1991) Revision of Indo-Pacific groupers (Perciformes: Serranidae: Epinephelinae) with descriptions of five new species, Indo-Pacific Fishes No. 20. Bernice Panahi Bishop-Museum, Hawaii

Rattanachot A, Pakdee K, Tantavanit C, Tarmasawart T, Pimoljinda T (1985) Experiment on propagation and nursing of brown spotted grouper (*Epinephelus tauvina*). In: Proceedings of fisheries research seminar. Department of Fisheries, Bangkok, pp 131–139

Reddy PK, Lam TJ (1992) Effect of thyroid hormones on morphogenesis and growth of larvae and fry of telescopic-eye black goldfish *Carassius auratus*. Aquaculture 107:383–394

Ren Z, Zhou W (2001) Review on status and future perspective of Chinese aquaculture feed. Feed China 8:1–4

Renaud FG, Kuenzer C (2012) The Mekong Delta system interdisciplinary analyses of a river delta. Springer

Rico A, Van den Brink PJ (2014) Probabilistic risk assessment of veterinary medicines applied to four major aquaculture species produced in Asia. Sci Total Environ 468–469:630–641

Rico A, Phu TM, Satapornvanit K, Min J, Shahabuddin AM, Henriksson PJG, Murray FJ, Little DC, Dalsgaard A, Van den Brink PJ (2013) Use of veterinary medicines, feed additives and probiotics in four major internationally traded aquaculture species farmed in Asia. Aquaculture 412–413:231–243

Rimmer, M. and Sugama K (2005). Sustainable marine finfish aquaculture in Indonesia and Australia. In A. Sudrajat, A.I. Azwar, L.E. Hadi, Haryanti, N.A. Giri and G. Sumiarsa (eds). Buku Perikanan Budidaya Berkelanjutan. Pusat Riset Perikanan Budidaya. Jakarta: 12-27

Rini WDE, Rahayu ES, Harisudin M, Supriyadi S (2021) Management of Gogo Rice production in realizing the commercialization of marginal land households in Yogyakarta. Int J Sustain Dev Plan 16(2):373–378

Ritvo G, Kochba M, Avnimelech Y (2004) The effects of common carp bioturbation on fishpond bottom soil. Aquaculture 242:345–356

Ross LG (2000) Environmental physiology and energetics. In: Beveridge MCM, McAndrew BJ (eds) Tilapias: biology and exploitation. Kluwer, London, pp 89–128

Ruangpan L, Tabkaew R (1993) Parasites of the cage cultured grouper *Epinephelus malabaricus* in Thailand. In: The Proceedings of grouper culture, November 30–December 1. National Institute of Coastal Aquaculture, and Japan International Cooperation Agency, Department of Fisheries, Songkhla, Thailand and Bangkok, Thailand, pp 106–111

Ruangpanit N (1993) Technical manual for seed production of grouper (Epinephelus malabaricus). National Institute of Coastal Aquaculture, Songkhla

Ruangpanit N, Yashiro R (1995) A review of grouper (Epinephelus spp.) and seabass (Lates calcarifer) culture in Thailand. In: Main KL, Rosenfeld C (eds) The culture of high-value marine fishes in Asia and the United States. The Oceanic Institute, Hawaii, pp 167–183

Ruangpanit N, Manuwotrg S, Tattanon T, Kraisingdecha P, Arkayanont P, Rojanapitayagul S (1986) Preliminary study on rearing fry of grouper Epinephelus malabaricus. Report of Thailand and Japan Joint Coastal Aquaculture Research Project 2:35–38. National Institute of Coastal Aquaculture, Songkhla, Thailand

Ruangpanit N, Bunliptanon P, Pechmanee T, Arkayanont P, Vanakovat J (1988) Propagation of grouper, *Epinephelus malabaricus*, at the National Institute of Coastal Aquaculture Songkhla. Technical Paper No. 5. National Institute of Coastal Aquaculture, Songkhla, Thailand

Ruangpanit N, Boonliptanon P, Kongkumnerd J (1993) Progress in the proportion and larval rearing of the grouper, *Epinephelus malabaricus*. In: The Proceedings of grouper culture, 30 November–December 1, 1993. National Institute of Coastal Aquaculture and Japan International Cooperation Agency, Songkhla, Thailand and Bangkok, Thailand

Ruddle K, Christensen V (1993) An energy flow model of the mulberry dike-carp pond farming system of the Zhujiang Delta, Guangdong Province, China. In: Christensen V, De P (eds) Trophic models of aquatic ecosystems. ICLARM conference proceedings 26. International Center for Living Aquatic Resources Management, Manila

Sakaras W, Kumpang P (1987) Effect of stocking density on growth and production of estuary grouper (Epinephelus tauvina Forskal) cultured in cages. Technical Paper No. 3. Rayong Brackish Water Fisheries Station, Brackish Water Fisheries Division, Department of Fisheries, Thailand

Sakaras W, Kumpang P (1988) Growth and production of brown spotted grouper (Epinephelus tauvina Forskal) cultured in cages. Technical Paper No. 2. Rayong Brackish Water Fisheries Station, Brackish Water Fisheries Division, Department of Fisheries, Thailand

Sakaras W, Sukbantaung S (1985) Experiment on grouper (Epinephelus tauvina Forskal) in cages with different stocking densities. Technical Paper No. l. Rayong Brackish Water Fisheries Station, Brackish Water Fisheries Division, Department of Fisheries, Thailand

Sarker MT (2000) Pangus Chash Babosthapana (Management of Pangus culture). Department of Fisheries, Bangladesh

Schad I, Schmitter P, Saint-Macary C et al (2012) Why do people not learn from flood disasters? Evidence from Vietnam's northwestern mountains. Nat Hazards 62(2):221–241

Shapiro DY (1987) Reproduction in groupers. In: Polovina JJ, Rabton S (eds) Tropical snappers and groupers: biology and fisheries management. Westview Press, Boulder, CO, pp 295–327

Shireman JV, Colle DE, Rottmann RW (1978) Growth of grass carp fed natural and prepared diets under intensive culture. J Fish Biol 12:457–463

Sirimontaporn P (1993) Species of groupers for aquaculture in Thailand. In: The Proceedings of grouper culture, November 30–December 1, Songkhla. National Institute of Coastal Aquaculture and Japan International Cooperation Agency, Department of Fisheries, Songkhla, Thailand and Bangkok, Thailand, pp 126–129

Soltani M, Sheikhzadeh N, Ebrahimzadeh-Mousavi HA, Zargar A (2010) Effects of Zataria multiflora essential oil on innate immune responses of common carp (*Cyprinus carpio*). J Fish Aquat Sci 5:191–199

Songsangjinda P, Na-anan P, Tunvilai D (1993) Water and sediment quality in grouper cage culture area at Khlong Pakhara, Langu District, Satun Province. In: The Proceedings of grouper culture,

November 30–December 1, Songkhla. National Institute of Coastal Aquaculture and Japan International Cooperation Agency, pp 112–119

Sorgeloos P (2001) Technologies for sustainable aquaculture development, Plenary Lecture II. In: Subasinghe RP, Bueno P, Phillips MJ, Hough C, McGladdery SE, Arthus JR (eds) Aquaculture in the third Millennium. Technical Proceedings of the Conference on aquaculture in the third millennium, Bangkok, Thailand, 20–25 Feb 2000. NACA and FAO, Bangkok and Rome, pp 23–28

Soundarapandian P, Kannan A (2008) South Indian technology of nursery farming for better survival and production of Macrobrachium rosenbergii (De Man). J Fish Aquat Sci 3:137–144

Statistics Indonesia (2014) Rice household survey

Steinbronn S (2009) A case study: fish production in the integrated farming system of the Black Thai in Yen Chau district (Son La province) in mountainous North-Western Vietnam – current state and potential. Dissertation, University of Hohenheim, Stuttgart

Stewart TW, Downing JA (2008) Macroinvertebrate communities and environmental conditions in recently constructed wetlands. Wetlands 28:141–150

Su MS, Liao IC (1996) Status and prospects of grouper culture in Taiwan. In: Book of abstracts - World Aquaculture '96; January 29–February 2, Bangkok, Thailand, p 391

Sugama K (2005) Status Teknologi Perikanan Budidaya untuk Mendukung Budidaya Berkelanjutan. In: Sudrajat A, Azwar AI, Hadi LE, Haryanti NAG, Sumiarsa G (eds) Buku Perikanan Budidaya Berkelanjutan. Pusat Riset Perikanan Budidaya, Jakarta, pp 1–5

Sukhanongs S, Tanakumchup N, Chungyampin S (1978) Feeding experiment on artificial diet for greasy grouper, Epinephelus tauvina in nylon cages. Annual report. Songkhla Fisheries Station, Department of Fisheries, Thailand, pp 103–117

Sumiono B (2006) Fish stock assessment in Lalaka Straits (In Indonesian). Technical paper. Fish stock assessment seminar. Central Research Institute for Capture Fisheries, Jakarta

Sutanto I (2005) Kesuksesan Budidaya Udang Vannamei (*Litopenaeus vannamei*) di Lampung. In: Sudrajat A, Azwar AI, Hadi LE, Haryanti, Giri NA, Sumiarsa G (eds) Buku Perikanan Budidaya Berkelanjutan. Pusat Riset Perikanan Budidaya, Jakarta, pp 67–72

Szucs I, Stundi L, Varadi L (2007) Carp farming in Central and Eastern Europe and a case study in multifunctional aquaculture. In: Leung PS, Lee CS, O'Bryan PJ (eds) Species and system selection for sustainable aquaculture. Blackwell Publishing, Ames, pp 389–413

Tan YT (1970) Composition and nutritive value of some grasses, plants and aquatic weeds tested as diets. J Fish Biol 2:253–257

Tan SM, Tan KS (1974) Biology of tropical grouper *Epinephelus tauvina* (Forskal). I. A preliminary study on the hermaphroditism in E. tauvina. Singap J Prim Ind 2:123–133

Tan-Fermin JD (1992) Withdrawal of exogenous 17-alpha methyltestosterone causes reversal of sex inversed male grouper, *Epinephelus suillus* (Val.). The Philippine Scientist 29:33–39

Tan-Fermin JD, Nagai A, Javellana D (1993) Induction of sex inversion in juvenile grouper, *Epinephelus suillus* (Val.) by injections of 17A-methyltestosterone. Japanese J lchthyol 40: 413–420

Tan-Fermin JD, Garcia LMB, Castillo AR Jr (1994) Induction of sex inversion in juvenile grouper, *Epinephelus suillus* (Valenciennes) by injection of 17A-methyltestosterone. Japanese J lchthyol 40:413–420

Tang Q (2012) Strategy study on sustainable development of aquaculture. China Poultry 34(11): 13–15

Tauhid A, Sunarto I, Koesharyani H, Supriyadi, Gardenia L (2005) Koi Harpes Virus Disease (KHVD) on common carp culture and it's control strategies. In: Supriyade H, Priono B (eds) Strategi Pengelolaan dan Pengendalian Penyakit KHV: Suatu Upaya Pemecahan dalam Pembudidayaan Ikan Air Tawar. Serial Bunga Rampai, vol 105. CRIA, Jakarta, pp 41–60

Tauhid, Partasamita S, Haryadi J, Sudrajat A (2006) Status of white shrimp (*Litopenaeus vannamei*) culture in East Java (In Indonesian). CRIA Report

Tay HC, Goh J, Yong AN, Lim HS, Chao TM, Chou R et al (1994) Effect of thyroid hormone on metamorphosis in greasy grouper, *Epinephelus tauvina*. Singap J Prim Ind 22:35–38

Teng SK, Chua TE (1979) Use of artificial hides to increase the stocking density and production of estuary grouper, Epinephelus salmoides, Maxwell, reared in floating net cages. Aquaculture 16: 219–232

Teng SK, Chua TE, Lim PE (1978) Preliminary observation on the dietary protein requirement of estuary grouper, *Epinephelus salmoides* Maxwell cultured in floating net cages. Aquaculture 15: 257–271

Toan PV, Sebesvar Z, Bläsing M, Rosendahl I, Renaud FG (2013) Pesticide management and their residues in sediments and surface and drinking water in the Mekong Delta, Vietnam. Sci Total Environ 452–453:28–39

Toledo JD, Nagai A, Javellana D (1993) Successive spawning of grouper, *Epinephelus suillus* (Valenciennes) in a tank and a floating net cage. Aquaculture 115:361–367

Tookwinas S (1989) Review of grow out techniques under tropical conditions: experience in Thailand on seabass (Lates calcarifer) and grouper (Epinephelus malabaricus). In: Advances in tropical aquaculture, Aquacap. IFEMER, pp 737–750

Tran N, Rodriguez U-P, Chan CY, Phillips MJ, Mohan CV, Henriksson PJG, Koeshendrajana S, Suri S, Hall S (2017) Indonesian aquaculture futures: an analysis of fish supply and demand in Indonesia to 2030 and role of aquaculture using the AsiaFish model. Mar Policy 79:25–32

Ukawa M, Higuchi M, Mito S (1966) Spawning habits and early life history of a serranid fish, *Epinephelus akaara.* Temmineket Schelegel. Japanese J lchthyol 13:156–161

United States Department of Agriculture (USDA) (2019) Indonesia's rice milled data

Usfar AA, Fahmida U (2011) Do Indonesians follow its Dietary Guidelines?: Evidence related to food consumption, healthy lifestyle, and nutritional status within the period 2000–2010. Asia Pac J Clin Nutr 20(3):484

Van Dam AA, Chikafumbwa FJKT, Jamu DM, Costa-Pierce BA (1993) Trophic interactions in a Napier grass (*Pennisetum purpureum*)-fed aquaculture pond in Malawi. In: Christensen V, Pauly D (eds) Trophic models of aquatic ecosystems. ICLARM conference proceedings 26. International Center for Living Aquatic Resources Management, Manila

Van Senten J, Engle CR, Smith MA (2021) Effects of COVID-19 on US aquaculture farms. Appl Econ Perspect Policy 43(1):355–367

Vu NS, Yang Y, Nguyen TP (2005) River pen culture of giant freshwater prawn *Macrobrachium rosnebergii* (De Man) in Southern Viet Nam. Aquac Res 36:284–191

Wahab MA, Rahman MM, Milstein A (2002) The effect of common carp, *Cyprinus carpio* (L.) and mrigal, *Cirrhinus mrigala* (Hamilton) as bottom feeders in major Indian carp polycultures. Aquac Res 33:547–556

Wang Q, Li Z, Gui G-F, Liu J, Ye S, Yuan J, De Silva SS (2017) Paradigm changes in freshwater aquaculture practices in China: moving towards achieving environmental integrity and sustainability. Ambio 47(4):410–426

Weber MJ, Brown ML (2009) Effects of common carp on aquatic ecosystems 80 years after 'carp as a dominant': ecological insights for fisheries management. Rev Fish Sci 17:524–537

Weber MJ, Brown ML (2011) Relationships among invasive common carp, native fishes, and physicochemical characteristics in upper Midwest (USA) lakes. Ecol Freshw Fish 20:270–278

Wienbeck H (1983) Reflections on ammonia content in fish culture. Archie Dtsch Fisch Verb 39: 21–33

Wongsomnuk S, Parnichsuka P, Danayadol Y (1978) Experiments on nursing of grouper, *Epinephelus tauvina* (Forskal) with various mixed feeds, Annual report. Songkhla Fisheries Station, Department of Fisheries, Thailand, pp 79–102

Woynarovich A, Moth-Poulsen T, Peteri A (2010) Carp polyculture in Central and Eastern Europe, the Caucasus and Central Asia, FAO Fisheries and Aquaculture Technical Paper No 554. http://www.fao.org/docrep/013/i1794e/i1794e00.htm

Wu RSS (1995) The environment impact of marine fish culture: towards a sustainable future. Mar Pollut Bull 31:4–12

Wu RSS (1999) Environmental impacts of marine fish farming and their mitigation. In: Garcia LMB (ed) Responsible aquaculture development in Southeast Asia. Proceeding of the Seminar-workshop on aquaculture development in Southeast Asia, Iloilo City, 12–14 Oct 1999, SEAFDEC, Iloilo, Philippines. 274p, pp 157–172

Wu RSS (2001) Environmental impacts of Marine Fish farming and their mitigation. In: Garcia LMB (ed) Responsible aquaculture development in Southeast Asia. Proceedings of the seminar-workshop on aquaculture development in Southeast Asia organized by the aquaculture department, Southeast Asian Fisheries Development center, Iloilo City, 12–14 Oct 1999, Aquaculture Department, Southeast Asian Fisheries Development Center, Tigbauan, Iloilo Philippines. pp. 157–172

Wu RSS, Lim KS, Mackay DW, Lau TC, Yam V (1994) Impact of marine fish farming on water quality and bottom sediment: a case study in the sub-tropical environment. Mar Environ Res 38: 115–145

Wu B, Xia S, Rahman MM, Rajkumar M, Fu Z, Tan J, Yang A (2015) Substituting seaweed with corn leaf in diet of sea cucumber (*Apostichopus japonicus*): effects on growth, feed conversion ratio and feed digestibility. Aquaculture 444:88–92

Xie J, Hu L, Tang J, Wu X, Li N, Yuan Y et al (2011) Ecological mechanisms underlying the sustainability of the agricultural heritage ricefish coculture system. Proc Natl Acad Sci USA 108:E1381–E1387

Xu BT, Li E, Zhou HT (1985) Observations on the development of egg and larvae of red spotted grouper. J Fish China 9:369–374

Yang X, Zuo W (1997) Relationship between water temperature and immune response of grass carp (*Ctenopharyngodon idellus* C. et V.). Asian Fish Sci 10:169–177

Yashiro R, Vatanakul V, Kongkumnard V, Ruangpanit N (1993) Variation of sex steroids level in maturing grouper. In: The Proceedings of grouper culture, November 30–December 1, 1993, Songkhla. National Institute of Coastal Aquaculture and Japan International Cooperation Agency, Department of Fisheries, Songkhla, Thailand and Bangkok, Thailand, pp 8–15

Yathavamoorthi R, Surendraraj A, Sabeena Farvin KH (2010) Enteric bacteria and water quality of freshwater prawn *Macrobrachium rosenbergii* (De Man) in culture environment from Kerala, India. J Fish Aquat Sci 5:282–292

Yuan Y, Yuan Y, Dai Y, Gong Y (2017) Economic profitability of tilapia farming in China. Aquac Int 25(3):1253–1264

Yuan Y, Miao W, Yuan X, Dai Y, Yuan Y, Gong Y (2022) The impact of COVID-19 on aquaculture in China and recommended strategies for mitigating the impact. J World Aquacult Soc 53(5):933–947

Zambrano L, Perrow MR, Macias-Garcia C, Aguirre-Hidalgo V (1999) Impact of introduced carp (*Cyprinus carpio*) in subtropical shallow ponds in Central Mexico. J Aquat Ecosyst Stress Recovery 6:281–288

Zambrano L, Scheffer M, Martınez-Ramos M (2001) Catastrophic response of lakes to benthivorous fish introduction. Oikos 94:344–350

Zhang F (2013) Review and outlook of modern aquaculture industry in China. World Sci Res Technol Dev 3:10–20

Zhu Z (2014) Agriculture multilateral South-South Cooperation by China. Manag Observer 1:93–96

Further Reading

Khoi LN (2011) Quality management in the pangasius export supply chain in Vietnam: the case of small-scale pangasius farming in the Mekong River Delta. PhD Thesis. ISBN 978-90-367-4332-7. 283 p

Phu TM, Hien TTT, Tien T, Dao NLA (2014) Assessment of striped catfish fillet quality at different rearing areas. Sci J Cantho Univ 1:15–21

Sang NV, Thomassen M, Klemetsdal G, Gjøen HM (2009) Prediction of fillet weight, fillet yield, and fillet fat for live river catfish (Pangasianodon hypophthalmus). Aquaculture 288:166–171

Sharma VP (2016) Marketable and marketed surplus of rice and wheat in India: distribution and determinants. Indian J Agric Econ 71(2):137–159

5 Use of Deep Tech in Integrated Aquaculture Systems

There are various types of IAS in which aquaculture and hydroponics are the two major combining techniques, namely called aquaponic systems. The system continuously recirculates water from the fish tank containing the metabolic waste of fish. The water passes through a mechanical filter that captures solid waste, and then goes through a biofilter, oxidizing ammonia to nitrate. The media beds, where plants and vegetables grow, are the next destination of the water, so the plants can absorb the nutrients, and finally, the water returns, purified to the fish tank (Fig. 5.1).

The task of the biofilter is to provide a habitat for bacteria that convert fish waste into dissolved nutrients for the plants. The nutrient removal process will remove the harmful toxic nitrite and ammonia; thus, fish, plants, and bacteria can effectively grow. The most important aspect of an aquaponic system is the nitrification process. Nitrogen is an essential block for all living creatures, and it is present in all amino acids. Additionally, nitrogen is an important inorganic nutrient for all plants.

Nitrogen in gas form is the most abundant element present in the earth's atmosphere, yet it is only present in the atmosphere as a very stable triple bond of nitrogen atoms. However, the molecular nitrogen (N_2) is not accessible for plants' intake and growth. Therefore, nitrogen must go through a process called nitrogen fixation to become absorbable by the plants. This process is facilitated by bacteria that chemically alter the molecular nitrogen by adding hydrogen and oxygen, creating ammonia (NH_3) and nitrate (NO_3-). Additionally, the Haber process can be used to convert atmospheric nitrogen into synthetic fertilizers.

Nitrifying bacteria are the type of bacteria converting ammonia into nitrite compounds (NO_2-) and finally into nitrate compounds (NO_3-). Plants use both ammonia and nitrates for their growth, but nitrates are easier to assimilate by the roots. These bacteria live in diverse environments such as soil, water, and air and are essential for the nitrification process to convert plant and animal wastes into accessible nutrients for plants. For this reason, nitrifying bacteria are essential for the aquaponic systems since they eliminate ammonia and nitrite, which are toxic and provide nutrients for the plants.

© The Author(s), under exclusive license to Springer Nature Singapore Pte Ltd. 2023

A. Rahimi-Midani, *Deep Technology for Sustainable Fisheries and Aquaculture*,
https://doi.org/10.1007/978-981-99-4917-5_5

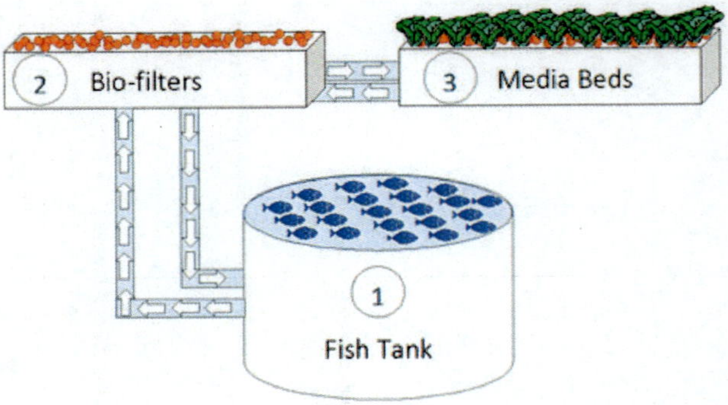

Fig. 5.1 Simple diagram of aquaponic systems

Nitrifying bacteria involved in the nitrification process are divided into two categories, first the ammonia-oxidizing bacteria and second the nitrite-oxidizing bacteria. The process starts when ammonia-oxidizing bacteria add oxygen to the NH_3 and create NO_2- and the nitrite-oxidizing bacteria oxidize the NO_2- into NO_3-. The most common ammonia-oxidizing bacteria in aquaponics is the genus *Nitrosomonas*, and the genus *Nitrobacter* is the most common nitrite-oxidizing bacteria. The aquaponic unit is totally reliant on these bacteria. In case of absence of bacteria or not functioning properly, ammonia concentrations in the water will increase to the level that suffocates the fish.

Bactria can live and grow on any substance, including the interior of grow pipes. The total area that can be used by these bacteria determines the amount of ammonia that they can metabolize. Plant roots and tank walls may be able to provide an adequate area, depending on the system design, fish biomass, and volume of circulating water.

The pH of the water also affects how the nitrifying bacteria function biologically. The pH range between 6 and 8.5 is the most ideal range for ammonia-oxidizing and nitrite-oxidizing bacteria, but in aquaponic systems, the range between 6 and 7 is better for both plants and fish.

Another important parameter for bacteria is the water temperature. The ideal water temperature for bacteria growth and productivity is 17 and 34 °C. However, temperature below 17 °C will drop the productivity of the mentioned bacteria. Additionally, temperatures below 25 °C will impact the growth of the fish species and increase the mortality level. Increasing the temperature of the system by locating the system under direct sunlight is common; however, photosensitive organisms such as nitrifying bacteria react to sunlight. This is because of ultraviolet (UV) light from the sun that kills these bacteria or reduce their colonization rate.

5.1 Impact of Water Quality in IAS

Water in an aquaponic system is the medium in which all macro- and micronutrients are transported to the plants and fish receive oxygen. For this reason, it is important to understand and analyze the water quality parameters such as D.O., pH, and total nitrogen. The water parameters can impact the fish, plants, and bacteria.

5.1.1 Oxygen

D.O. level shows the amount of oxygen molecules within waters. In case of low D. O., fish will suffocate and die within hours. Monitoring D.O. levels is very complicated and expensive while using a small air pump can ensure aeration of the water.

Water surface helps the oxygen from the atmosphere to be dissolved in the water. In normal conditions, fish can survive in such waters, but with high-density D.O. diffusion, this is not sufficient. To face this problem in aquaponic systems, creating dynamic water flow and use of aerators to produce air bubbles are two common ways. As it is shown in Fig. 5.2, the optimum D.O. levels for each organism are between 5 and 8 mg/L. Carp and tilapia can tolerate D.O. levels as low as 2–3 mg/L.

As water temperature rises, the solubility of oxygen decreases. This is mainly because the water capacity to hold D.O. decreases as the temperature increases.

5.1.2 Power of Hydrogen (pH)

The pH is a measure of acidity or basic of the solution on a scale of 1–14. pH 7 is neutral, with anything below 7 being acidic and above 7 being basic. Figure 5.3 shows that the negative pH scale means that a pH 7 has less hydrogen than a pH of 6; also, it shows logarithmic scale of pH, where a pH of 7 has 10 times fewer hydrogen ions than a pH of 6, 100 times fewer than a pH of 5.

In aquaponic, pH has a major impact on plants and bacteria. The pH controls the plants' access to nutrients. It is believed that at a pH of 6–6.5, all or most of the nutrients are available for the plants, but outside this range, the nutrients become difficult to access. For example, at 7.5 pH level, iron, phosphorus, and manganese deficiencies are very common.

In case of nitrifying bacteria, below 6, the bacteria face difficulty to convert ammonia into nitrate. This unbalanced system will cause stressful conditions for other organisms. Fish also have specific pH tolerance levels ranging from 6 to 8.5, but ammonia toxicity is caused with higher pH levels in the system.

5.1.2.1 Impact of the Nitrification Process on pH

In normal conditions in an aquaponic system, the nitrification process will lower the pH. This is because in the nitrification process, weak concentrations of nitric acid are produced since the bacteria liberate hydrogen ions during the conversion of

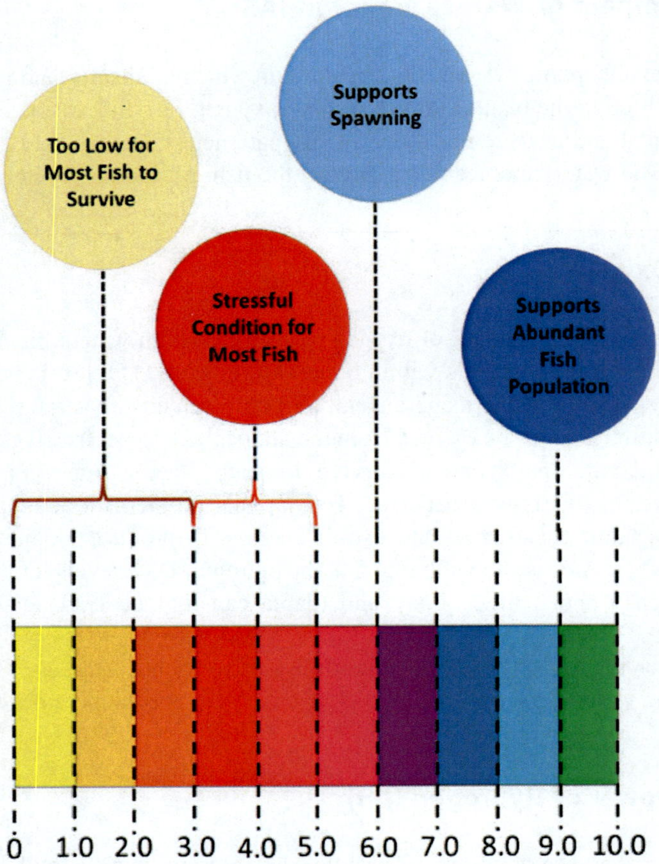

Fig. 5.2 D.O. tolerance for fish species

ammonia to nitrate. Hence, over the period of time, the aquaponic system will gradually become more acidic due to bacterial activity.

5.1.2.2 Impact of pH on Phytoplankton Activities

Photosynthesis of plankton, algae, and aquatic plants removes CO_2 from the water and raises the pH, while respiration by fish lowers the pH due to release of CO_2. The aquatic plants' ability to photosynthesize and remove carbonic acid (H_2CO_3) causes the pH of the water to rise early in the morning. Overnight, pH falls because of the plants' ability to respire and release H_2CO_3. Because of this, the pH is lowest at sunrise and highest at sunset.

5.1.2.3 Fish Density

Breathing fish releases CO_2 into the water. CO_2 lowers pH because CO_2 upon contact with water naturally converts into H_2CO_3. Therefore, the higher the fish stocking density, the more CO_2 will be released, lowering the overall pH level. This

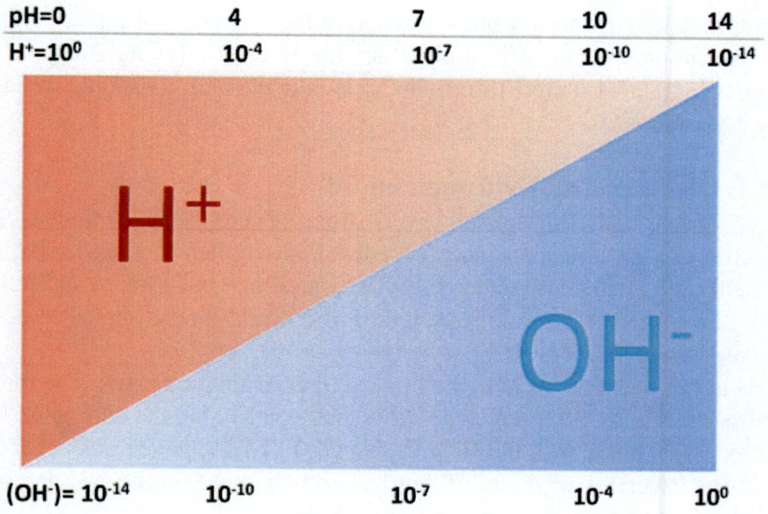

Fig. 5.3 pH scale

effect will get greater as a result of higher fish stock respiration as the fish become more active, which occurs in warmer temperatures.

5.1.3 Temperature

Temperature influences various aspects of the aquaponic systems. Temperature influences D.O. and ionization of ammonia. Higher temperature reduces the D.O. and more ammonia toxicity. Additionally, temperature will restrict the intake of some elements such as calcium.

Fish and plants in an aquaponic system should match the temperature as changing the water's temperature can be very expensive and energy intensive. Higher water temperatures are beneficial for nitrifying bacteria, popular vegetables, and fish like tilapia, carp, and catfish.

5.1.4 Ammonia, Nitrite, and Nitrate

Nitrogen is part of all proteins. Aquaponic systems initially receive nitrogen from fish feed that contains crude protein. The protein in the feed is used for growth, and some that cannot be absorbed by the fish is released in the water through excretion of the fish. The urine that the fish excretes from its gills contains a significant amount of NH_3. The NH_3 is then nitrified by bacteria and will be converted into NO_2- and NO_3-.

NH_3, NO_2-, and NO_3- are poisonous for fish, and comparing among these three, NH_3 and NO_2- are 100 times more poisonous than NO_3-. However, these compounds are nutritious for plants, and all of them can be consumed by the plants. In the aquaponic system with biofiltration, the level of NH_3 and NO_2- should be close to zero (0.25–1.0 mg/L) since the bacteria in the biofilter should be converting all the NH_3 and NO_2-.

5.1.4.1 Effects of Total Nitrogen on IAS

NH_3 is toxic for all fish species; even strong species such as tilapia can show symptoms of ammonia poisoning at a level lower than 1.0 mg/L. Exposure to above 1.0 mg/L will cause damage to fish central nervous system and gills. Additionally, lower levels over a long period of time can also result in fish stress, spread of disease, and mortality.

NH_3 in the water can exist in the form of ionized and unionized. The ionized and unionized forms of NH_3 together are called total ammonia nitrogen. In acidic condition, NH_3 binds with the H^+ and changes to NH_4+, which is less toxic. When NH_3 levels exceed 4 mg/L, nitrifying bacteria become significantly less effective because they are less active at high levels of NH_3.

NO_2- can also create health issues for the fish with concentrations as low as 0.25 mg/L. Small amounts of NO_2- can prevent the transport of O_2 within the bloodstream of fish, causing the blood to turn brown, and it is known as brown blood disease. The effect can be seen in the gills, and fish will show symptoms similar to ammonia poisoning.

NO_3- is less toxic than the other two nitrogen compounds. It is the most obtainable type, and fish can withstand concentrations of up to 300–400 mg/L. However, high levels of NO_3- will have negative impact on plants leading to hazardous accumulation of NO_3- in leaves. It is recommended to keep the NO_3- levels at 5–150 mg/L in the aquaponic systems.

5.1.5 Water Hardness

General hardness and carbonate hardness are the two major types of water hardness. General hardness measures the positive ions in the water, while carbonate hardness measures the buffering capacity of the water. General hardness does not have an impact on aquaponic process, but carbonate hardness has a direct relationship with the water pH.

General hardness is the amount of calcium (Ca^{2+}), magnesium (Mg^{2+}), and iron (Fe^+) ions in the water. Waters with high source of Ca^{2+}, Mg^{2+}, and, to a lesser extent, Fe^+ are found in riverbeds or limestone-based aquifers. Plants enjoy the waters with Ca^{2+} and Mg^{2+}, which are essential ions for plant nutrients. Water with general hardness can be a good source of micronutrients for aquaponics with no health effect on the organisms. Since these ions are considered heavy ions, they cannot be found in the atmosphere, and so, rainwater has low water hardness.

The total amount of carbonates (CO_3^{2-}) and bicarbonates (HCO_3-) dissolved in the water is called carbonate hardness. Water with levels between 121 and 180 mg/L are considered to have high carbonate hardness. Carbonate hardness has an impact of pH level since it acts as a resistance to the lowering of pH. CO_3^{2-} and HCO_3- in the water will bind with the H^+ released by any acid, removing these free H^+ ions from the water. For this reason, the pH will stay constant even as new H^+ ions from the acid are added to the water. Alkalinity is important since rapid changes in pH are stressful to the entire aquaponic system. If there are no CO_3^{2-} and HCO_3- present, then the pH of the water would quickly drop in the system. The higher the carbonate hardness in the water, the longer it will act as a buffer for pH to keep the system stable against acidification.

It is essential for the aquaponic systems to always have certain concentrations of carbonate hardness in the water, as it can neutralize the acids and keep the pH constant. Without this, the systems will have rapid pH changes that would have negative impacts on the fish.

5.1.6 Activity of Algae

Water quality parameters are also being affected by algae growth in aquaponic systems. Algae are a class of photosynthetic organisms that grow in any waterbody with sufficient nutrients and sunlight. Microscopically small, single-celled algae are a kind of phytoplankton. On the other hand, macroalgae are much larger and form filamentous mats in the aquaponic systems. Algae will compete with vegetables in the aquaponic systems, consuming the nutrients in the water. Furthermore, they will reduce the D.O. levels in the water at night by production of CO_2. This will also cause a daily shift in the water's pH level.

Filamentous algae can block filters leading to interruption in water circulation, while brown algae can negatively impact plants' growth. However, some operations, such as those that raise tilapia and shrimp, profit from the cultivation of algae for feed (green water culture).

5.1.7 Small Organisms Living in the Water

Over time, many small organisms will grow in the aquaponic systems, which will contribute to the ecosystem. These organisms can be divided into three categories. First are the helpful organisms that help decomposition of fish wastes. Second are the ones that are neither helpful nor harmful for the systems. Finally, there are the small organisms that are threats and harmful for the systems such as parasites, pests, and bacteria. It is best to prevent these small organisms from becoming dangerous infestations, putting fish and plants through stressful conditions by ensuring highly aerobic conditions with access to all essential nutrients. This will allow organisms to stave off diseases using their own healthy immune systems.

5.1.8 Water Sources

Depending on the plant types and the location of the systems, an aquaponic system uses up to 3% of its water volume per day. Water is lost from direct splashing as well as through natural evapotranspiration and retaining within the plants. For this reason, water availability to refill these usages is important. However, the new water sources need to be checked for pH, hardness, salinity, chlorine, and any pollutants to ensure that the water is safe to use.

Salinity level in the water needs to be considered since it indicates the concentration of salts in the water. Water salinity can be monitored with electrical conductivity (EC) and/or total dissolved solids (TDSs). EC is used to monitor the electricity amount that goes through water, and the unit is micro-Siemens per centimeter. TDS is also used to monitor the salt particles in the water, and the unit is milligrams per liter.

5.1.8.1 Aquifer Water

The water drawn from the wells varies in quality according to the bedrock and cistern compositions. If the bedrock is limestone, then the water will have high concentrations of hardness. In the aquaponic systems, the water hardness will not cause any problem since alkalinity is consumed by the nitric acid produced by the nitrifying bacteria. If fish biomass density is too low and hardness levels are very high, this will change; in such case, the water will stay basic and resist turning acidic due to the nitrification cycle. In coral islands, aquifers often have saltwater intrusion into the sweet water (freshwater) and have high salinity level for the aquaponic systems.

5.1.8.2 Filtered Water

When water is filtered, depending on the filter used, most of the metals and ions are removed. This deionization from reverse osmosis will significantly reduce the hardness, and it must be buffered.

5.1.8.3 Rainwater

The most suitable water for the aquaponic systems is rainwater; thus, collected rainwater can be a perfect source. Rainwater pH is usually neutral and has a very low concentration of hardness with zero salinity. There are reports of acidic rains in Europe, the USA, and Asia. In this case, if the pH level drops below 6, it is necessary to add a base or increase the carbonate hardness. Commonly used bases are potassium hydroxide (KOH) and calcium hydroxide ($Ca(OH)_2$). It is always safer to add calcium carbonate ($CaCO_3$) or potassium carbonate (K_2CO_3) to increase both hardness and pH. There are many natural sources of $CaCO_3$ such as eggshells, crushed seashells, and crushed chalk. The choice of the bases and buffers can also be driven by the type of plants growing in the system, as each of these compounds adds an important macronutrient. Leafy vegetables can be favored by calcium bases to avoid tip burns on leaves, while potassium is optimal in fruit plants to favor flowering, fruit settings, and optimal ripening. Usually in RAS, baking soda (sodium bicarbonate) is used to increase the hardness, but it cannot be used in the aquaponic systems because it increases the sodium level, which is dangerous to the plants. Rainwater is more sustainable and will reduce the operational costs of the aquaponic systems.

5.1.8.4 Tap Water

Water supplied in urban areas is often treated with different chemicals to remove the pathogens. Chlorine and chloramines are the most common chemicals used for water treatment, which are toxic to fish, plants, and bacteria. The most common way to get rid of these chemicals is to store water and allow chlorine to dissipate into the atmosphere. Chloramines are reliable and do not readily off-gas. If the municipality uses chloramines, it may be necessary to use chemical treatment techniques such as charcoal filtration or other dechlorinating chemicals.

5.2 Impact of Bacteria in IAS

As discussed, nitrification and a healthy bacterial colony are essential to a functioning aquaponic system. Since nitrification bacteria are slow in reproducing and establishing colonies, it is necessary to maintain the environmental parameters to create a suitable environment for these bacteria. In general, bacteria need large, dark places with clean water, food, and oxygen to colonize.

High surface area with sufficient biofiltration material is optimal to develop colonies of bacteria. Smaller and more porous materials in the biofiltration media will provide larger surfaces for bacteria to colonize.

For monitoring the health of bacteria, measuring ammonia, nitrate, and nitrite can provide useful information. When the ammonia and nitrite levels are between 0 and 1 mg/L, the system is functioning properly. If the ammonia and nitrite levels are higher, then this shows problems whether in the surface size of the biofilters, number of fish, feeding amount, or the bacteria themselves that may not be functioning properly due to water quality.

Heterotrophic bacteria are a different class of bacteria found in aquaponic systems that use organic carbon as a food source. These bacteria are involved in the decomposition of fish and plant wastes. Nearly 70% of the feed consumed by fish is dumped as waste, where 50% dissolved waste is released as ammonia. In a process called mineralization, heterotrophic bacteria metabolize these solid wastes, making essential micronutrients for the plants.

In the process of decomposition of the solid portion of the fish waste, heterotrophic bacteria release nutrients locked in these solid wastes. This is essential because plants cannot take up nutrients in solid forms. Heterotrophic bacteria feed on any form of organic material, such as solid fish waste, uneaten fish food, dying plants, dying plant leaves, and even dead bacteria.

The heterotrophic bacteria grow faster than the nitrifying bacteria, reproducing in hours rather than days. The bottom part of the media beds, where the waste collects on the bottom, is the best zone for these bacteria. Without mineralization, plants may experience nutrient deficiencies.

Within the media beds, beside heterotrophic bacteria, earthworms, isopods, amphipods, larvae, and other small animals can be found in aquaponic systems. These organisms work together with the bacteria to decompose the solid waste, and having this community can prevent accumulation of solids.

There are harmful bacteria that also grow in the aquaponic systems. One of these bacteria is sulfate-reducing ones that are found in no-oxygen conditions, where they obtain energy through a redox reaction using sulfur. This process will produce hydrogen sulfide (H_2S), which is toxic to fish. The rotten eggs smell in lakes, salt marshes, and estuaries because of the H_2S. When the solid wastes accumulate faster than what heterotrophic bacteria can process, this can produce a condition that supports sulfate-reducing bacteria. In high fish density systems, the fish produce so much solid waste that the mechanical filters cannot be cleaned fast enough, which encourages these bacteria to multiply and produce their noxious metabolites. These bacteria only grow in anoxic conditions, so providing adequate aerations can prevent accumulation in the aquaponic systems.

Another group of harmful bacteria in the aquaponic systems is the denitrification ones. These bacteria live in anaerobic conditions, and their job is to convert nitrite into atmospheric nitrogen. These bacteria reduce the efficiency of the aquaponic systems by removing the nitrogen fertilizers, and plants will show signs of nitrogen deficiencies.

Another unwanted group of bacteria in aquaponic system is those causing diseases in fish, plants, and even humans. In the aquaponic systems, the pathogens can be prevented from entering the systems by ensuring good worker hygiene; preventing rodents from defecating in the system; keeping wild mammals (and dogs and cats) away from aquaponic systems; avoiding using water that is contaminated; and being aware that any live feed can be a vector for introducing alien microorganisms into the system. It is especially important not to use rainwater collection from roofs with bird feces unless the water is treated first. Warm-blooded animals can introduce *Escherichia coli*, while birds carry *Salmonella* spp., entering the systems with animal feces. Additionally, aquaponic water should never come into contact with the leaves of the plants.

5.3 Crop Production in IAS

There are many similarities between earth-grown agriculture and soilless production. The differences between traditional land practices and soilless techniques are in the use of fertilizers, water consumption, and overall productivity.

5.3.1 Fertilizer Use

In an intensive crop cultivation where fertilizers are required, farmers cannot control the amount of nutrients absorbed by the plants. This is because the process in the soil is very complex, and the abiotic and biotic interactions will directly determine the availability of nutrients to the plant roots. In soilless culture, the nutrients are dissolved in water that moves through the plants' roots and can be made according to plants' needs.

Additionally, certain fertilizers are consumed by weeds, which lowers their effectiveness and results in environmental problems. Fertilizers are also expensive, and excess usage will have short- and long-term impacts on the soil and water.

The benefits of a soilless aquaponic system include little fertilizer loss due to biological, physical, and chemical processes and exact control of nutrient concentrations based on plant needs. Monitoring and control can help boost output and improve the quality of the produced goods.

5.3.2 Water Use

There are studies proving that aquaponic systems use less water than in soil-based crop production. Water will evaporate from the soil's surface, perspire via leaves, seep into the subsoil, run off, and support weed development in soil-based agriculture. On the other hand, in soilless aquaponic systems, the water goes through crops to help in their growth while transpiration happens through the leaves. The water used is the absolute minimum needed to grow the plants, and only a negligible amount of water is lost for evaporation from the soilless media. Overall, aquaponics uses only about 10% of the water needed to grow the same plant in soil.

5.3.3 Urban Farming with IAS

Aquaponic systems are best fit for urban areas where traditional agriculture cannot be supported and/or implemented. There will be less production footprint because aquaponic systems require less transportation. Additionally, aquaponic systems can be used in areas that agriculture cannot be deployed such as extreme dry lands like deserts, lands with high salinity such as coastal areas, and lands with low-quality soil due to overuse of fertilizers. The need for aquaponic systems is increasing due to decrease in arable lands suitable for traditional agriculture, and thus, these systems allow the urban population to grow food intensively.

5.3.4 Productivity in IAS

It is proven that hydroponic culture can achieve 25% higher yields than traditional soil-based agriculture. According to experts, hydroponic can produce 2–5 times more than traditional agriculture. However, this production rate can only be achieved when systems are used in greenhouses with expensive inputs such as fertilizers. Aquaponic systems can also produce higher than soil-based agriculture. This is due to the monitoring and maintenance ability of farmers in these systems that can ensure the growing conditions of the plants, ensuring optimal real-time nutrient balances, water delivery, pH, and temperature. In addition, in soilless culture, there is no competition with weeds, and plants benefit from higher control of pests and diseases.

5.3.5 More Efficient with Less Tasks

In aquaponic systems, there are no requirements for ploughing, tilling, mulching, or weeding. In agriculture fields, these activities will need large agricultural machinery that works with fossil fuels. Aquaponic systems are less labor intensive, and less tasks are required compared to traditional agriculture. Harvesting is also a simple procedure compared with soil-based agriculture, and the products do not need extensive cleaning to remove soil contamination. Aquaponic systems are suitable for any gender, and many age classes with different capabilities can work with these systems.

Soilless culture is sustainable since same crop can be cultured annually, compared to soil culture activities in which soil degrades significantly and loses its fertility, while use of pests and spread of diseases will increase.

Aquaponic systems are expensive and require initial setup and installation. Also, the high initial cost can prevent farmers from adopting and using such systems. These systems are more expensive than hydroponics since the system needs aquaculture installation. The system is very labor oriented and requires daily management to balance the interaction between three groups of organisms. Aquaponic systems require stable and reliable electricity and are much more complex than normal soil-based agriculture. There is a learning curve, as with many new technologies, and any new aquaponic farmer needs to be fully dedicated to learning. Aquaponics are not appropriate for every situation, and the benefits should be weighed against the costs before embarking on any new venture.

5.4 Plants in IAS

The proper operation of any aquaponic system depends on the plants. Plants can produce their own food using photosynthesis, and the process requires O_2, CO_2, and sunlight. Chlorophyll is produced by micro-organelles called chloroplasts with the task of using the energy from sunlight to break CO_2 and create sugar such as glucose. During this process that happens in the presence of water, O_2 is released. The produced sugar molecules are transported throughout the plant and used later for growth, reproduction, and metabolism. At night, plants use these same sugars, as well as oxygen, to generate the energy needed for growth. This process is called respiration. For this reason, it is essential for plants to have access to sunlight to ensure photosynthesis. CO_2 is naturally available in the atmosphere, and plants can use all the necessary amount from the enclosed area with ventilation.

The plants use photosynthesis to gain nutrients but require more such as inorganic salts. These nutrients are required for the enzymes that facilitate photosynthesis, for growth and reproduction. In the absence of soil, these nutrients need to be supplied in another way; in aquaponic systems, all these essential nutrients come from the fish waste.

The nutrients provided to the plants are micro- and macronutrients. The macronutrients are nitrogen (*N*), phosphorus (*P*), potassium (*K*), calcium (*Ca*), magnesium (*Mg*), and sulfur (*S*). In the basis of all proteins, there is *N*. The micronutrients are boron (*B*), copper (*Cu*), iron (*Fe*), manganese (*Mn*), molybdenum (*Mo*), and zinc (*Zn*).

B is a molecular catalyst involved in structural polysaccharides and glycoproteins, carbohydrate transport, and regulation of some metabolic pathways in plants. This element helps in reproduction and water uptake by plants' cells. Deficiencies may be seen as incomplete bud development and flower set, growth interruption and tip necrosis, and stem and root necrosis.

Ca is involved in strengthening the stems and helps in root development. It is used as structural component of both cell walls and cell membranes. The calcium in hydroponic systems and the most recent growth are a result of calcium immobilization in the plants. Examples of deficiencies are tip burn of lettuces and blossom-end rot of tomatoes and zucchinis. *Ca* can only be transported through active xylem transpiration, so when conditions are too humid, calcium can be available but locked-out because the plants are not transpiring. Increasing airflow with vents or fans can prevent this problem. In aquaponics, calcium can be supplemented with coral sand or calcium carbonate, which also has the advantage of balancing pH.

Cu helps in strengthening stems and is used by some enzymes. Deficiencies may include chlorosis and brown or orange leaf tips, reduced growth of fruits, and necrosis. Copper deficiency can occasionally manifest as very dark green growth.

Mg is a key element in photosynthesis and is the electron acceptor in chlorophyll molecules. Deficiencies can be seen as yellowing of leaves between the veins especially in older parts of the plant.

Mo catalyzes the oxidation reduction with different forms of nitrogen. Insufficient *Mo* plants show nitrogen deficiencies although nitrogen exists in the water. *Mo* is biologically unavailable at pH less than 5.

N is essential for photosynthesis, building structures, cell growth, metabolic process, and production of chlorophyll. After *C* and *O*, *N* is the most common element in a plant. In aquaponic systems, plants can absorb dissolved nitrogen in the form of nitrate, but plants can also utilize ammonia and give back some amino acids. Nitrogen deficiencies can be distinguished in a plant with yellowing older leaves, thin stems, and poor vigor. An overabundance of nitrogen can cause excess vegetative growth, resulting in lush, soft plants susceptible to disease and insect damage, as well as causing difficulties in flower and fruit set.

P is a backbone of deoxyribonucleic acid (DNA) in plants as a component of phospholipid membrane as well as the component to store energy in the cells (adenosine triphosphate (ADP)). *P* is an important element for formation of oils and sugars in the plants. Phosphorus deficiencies commonly cause poor root development because energy cannot be properly transported through the plant; older leaves appear dull green or even purplish brown, and leaf tips appear burnt.

S helps in the production of some proteins such as photosynthetic enzymes. The amino acids methionine and cysteine both contain sulfur, which contributes to some proteins' tertiary structure. Deficiencies are rare but include general yellowing of the entire foliage in new growth. Also, leaves may become yellow, stiff, and brittle and fall off.

Zn is used by enzymes helping in plant growth and maturation. Deficiencies may be noticed as poor vigor, stunted growth with reduced internodal length and leaf size, and intravenous chlorosis that may be confused with other deficiencies.

5.4.1 Plant Selection for IAS

Until today, more than 150 different plants have been successfully grown in aquaponic systems. Based on demand, aquaponic plants can be divided into two primary groups. First are low-nutrient plants such as lettuce, chard, salad rocket, basil, mint, parsley, coriander, chives, pak choi, and watercress. Many of the legumes such as peas and beans also have low-nutrient demands. Second are high-nutrient plants or nutrient-hungry plants. These include botanical fruits, such as tomatoes, eggplants, cucumbers, zucchini, strawberries, and peppers. Other plants with medium nutrient demands are cabbages, such as kale, cauliflower, broccoli, and kohlrabi. Bulbing plants such as beets, taro, onions, and carrots have medium to high requirements, while radish requires less nutrients.

5.4.1.1 Basil

Basil is one of the most popular herbs to grow in aquaponic systems because of its high value and high demand in urban zones. There are many different types of basil that grow well in aquaponic systems including the Italian Genovese basil (sweet basil), lemon basil, and purple passion basil. However, due to high nitrogen uptake, excessive nutrient depletion can occur in aquaponic systems.

Basil seeds require high and stable temperatures between 20 and 25 °C. Once planted into the aquaponic system, basil grows best in warm conditions and with full exposure to the sun. However, better quality leaves are obtained through slight shading. The suitable pH for healthy growth of this plant is 5.5–6.5 with plant spacing of 15–25 cm (8–40 plants/m^2).

When plants are 15 cm tall, leaf harvesting begins and lasts for 30–50 days. Flowering tips during plant growth need to be removed in order to avoid bitter tastes in leaves. However, basil flowers are attractive to pollinators and beneficial insects, so leaving a few flowering plants can improve the overall garden and ensure a constant supply of basil seeds.

5.4.1.2 Beans

Aquaponic systems can support both climbing and bush bean varieties. Climbing varieties can yield three times more pods than bush varieties. Beans have low NO_3- needs, but have a moderate demand in terms of *P* and *K*. These requirements make beans an ideal choice for aquaponic production, although excess NO_3- will delay flowering.

Plants that do not grow in temperatures under 14 °C and above 35 °C will cause floral abortion and poor fruit set. Optimal relative humidity for plants is 70% or 80%. Beans are sensitive to the photoperiod; thus, it is important to choose the right varieties according to the location and season. In general, climbing varieties are cultivated in summer, while dwarf varieties are adapted to spring and autumn conditions. Plants need 10–30 cm of space, and pH can be between 5.5 and 7.

Green or yellow wax beans should be firm and crisp at harvest, and the seeds are small. Black beans are ready to harvest when pods change color and the beans inside are fully formed but not dried out. Before the weather becomes cooler and the plants turn brown or lose their leaves, kidney beans are ready to be harvested.

5.4.1.3 Broccoli

Broccoli is a winter vegetable. In aquaponic system, the media bed method is recommended since the broccoli is large and heavy at harvest. It is a nutrient-demanding plant and is very susceptible to warm temperatures.

It grows best during daytime with temperatures between 14 and 17 °C, and for head formation, winter varieties need 10–15 °C temperature range. Hot temperatures cause premature bolting. Broccoli grows best in 6–7 pH, and plants can be cultured in 40–70 cm distance from one another. The seedlings can be transplanted to the media beds when plants are 15–20 cm tall. Broccoli, as well as cabbage, is susceptible to cabbage worms and other persistent pests. While some mechanical removal can have marginal effect, treatment with biological pesticides and repellents can control the infestations.

5.4.1.4 Cauliflower

This plant is a high-value, nutritious, and winter crop that needs adequate plant spacing. Cauliflower reacts well in environments with high *N* and *P* concentrations. Among other nutrients, *K* and *Ca* are important for the production of heads. The plant is particularly sensitive to climatic conditions, and the heads do not develop properly in hot, very cold, or very dry conditions.

Cauliflower growth optimal air temperature is 15–25 °C, but for the formation of the heads, the plants require a temperature of 10–15 °C or 15–20 °C. Plants can tolerate cold temperatures; however, heads can be damaged by frost. Light shade can be beneficial in warmer temperatures. The optimal pH for the vegetable is between 6 and 6.5, and plant spacing must be between 45 and 60 cm (3–5 plants/m^2). The plants can be transplanted to the aquaponic systems when they are 3–5 weeks old and have 4 or 5 leaves. The growth time is 2–3 months for spring crops and 3–4 months for autumn crops.

Too much sun, heat, or nitrogen uptake can cause "ricey" heads where the main flower separates into small, rice-like grains. Temperatures below 12 °C could instead produce "buttoning." Cauliflower is susceptible to some pests including cabbage worms, flea beetle, white maggots (larvae), and cabbage aphids, which can be removed manually or by using other pest management techniques. Cauliflower can be cut off when the heads are compact, white, and firm. Once the head is removed, the rest of the plant can be removed and replaced with a new plant.

5.4.1.5 Cucumbers

Members of the Cucurbitaceae family including squash, zucchini, and melons and cucumbers are high-value summer vegetables. With their large root structure, they are ideal plants for aquaponic systems. Cucumbers require large amounts of *N* and *K* and grow best with long hot humid days with ample sunshine and warm nights.

Optimal growth temperatures are between 24 and 27 °C during daytime with 70–90% humidity. Temperature of the substrate of about 21 °C is also optimal for production. Plants stop their growth and production at 10–13 °C. It is recommended to have higher potassium concentration to favor higher fruit settings and yields. Optimal pH for cucumbers is between 5.5 and 6.5. Cucumbers need to be planted 30–60 cm, which are approximately 2–5 plants/m^2.

Cucumber can be transplanted to the aquaponic systems at 2–3 weeks at four- or five-leaf stage. Plants grow very quickly, and it is a good practice to limit their vegetative vigor and divert nutrients to fruits by cutting their apical tips when the stem is 2 m long; removing the lateral branches also favors ventilation. Further plant elongation can be successively secured by leaving only the two farthest buds coming out from the main stem. The presence of pollinating insects is necessary for good fecundation and fruit set. Cucumber plants need support for their growth, which will also provide plants with adequate aeration to prevent foliar diseases. Owing to the high incidence of pest occurrences in cucumber plants, it is important to plan appropriate integrated pest management strategies. Cucumbers can start production after 2–3 weeks, and in optimal conditions, plants can be harvested 10–15 times.

5.4.1.6 Eggplants

These plants are summer fruiting vegetables that grow well in media beds with deep root growth systems. Each plant produces 10–15 fruits for a total yield of 3–7 kg. Eggplants have high *N* and *K* requirements, which shows the necessity for a careful selection of number of plants to grow in each aquaponic system.

Eggplants grow in warm temperatures with full sun exposure. Plants grow best in the temperature between 22 and 26 °C with a humidity of 60–70%. Temperatures below 10 °C and above 32 °C limit plant growth. Eggplants enjoy pH between 5.5 and 7, and plants need to be distanced 40–60 cm (3–5 plants/m^2).

Plant seedlings can be transplanted with 4–5 leaves. These plants toward the end of the summer begin pinching off new blossoms to favor the ripening of the existing fruit. At the end of the season, plants can be drastically pruned at 20–30 cm by leaving just three branches. Plants can be grown without pruning; however, in limited spaces or in greenhouses, management of the branches can be facilitated with stakes or vertical strings.

Harvesting can start when eggplants are 10–15 cm and when the skin is shiny. If the eggplants are dull or yellow skin, it is a sign that it is overripe.

5.4.1.7 Lettuce

Aquaponic systems are most suitable for lettuce production with the optimal nutrient concentration in the water. Four main lettuce species can grow well in aquaponic systems: crisp head lettuce (iceberg), which has tight head with crispy leaves, ideal for cooler conditions; butterhead lettuce, which shows leaves that are loosely piled on one another and have no bitter taste; Romaine lettuce, which has upright and tightly folded leaves that are slow to bolt and are sweet in taste; and loose-leaf lettuce, which comes out in a variety of colors and shapes with no head and can be directly sowed on media beds and harvested by picking single leaves without collecting the whole plant. Lettuce is in high demand and has a high value in urban zones.

For head growth, the night temperature should be 3–13 °C with a day temperature of 17–28 °C. The generative growth is affected by photoperiod and temperature—extended daylight and warm conditions at night cause bolting. Water temperature above 26 °C may also favor bolting and leaf bitterness. The plant has low nutrient demand; however, higher calcium concentrations in water help to prevent tip burn in leaf in summer crops. The ideal pH is 5.8–6.2, but lettuce still grows well with a pH as high as 7, although some iron deficiencies might appear owing to reduced bioavailability of this nutrient above neutrality.

Lettuce can be transplanted in the system at 3 weeks when it has three leaves. Plants need to be cultured with 18–30 cm (20–25 heads/m^2). Supplemental fertilization with P to the seedlings in the second and third weeks favors root growth and avoids plant stress at transplant. Harvesting can start when the leaves are large to eat.

5.4.1.8 Mangold

One of the popular leafy green vegetables that grow well in the aquaponic systems is mangold. It is a moderate NO_3- feeder and requires lower concentrations of P and K than fruiting vegetables. Thanks to high market value, fast growth rate, and its rich nutritional contents, mangold is common in aquaponic systems. The foliage is green to dark green, but the stems can have attractive colors of yellow, purple, or red.

The optimal temperature for these plants is between 16 and 24 °C, while minimum temperature is 5 °C. Mangold grows as a late-winter crop and can grow in full sun during summer seasons. The plant has a moderate tolerance to salinity. Acceptable pH for mangold is 6–7.5, and plants need 30 cm space from one another.

Mangold seeds can produce more than one seedling, and thinning is required for the seedling to grow. As plants become senescent during the season, older leaves can be removed to encourage new growth. The leaves can be continuously cut whenever reaching harvestable sizes, and removal of larger leaves favors the growth of the new ones.

5.4.1.9 Parsley

Although parsley typically grows once a year, it is actually a biennial herb if the winters are mild with little to no frost. Although the plant can withstand temperatures as low as 0 °C, 8 °C is the absolute minimum for growth. In the first year, the plants produce leaves, while in the second year, the plants will begin sending up flower stalks for seed production.

The main problem with parsley is initial germination that takes 2–5 weeks. Water temperatures between 20 and 23 °C can help in accelerating germination. Emerging seedlings will have the appearance of grass, with two narrow seed leaves opposite to each other. After 6 weeks, seedlings can be transplanted into the aquaponic unit during early spring. Acceptable pH for parsley is between 6 and 7, and plants need 15–30 cm of space for efficiently growing in the aquaponic systems. Harvesting begins once the individual stalks of the plant are at least 15 cm long. If only the top leaves are cut, the stalks will remain, causing the plant to be less productive.

5.4.1.10 Peppers

Many varieties of peppers depending on the color and degree of spice can be grown in aquaponic systems. They are summer fruiting vegetables that prefer warm weather and sun exposure. Temperatures between 22 and 34 °C are suitable for seed germinations, and seeds will not germinate below 15 °C. Daytime temperatures of 22–28 °C and nighttime temperatures of 14–16 °C favor best fruiting conditions under a relative humidity of 65–60%. A comfortable pH for peppers to grow is 5.5–6.5, and plant needs 30–60 cm spacing which is 3 or 4 plants/m^2. Temperature suitable for the root is 15–20 °C. When the air temperature goes below 10 °C, plants will stop growing, and it causes abnormal deformation of the fruits. Temperatures above 30–35 °C will cause floral abortion. NO_3- must be between 20 and 120 mg/L and supports the initial vegetative growth, but higher concentrations of *K* and *P* are needed for flowering and fruiting.

Seedlings can be transplanted to the systems when they have 6–8 leaves, and the best time is at night when the temperature is above 10 °C. The heavy plants need support with vertical strings hanging from iron pulled above the unit. Harvesting can start when peppers reach a market size. Peppers will change color and ripen while improving their levels of vitamin C.

5.4.1.11 Tomatoes

Tomatoes can grow in an aquaponic system with physical supports. Tomatoes need high *K* level, and so plants per system must be planned according to the fish biomass. Higher *N* concentration is preferable during early stages to favor plants' vegetative growth; however, *K* should be presented from the flowering stage to favor fruit settings and growth.

Tomatoes prefer warm temperatures with sun exposure. Temperatures below 8–10 °C will stop the plants from growing, and night temperatures of 14 °C will help in fruit set. Temperatures above 40 °C cause floral abortion and poor fruit setting. Tomato plants are two types: seasonal production or determinate, and indeterminate with continuous production of floral branches. In the first type, the plant can be left like bushes with only 3–4 branches and removing all the auxiliary suckers to divert nutrients to fruits. Both determinate and indeterminate varieties should be grown with a single stem (double in case of high plant vigor) by removing all the auxiliary suckers. Tomatoes rely on supports that can be either made of stakes (bush plants) or bound to vertical plastic/nylon strings that are attached to iron wires pulled horizontally above the plant units.

Tomatoes have a moderate tolerance to salinity, which makes them suitable for areas where pure freshwater is not available. Higher salinity at fruiting stage improves quality of the products. The best pH for tomatoes is between 5.5 and 6.5, and plants need to have 40–60 cm spacing.

Three to six weeks after germination, seedlings can be transplanted into the system. Once the plants are 60 cm tall, the growing method can be determined to be either bush or single stem. To prevent fungal incidence, lower leaves need to be removed, and before ripening the fruit, leaves need to be removed to avoid covering the fruits to accelerate maturation.

5.4.2 Plant Pests and Pest Management in IAS

One of the problems faced in IAS is insect pests because they carry diseases that plants can contract. Pests can extract liquids leading to stunted growth. Pest management for outdoor conditions differs from the protected cultivation due to physical separation of the plants from the surrounding areas. Additionally, insect pest prevalence is highly dependent on climate and environment. For example, pest management in temperate or arid zones is easier than in tropical regions, where higher incidence and competition among insects make pest control a far more difficult task.

In an aquaponic system, it is typical for the media bed to contain microorganisms, tiny insects, and spiders (Fig. 5.4). However, other harmful insect pests, such as whiteflies, thrips, aphids, leaf miners, cabbage moths, and spider mites feed upon and damage the plants.

In soil agriculture, pesticides and insecticides are typically used, but in aquaponic systems, this is not practical. Fish and the good bacteria in the system may both die when exposed to any strong chemical pesticide; therefore, commercial chemical pesticides must never be used. There are, however, other practical physical, environmental, and cultural controls that can be used to lessen the danger from pests in aquaponic systems.

One of the solutions is to use integrated production and pest management approach, which is an ecosystem approach to both soilless and soil-based plant production. Together with host-plant resistance and cultural practices, it combines mechanical, physical, chemical, biological, and microbiological controls. Not all these controls are applicable for aquaponics as some may be fatal for fish and bacteria, while others may not be economically justified for small-scale aquaponics.

Fig. 5.4 Poseidon-AI® IAS hosting a spider

5.4.2.1 Physical and Mechanical Controls

Prevention is the most fundamental part of pest management in aquaponic systems. Regular and thorough monitoring for pests is vital, and, ideally, minor infestations can be identified and managed before the insects damage the entire crop. Physical exclusion refers to keeping the pests away. Mechanical removal is when the farmer actively takes the pests away from the plants. Cultural controls are the choices and management activities that the farmer can undertake to prevent pests. These controls should be used as a first line of defense against insect pests before other methods are considered.

5.4.2.2 Setting Up Screens and Nettings

Whenever organic horticulture is practiced or pesticides are not effective, setting up screens are common to prevent pest damage. Targeted pest can determine the netting mesh size, for example, nets with a mesh size of 0.15 mm to exclude thrips, 0.35 mm to exclude whitefly and aphids, and 0.8 mm to keep out leaf miners. Screens do not suppress or eradicate pests; they only exclude most of them; therefore, they must be installed prior to pest appearance, and care should be taken not to let pests enter the protected environment.

5.4.2.3 Manual Inspection and Removal

Manual inspection and removal can be done using a high-pressure stream of water on the heavily infested leaves to delay the spread of insects to the surrounding plants. Larger pests and larvae may also be used as supplementary food for the fish. Water sprayed from a hose directed at the underside of the leaves is an extremely effective management technique on many types of sucking insects. This is one of the most effective methods on small-scale systems, but it can be just a temporary remedy as the displaced pests can return to the plants.

5.4.2.4 Sticky Traps

Sticky traps are effective in protected environments such as greenhouses. These traps are less effective in outdoor conditions since new insects can easily come from the surrounding area. Continuous monitoring of the insects captured can help farmers to adopt in accordance with the specific pests.

5.4.2.5 Environmental Management

To favor healthier plant growth and to build unfavorable conditions for pests, maintaining optimal light, temperature, and humidity conditions can help in protecting the cultivation. Some pests are attracted to specific plants, and different plant varieties from the same species have different resistance/tolerance to pests. Moreover, some plants attract and retain more beneficial insects to help manage pest populations.

Some plants, such as cucumber and legumes, are more prone to aphids or red mite infestations and thus can be used to detect pest prevalence early. Another strategy that can be adopted in aquaponic systems is the use of biological insecticides on sacrificial plants near or within these systems. This strategy would not affect the aquaponic ecosystem or beneficial insects present around the unit. Fava beans and petunias (flowers) can be used to catch thrips, aphids, and mites. Cucumbers are also used to catch aphids and hoppers, while succulent lettuce seedlings are used to capture other leaf-eating insects.

5.4.2.6 Spacing

Insect infestations are encouraged by high-density planting because it enhances the competition for light. This will make the plant tissue more succulent for pests to bore through or for pathogens to penetrate. As mentioned, many plants have special needs for sunlight or a lack of it. By combining full-sun with shade-tolerant plants, it is possible to intensify production without the risk of raising competition and weakening the plants.

5.4.2.7 Biological Control

Some pesticides that are obtained from microorganisms are safe for aquatic animals because they act specifically on insect structures and do not harm fish species nor are harmful for humans. *Bacillus thuringiensis* and *Beauveria bassiana* are the two organisms widely used in aquaponics and organic agriculture. *Beauveria bassiana* is a fungus that germinates and penetrates the insect's skin (chitin), killing the pest through dehydration. *Bacillus thuringiensis* is a toxin extract from a bacterium that damages the insect's digestive tract and kills it. It can be sprayed on leaves and specifically targets caterpillars, leaf rollers, moths, or butterfly larvae without damaging other beneficial insects.

5.4.2.8 Pest Predators

In greenhouses and controlled environment, pest predators and/or beneficial insects are effective to control pests. Predator insects are introduced into the media beds to control any further infestation. The advantages of using pest predators are:

- Absence of pesticide residue or pesticide-induced resistance in pests
- In long run and for large-scale aquaponic systems, it is economically feasible
- Ecologically sound

Many of these beneficial insects feed on nectar in their adult stages, so a selection of flowers near the aquaponic unit can maintain a population that can keep pests in balance. This method will never fully eradicate pests. Instead, pests are suppressed under a tight prey-predator relationship.

5.4.3 Plant Disease in IAS

Aquaponic systems take advantage of microscopic ecosystems, which include bacteria, fungi, and others. The presence of these microorganisms makes systems more resilient toward spread of pests and diseases. For this reason, successful plant production in aquaponic systems is the result of well-balanced environmental management. Similarly, integrated disease management relies on prevention, plant choice, and monitoring as a defense against disease.

Environmental conditions such as pH, temperature, and humidity play an important role in the health of the plants. Each bacterium, fungus, or parasite has optimal growth temperatures that can have different impact on the plants; thus, diseases occur in certain areas and periods when conditions are favorable to the pathogens more than its host. Humidity and moisture play key roles in the germination of fungal spores. The activation of some bacterial and fungal diseases also has direct correlations with the presence of water. Hence, monitoring the moisture and humidity can reduce the risks of disease outbreaks.

Through ventilation that minimizes temperature differential and creates horizontal airflow, ventilation may be controlled. This flow of air will prevent the temperature from dropping below the dew point, and so, water will not condense on the vegetables. Water temperature is important in avoiding fungal outbreaks. *Pythium* spp. is a common disease in aquaponic systems, which is soilborne pathogen that can be introduced into the systems from contaminated materials. However, in aquaponic system, the fungus does not cause damage below certain temperature due to the presence of other competitive microorganisms. Maintaining the temperature below 28–30 °C can avoid the exponential germination of spores that eventually will cause a disease outbreak.

High plant densities reduce the internal ventilation causing an increase in humidity among plants in the media beds. Under intense light competition among high-density plants, the plants grow without consolidating their cells, which leads to softer tissue walls. Tender tissues are more prone to disease because of their limited resistance to pests and/or pathogen penetration.

Some varieties of plants have higher resistances to pathogens; thus, in some cases, culturing resistant cultivars can successfully prevent diseases. It is vital to select plant varieties that are more adapted to grow in certain environments or have higher degree of resistance against certain pathogens. The use of local varieties that are naturally selected for a specific environment can ensure healthy plant growth.

Not all the diseases can be controlled with resistant varieties, and it is recommended to shift to other varieties during the critical season. In the case of *Pythium* spp., resistance varieties of lettuce are not able to control infestation and it is better to shift to other species such as basil, which is more tolerant to the pathogen and to high water temperature.

Seeds and seedlings must be brought from acceptable and certified nurseries to make sure that the plants are disease free. Additionally, plants with injuries such as broken branches, pest damage, and cuts should not be added or located in the same area as other plants in the media bed or greenhouse environment.

In aquaponic systems, soft chemicals and organic mixes can be used and applied to deter pests. Alcohol extracts, such as rosemary, hyssop, sag, and thyme, can be used like a repellent by mixing with water and spraying on the plants. Citrus can be dissolved in water and sprayed on plants thoroughly as a repellent, controlling a broad spectrum of pests. Essential oils such as sage or thyme can be mixed with water and sprayed on the plants. The solution can be used as a pest repellent for eliminating a broad range of pests. Garlic oil can be used as insecticides for pests such as aphids, cabbage worms, leafhoppers, whiteflies, some beetles, and nematodes. The mix of minced garlic, vegetable oil, and water can be sprayed on the plants. Hot peppers can remove maggots and ants when dust is sprinkled over the plants. Lime can also be used for removing a broad range of pests by blowing on the wet leaves like a repellent.

Tomato leaves diluted in water can remove aphids and corn earworms by attracting beneficial microbes and have toxic effect for alkaloids. Starch spray can eliminate aphids, spiders, mites, thrips, and whiteflies when sprayed on the leaves. Soups can also be used as pest repellents, but excess use of soap in aquaponic systems will damage the fish gills. This empirical knowledge is working in many systems, but there has been no systematic scientific research on their efficacy.

On the other hand, biological insecticides are classified for organic use, but most of them are toxic to fish. Copper is an inorganic insecticide and fungicide that is used as an insect repellent; however, over accumulation in water, it is toxic especially to crustaceans. Diatomaceous earth is an inorganic insecticide that absorbs lipids from the outer layer of insects' skeletons causing them to dehydrate. Nicotine is extracted from tobacco, which acts as a neurotoxic insecticide but is toxic for fish in the aquaponic systems. Neem (*Azadirachta indica*) is an organic antifeedant and fungicide, which is toxic to fish but can be used as foliar spray away from the water (Ellis and Bradley 1996).

Pyrethrum (*Chrysanthemum cinerariaefolium*) is a natural neurotoxic insecticide, which not only kills harmful insects but also wipes out the beneficial microorganisms. This organic insecticide is toxic for fish but has low persistence and easily gets destroyed with light in 3 days. Rotenone (*Derris elliptica*, *Lonchocarpus* spp., *Tephrosia* spp.) is a natural insecticide, which is extremely toxic for fish but can be used for plant nurseries before transplanting to the aquaponic systems. Ryania (*Ryania speciosa*) is an organic insecticide that disrupts the *Ca* channel in the cells of the pests with moderate toxic effect to fish. Sabadilla is a plant that interferes with the nerve membrane of the pests with no well-known effect on the fish. Sulfur is an inorganic insecticide and fungicide that can be used as a pest repellent against mites. Quassia (*Quassia amara*) is another organic insecticide that causes phagodeterrence in insect larvae with no toxic effect on fish (Copping 2004; Shour 2000; Soil Association 2011; IFOAM 2012).

Another way to control pests is to use beneficial insects. It is more applicable for large producers and industrial scale aquaponic systems. *Adalia bipunctata* (predatory beetle), *Aphelinus abdominalis* (parasitoid), *Chrysoperla carnea* (lacewings), and *Aphidius colemani* (predatory wasp) are beneficial pests that can be used for eliminating Aphids. *Cryptolaemus montrouzieri* (predatory beetle) and *Coccidoxenoides perminutus* (parasitoid wasp) are used for removing mealybugs. *Trichogramma* spp. (parasitoid) can kill and eliminate caterpillars. *Heterorhabditis megidis* (nematode) can remove chafer grub larvae. *Steinernema carpocapsae* (nematode) and *Cydia pomonella* (granular virus) are types of beneficial insects with the ability to eliminate codling moths.

Anagrus atomus (parasitic wasp) can kill leafhoppers, while *Dacnusa sibirica* and *Diglyphus* (parasitoid) remove leaf miners. *Chilocorus nigritus* (predatory beetle) eliminates scale insects. *Hypoaspis miles* (predatory mite) and *Steinernema feltiae* (nematode) can kill and eliminate sciarid fly and thrips. *Amblyseius cucumeris* and *Phytoseiulus persimilis* are predatory mites, and *Orius insidiosus* which is a predatory bug can be used for killing thrips.

Amblyseius californicus (predatory mite) and *Feltiella acarisuga* (mite midge) are used for removing spider mites. *Encarsia formosa* and *Eretmocerus eremicus* are parasitoids that remove greenhouse whitefly. *Heterorhabditis megidis* and *Phasmarhabditis hermaphrodita* are nematodes that can eliminate vine weevil and slugs (Olkowski et al. 2003; Soil Association 2011).

5.5 Fish Production in IAS

Fish species play an important role in the aquaponic systems, and so, understanding their life cycle, nutritional needs, and diseases can help in building sustainable farms, especially in urban and landlocked zones. An average fish will go through life stages of egg, larvae, fry, fingerling, adult, and maturity, which the duration of each stage depends on both environmental characteristics and physiology and biology of the species. Usually, egg stage is very short and depends on the water temperature. The water must have a high D.O. level with a gentle aeration system. With sterile procedures and good hatchery practices, bacterial and fungal diseases can be prevented in eggs' pre-hatched stage.

Larvae fish carry a yolk sac that is used as nourishment, which is absorbed through the larval stage. Once the yolk sac is absorbed, the fry stage will start where species can swim more actively. The fry and fingerling stage is when the fish will begin to eat solid food. In the wild, this food is generally plankton found in the water column and algae from the substrate. The fingerlings usually consume 10% of their body weight per day as feed. Generally, fry, fingerlings, and juvenile fish need to be kept separate to prevent the larger fish from eating the smaller individuals.

Aquaponic systems concentrate on the growth phase of the fish typically focusing on fish eating, growing, and excreting waste. The fish will be harvested during the growth stage since past this stage the species will reach sexual maturity. In this phase, fish growth will be slowed and most of the energy will be devoted into the development of sex organs. Tilapias are exceptionally easy breeders and can in fact breed too much for a small-scale system. Catfish, carp, and trout require more careful management, and it may be better to source fish from a reputable supplier.

5.5.1 Feeding in IAS

Feed consists of nutrients that are required for healthy fish growth. Most species need proteins, amino acids, carbohydrates, lipids, energy, minerals, and vitamins. In aquaponic systems, commercially available feed pellets are used especially at the beginning. It is possible to create fish feed in locations that have limited access to manufactured feeds. However, these home-made feeds need special attention because they are often not whole feeds and may lack essential nutritional components.

Growth and metabolism of animals have a direct relation to protein level in the feeds. They are made of 20 different amino acids, reassembled in innumerable combinations to provide all the indispensable proteins for life and growth. For aquatic animals, there are then essential amino acids which are arginine, histidine, isoleucine, leucine, lysine, methionine, phenylalanine, threonine, tryptophan, and valine. For this reason, feed formulation needs to find an optimal balance between these amino acids to meet the required needs of each species. Protein intake for every species varies through their life cycle. Herbivorous species need protein ranging from 30% to 40%, while carnivorous species require 40% to 50%. Additionally, juvenile fish require higher protein diets than adults.

The ideal dietary balance between proteins and energy-producing carbs and fats is another important consideration. This is to maximize growth and reduce costs and waste from using proteins as fuel. Although they can serve as an energy source, proteins are much more expensive than the preferred sources of energy such as lipids and carbohydrates. In aquaponic systems, an increase in dietary protein intake will have an impact on the amount of nitrogen in the water; therefore, it should be balanced either by a growth in plants or a choice of vegetables with higher *N* requirements.

As mentioned, carbohydrates are the cheapest but most important source of energy for fish. While cellulose and hemicellulose are not digestible in fish, carbohydrates are mainly composed of simple sugars and starch. The maximum tolerated amount of carbohydrates should be included in the diet in order to lower the feed costs. Omnivorous and warmwater fish can easily digest quantities up to 40%, but the percentage falls to about 25% in carnivorous and cold-water fish. Starch is the most used product in pelleted feed, which undergoes a gelatinization process of 60–85 °C, preventing pellets from easily dissolving in the water.

Lipids are another essential element necessary for the growth and other biological functions of fish. Fish cannot synthesize essential fatty acids, and so, deficiency in the supplement of fatty acids will reduce the growth and reproductive efficiency. In general, freshwater fish require a combination of both omega-3 and omega-6 fatty acids, whereas marine fish need mainly omega-3. Tilapias mostly require omega-6 in order to secure optimal growth and high feed conversion efficiency. Lipid inclusion in the feed needs to follow optimal protein/energy ratios to secure good growth, to avoid misuse of protein for energy purposes, and to avoid fat accumulation in the body. By oxidation of carbohydrates, lipids, and proteins, fish can absorb energy. Each species requires an optimum amount of protein and energy for efficient growth. Feed needs to be selected to meet the required level of digestible energy.

Fish body cannot produce vitamins, and so it is important to appropriately be supplied through the diet of the species. Minerals are the core part of many enzymes, and fish require *Ca*, *P*, *K*, *Na*, *CL*, *Mg*, and *S* plus 15 other minerals.

In aquaponic systems, feed waste from overfeeding is consumed by heterotrophic bacteria, which consume large amounts of O_2. Additionally, decomposition of excessed feed can increase the amount of NH_3 and NO_2- to toxic levels. The uneaten pellets can clog the mechanical filters, leading to decreased water flow and anoxic areas.

It is essential to track the FCR in aquaponic systems since the cost required for growth can be optimized. It is necessary to calculate the FCR as part of the business plan and/or financial analysis. FCR will also provide a more accurate growth rate expectation for harvest timing and production.

5.5.2 Fish Species Growing in IAS

There are several freshwater species that can grow well in aquaponic systems. These species are barramundi, catfish, common carp, grass carp, largemouth bass, Murray cod perch, silver carp, tilapia, and trout. Although all these species can grow in an aquaponic system, it is important not to introduce new species into the local waterbodies.

5.5.2.1 Carp

Carp is the most cultured species in the world. Like Tilapia, carp can tolerate low D.O. and poor water quality. However, carp can survive in lower temperatures than tilapia and survive in as low as 4 °C and as high as 34 °C. This makes carp an ideal selection for the aquaponic systems in tropical conditions. Optimal growth temperatures are between 25 and 30 °C, and in 10 months, these species can grow from fingerlings to a half kilogram. The growth rate will significantly reduce in temperatures less than 12 °C.

In the wild, carps can grow up to 40 kg and more than a meter long. They are bottom feeders that eat a wide range of feeds. As mentioned, carp prefer feeding on insects, larvae, worms, mollusks, and zooplankton, while some herbivores eat leaves, stalks, and seeds of aquatic and terrestrial plants.

In aquaponic systems, carp fingerlings are bought from licensed hatcheries and dedicated breeding facilities while obtaining juveniles are more complex due to necessity for hormone injection in female carps. In order to have better use of food, herbivorous, planktivorous, and omnivorous species can be cultured together. For this reason, using common carp and grass carp together can help in more efficient use of feeds.

To add more to the nutrient pool of the aquaponic systems, roots and other crop residues can be extremely beneficial since carp excrement can add most of the micronutrients back to the plants.

5.5.2.2 Catfish

Catfish are a robust species that can adapt well to changes in pH, temperature, and D.O. These species are resistant toward many diseases and parasites and can easily be stocked at a very high density. High stocking densities require mechanical filtration and solid removals. African catfish (*Clarias gariepinus*) is one of the air breathers, making them ideal for aquaponic systems since sudden drop in D.O. will not cause high mortalities. With high mechanical filtration, catfish can overcome the low D.O. and high NH_3 levels in highly dense aquaponic systems.

Waste management for catfish is different from tilapia since catfish excrement is less voluminous and more dissolved. Catfish grow best in warm waters and prefer temperature of 26 °C, but in the case of African catfish, growth will stop below 22 °C. Physiology of the catfish species is different from other fish since it can tolerate high levels of NH_3, but high nitrate concentration will reduce their appetite due to an internal regulatory control trigged by high levels of NO_2- in their blood.

Catfish will occupy only the bottom of the aquaponic tanks since they are benthic fish. This causes difficulties in raising them at high densities since they do not spread out through the water column. If aquaponic systems are overcrowded by catfish, they will hurt each other with their spines. Catfish can be raised with another species that utilize the upper portion of the tank such as bluegill sunfish, perch, or tilapia.

5.5.2.3 Largemouth Bass

Native to North America, largemouth bass (*Micropterus salmoides*) belong to the order Perciformes, which includes striped bass, Australian bass, the black sea bass, and the European sea bass. The species of largemouth bass can tolerate wide temperature ranges, but the growth will stop in temperature below 10 °C and above 36 °C, with loss of appetite below 10 °C. The optimal condition for their growth is temperatures between 24 and 30 °C but can tolerate low D.O. and pH with high FCR level at D.O. level above 4 mg/L.

The water transparency for largemouth bass should be high with suspended solids less than 25 mg/L.

Largemouth bass is a carnivorous species that require large amounts of protein in their diets with most of the growth curve happening between moderate environmental conditions. Largemouth bass have higher omega-3 compared to other freshwater species such as carp and tilapia.

5.5.2.4 Prawns and Shrimps

Prawns and shrimps can be found feeding on the bottom of the oceans as well as in the freshwater. Shrimps are species living in saltwater, and prawns are sweet-water or freshwater species. Most species are omnivores and live up to 7 years.

Prawns are good to be raised in the aquaponic systems since they consume uneaten feeds, fish wastes, and other organic materials at the bottom of the aquaponic tanks. In other words, they help accelerate the decomposition process of the organic materials in the tanks. Prawns cannot produce sufficient amount of waste for the plants to grow, and thus, it is better to grow them with mid-water swimming fish simultaneously. Since prawns are territorial, they cannot be cultured in high densities.

Prawns need warm water with temperatures between 24 and 31 °C. In the aquaponic systems, prawns can grow in 4 months and so annually the systems can produce three times. The species need to grow in a good water quality and are very sensitive. The larvae cycle of prawns is complex, requiring carefully monitored water quality and special feed.

5.5.2.5 Tilapia

Tilapias are the most popular freshwater species cultured worldwide. The species are resistant to many parasites and pathogens and can handle stressful conditions. They can handle a wide range of water qualities but grow best in water temperatures. Tilapias can tolerate water temperatures under 14 °C and up to 34 °C; however, below 17 °C, the species will stop eating and growing and will die below 12 °C. The best water temperature range for tilapias is between 27 and 30 °C. In an ideal condition, tilapias can grow from fingerling size to maturity in approximately 6 months.

Tilapia is an omnivore species that eats both plant-based and animal-based feeds. Tilapias are aggressive species, and hence, they cannot be grown in a small area since the males are territorial. In aquaponic systems, it is better to only use male species since they grow larger and faster. Mono-sex male tilapia can be obtained through hormone treatment or hand sexing of fingerlings. In the first case, fry is fed a testosterone-enriched feed during their first 3 weeks of life. High levels of the hormone in the blood cause a sex reversal in female fry. This technique, widely used in Asia and America but not in Europe (owing to different regulations), allows farmers to stock same-size male tilapia in ponds in order to avoid any problems of spawning and growth depression by feed competition from newer juveniles. Hand sexing simply consists of separating males from females by looking at their genital papilla when fish are about 40 g or larger. The males can be distinguished by their behaviors since they are territorial and chase after other males. Also, male fish have larger heads with a more pronounced forehead region, a humped back, and more squared-off features.

Mix-sex culture can be reared in aquaponic tanks until sexual maturity is reached at 5 months. The females can be harvested at an earlier stage, leaving the males to grow in the aquaponic systems.

5.5.2.6 Trout

Trout are cold-water, carnivorous species that belong to the salmon family. All trout species require colder water ranging between 10 and 18 °C with optimum temperature of 15 °C. Aquaponic systems in the temperate climates or Nordic, with cold winters, can be a good solution for culturing trout. Above 21 °C, the growth significantly decreases, and fish cannot utilize D.O. properly.

Diet for trout is high-protein diet, and so, more NO_3- will be added into the aquaponic systems. This allows to cultivate more leafy vegetables in the media beds of aquaponic systems. Species have high tolerance toward salinity and can survive brackish water and freshwater. Successful culture of trout requires constant monitoring of the water quality such as D.O. level and NH_3.

Rainbow trout (*Oncorhynchus mykiss*) is the most common species cultured in the USA and Canada; in cage and tanks in Norway, the UK, Chile, and Peru; and in many upland areas in tropical and subtropical areas in Iran, Nepal, and Japan as well as Australia. Trout requires a high-protein diet with substantial amount of fats. Trout are considered an “oily fish,” a nutritional description indicating a high amount of vitamin A, vitamin D, and omega-3 fatty acid, making them an excellent choice to grow for domestic consumption. Trout command higher prices in some markets for the same reason, but they require diets comparatively rich in fish oil.

5.5.3 Fish Disease in IAS

Any imbalance in the aquaponic system can cause the spread of disease in the system. For this reason, it is important to build a healthy defense system to prevent the spread of disease in the system. Therefore, adequate environmental control is essential to avoid the incidence of pathogens. Fish disease can be caused by biotic or abiotic factors; thus, monitoring the water quality parameters can avoid metabolic disorders and mortality. Additionally, control of contaminants can offset many toxicities and infections.

Because aquaponic systems have recirculating parts, it is less prone to pathogen introduction and disease outbreaks. This is because it is easier to control the inputs, especially environmental parameters. Water from the nearby waterbodies can be filtered using slow sand filtration that will reduce the chances of possible parasite or bacteria from entering the system. Also, small crustaceans will be prevented from entering the system.

Three major groups of pathogens that can cause fish diseases are bacteria, fungus, and parasites, which can enter the system from the environment. It is easier to prevent introducing new pathogens than treating them; hence, prevention is always the better solution in aquaponic systems.

For preventing new pathogens, it is important to get healthy fingerlings from reliable and professional licensed hatcheries. Fish need to be checked and examined for signs of disease. In some cases, new fish need to be quarantined for 30–45 days before adding into the system. Fish always need to be fed with nutritious diet, and feed needs to be kept in dry and dark places. Pets and other domestic animals should not have access to the aquaponic systems, and access to birds, snails, and rodents that can be vectors of parasites should be prevented.

Despite all the prevention techniques, there are still chances of spreading diseases in the aquaponic systems. For this reason, it is important to monitor and observe the fish's behavior. Some diseases will have external signs such as ulcer on the body surface, ragged fins, gill necrosis, extended abdomen, and swollen eyes. In addition to these signs, fish behavior might change that can be seen as loss of appetite, lethargy, and odd position in the water.

5.5.3.1 Abiotic Diseases in IAS

Many mortalities in the aquaponic systems are caused by abiotic diseases. Abiotic diseases are mainly related to water quality and toxicity, which induce infection opportunities in stressed and unhealthy fish. Examples of biotic diseases are ammonia poisoning, food deficiencies, gas supersaturation, hydrogen sulfide, hypoxia, improper salinity, nitrite poisoning, pH, and temperature stress.

Ammonia poisoning occurs when fish enters a new tank, biofilters fail, or tank is overpopulated and excessive protein added through overfeeding; also, reduced water flow and low D.O. can cause increase in water ammonia levels. Due to ammonia poisoning, the fish will have abnormal movement, loss of appetite, large and dark gills, and redness around the fins and eyes. In case of seeing such symptoms, 20–50% of water needs to be exchanged and acid buffer can be added. Additionally, improving oxygenation and adding biofiltering media can be other solutions to ammonia poisoning.

Food lack or improper storage of feed can cause food deficiency in the aquaponic systems. Fish will grow slow and face abnormality in the skeleton while showing depression and anemia. Food deficiency can be solved by varying the fish diet and providing vitamins and minerals.

Rapid increase in temperature or rapid decrease of water pressure that reduces the gas solubility will cause gas bubble disease. As a result of this, fish will start floating on the surface with popped eyes and will have emboli in blood and/or other organs. To solve this disease, reduce gas excess and fish need to be in stress-free zone.

In case of solid waste accumulation, low aeration, and increase in temperature, fish can have abiotic disease caused by hydrogen sulfide. In this case, the system will smell like rotten eggs, and fish will have usual swimming behavior with purple-violet gills. Removal of organic wastes and an increase in D.O. in the water can help in facing this disease.

Another disease caused by insufficient aeration, high-density tanks, and low water flow is hypoxia. Fish with this disease will show signs of anorexia or depression, and all gather at the water inflow while large fish die with smaller fish alive. Feed reduction as well as removing fish while adding more aeration can help overcome this problem.

Improper salinity will cause skin lesions and depression in the species living in the aquaponic systems. This disease is due to salt concentrations beyond fish tolerance and can be overcome by adding more rainwater or freshwater to reduce the salinity level in the tanks.

Nitrite poisoning will cause difficulty in species breathing, brownish blood, and abnormal swimming. This can be due to failure of biofilter, low $Cl{:}NO_2$ ratio, and sudden temperature drop in the systems. Immediate water replacement and reducing fish density can be a good remedy for this abiotic disease.

pH will drop when nitrification occurs or reduction in the buffer happens in the water, while increase in the pH can be due to adding an improper buffer or the water being high in alkalinity or hardness. In low pH case, fish will have difficulty breathing and increase their mucus production. In case of high pH, fish will have opacity in skin and gills and, in rare cases, corneal damages. Adding proper buffer to adjust the pH can help in overcoming this issue.

Temperature change can cause stress for the species. Change in heating or insulation or lack of thermostat can cause sudden temperature changes in the aquaponic systems. Fish will face hypothermia or hyperthermia, and there are possibilities of dyspnea or spread of mold disease.

5.5.3.2 Biotic Diseases in IAS

Aquaponic systems are less affected by biotic diseases. However, in most cases, pathogens are already present in the system, but disease occurs because the fish immune system weakens or environment is favorable for thrive of the pathogen. Healthy management, stress avoidance, and quality control of water are thus necessary to minimize any disease incidence. In case of disease occurrence, the first act is to isolate the infected fish from the rest of the stock. If any cure is put into action, it is fundamental that the fish be treated in a quarantine tank, and that any products used are not introduced into the aquaponic system. This is in order to avoid any unpredictable consequences to the beneficial bacteria. Examples of bacterial diseases in aquaponic systems are columnaris, dropsy, fin rot, streptococcosis, tuberculosis, and vibriosis. Fungal and protozoan diseases are white cotton saprolegniasis, coccidiosis, hexamitosis, ich/white spot, Trichodina infection, and velvet/dust. Parasitic diseases are caused by anchor worm, flukes, leeches, and nematodes.

Columnaris is caused by a type of bacteria with the main agent *Flexibacter columnaris* due to acute stress, increase of temperatures, and low oxygen and nitrite. Symptoms are reddening and erosion of skin turning into shallow ulcers and necrosis, necrosis of gills, and release of yellowish mucus from the lesions. To overcome this bacterial disease, potassium permanganate is used to treat fish and increase appetite to let them eat medicated feed. Also, immersion in copper sulfate and oxytetracycline or nifurpirinol can be used for eliminating the bacteria.

Various bacteria and/or parasites or a virus can cause dropsy. This disease causes infection of internal organs leading to fluid accumulation in the body, making the fish appear bloated. Chloramphenicol or tetracycline can be mixed with feed to overcome the disease.

The most common agent for fin rot is *Pseudomonas* spp. Symptoms are damaged fins with fin ray exposed, erosion, loss of color, ulceration, and bleeding. Antibiotics such as chloramphenicol or tetracycline can help remove the bacteria.

Streptococcosis is caused by an agent called *Streptococcus* spp. The fish will have hemorrhage on their body with popped eyes. Additionally, sanguineous liquid in peritoneal cavity can be detected. The disease can be treated with oxytetracycline, erythromycin, and ampicillin.

The bacteria responsible for tuberculosis are *Mycobacterium* spp. However, high density, poor water quality, and susceptible fish species are supplementary causes. Ingestion is the most common transmission factor, and encysted bacteria can survive 2 years in the environment. Treatment is with erythromycin, streptomycin or kanamycin, and vitamin B6 or elimination of the fish.

Various types of *Vibrio* spp. can cause vibriosis, which is more common in brackish and tropical fish. Higher temperatures or organic pollution can cause this disease. Fish will have skin hemorrhage with reddening spots in the lateral and ventral parts. Swollen lesions turn into ulcers releasing pus. Additionally, there will be signs of systemic infection in kidney and spleen. Eye lesions such as eye cloudiness, ulceration, popped-out eyes, and eventually organ loss are very common in this disease. Oxytetracycline and sulfonamides need to be added to the water or feeds as soon as early signs appear. The disease can be transmitted to humans, and so fish or systems need to be handled with caution.

White cotton saprolegniasis is a fungal disease caused by *Saprolegnia* spp. Concurrent causes are acute stress, temperature drop, and transport stress. The main symptom is white, brown, or red cottonish growth on fish skin that keeps on expanding. Ocular lesions as cloudy eyes cause blindness and loss of organs. Remedies can be treatment of eggs with hydrogen peroxide or prolonged immersion in methylene blue. Lesions may be treated with cloth soaked with povidone-iodine or mercurochrome. Also, salt bath or formalin bath can be a good treatment.

Coccidiosis is caused by coccidia belonging to different families. The symptoms are intestinal infestation and enteritis, and epithelial necrosis. Lesions occur in internal organs such as liver, spleen, reproductive organs, and swim bladder. Use of coccidiostat monensin, sulfadimidine, or amprolium can treat this disease.

Hexamita spp. and *Spironucleus* spp. can cause hexamitosis in which flagellate protozoa attach to the intestinal tract. The occurrence of parasite in the intestine and gallbladder is the sign of this disease. Other symptoms are presence of abdominal distension and white, mucous excrements followed by behavioral disorders such as fish hiding in corners with head down and/or swimming backwards, progressive reduction of head volume above the eyes, and darkening of body. Treatment is by metronidazole in feed and in the water. Also, adding magnesium sulfate and cathartic is a helpful solution. Increase in temperature and improvement in environmental conditions can increase the treatment process.

White spot disease is caused by *Ichthyophthirius multifiliis*. The disease creates small white cysts covering the body of the fish giving an appearance of salt grains that emerge, mucous skin, and skin erosion. Additionally, behavioral disorders as lethargy, loss of appetite, and body rubbing to remove the parasite can be seen. The parasite is susceptible to treatment during the free-swimming stage of juveniles following the adult stage of the fish and the production of cysts that fall on the bottom. It is recommended to apply salt bath or formalin bath every week until the disease is cured. Meanwhile, water temperatures need to be above 30 °C for 10 days. Raising the temperature from 21 to 26 °C shortens the cycle of the parasite from 28 to 5 days making the treatment period in curative bath shorter.

Trichodina is saucer-shaped protozoan parasite that attaches to gills and the body surface of the host fish. The parasite is often found in poor water quality and high-density stocked systems. The parasite can be found by getting a wet mount of skin scraping. A gray film on skin and gills, along with an excess of white mucous secretion, is common in the host fish. Treatment is with formalin or potassium permanganate bath.

Piscinoodinium spp., which is a parasitic skin flagellate that binds to the host, causes velvet disease. This disease causes brownish dust and covers the body and the fins. Respiratory discomfort with quick gill movement due to presence of parasite on the gills and cloudy eyes are the symptoms. Additionally, formation of cysts that discharge free infective parasites is among the other symptoms. This disease is highly contagious and fatal. The system needs to be left with no fish for 2–3 weeks in order to remove the protozoan. For heavy infestation, a bath with 3.5% salt for 1–3 min is effective to remove the trophonts. Alternatively, treat with copper sulfate at 0.2 mg/L in a separate tank (copper can bioaccumulate and cause toxicity). Raising temperatures from 24 to 27 °C can speed up the treatment cycle.

Lice is a parasitic disease which was introduced from the wild. The parasite can be seen on the skin, gill, and mouth. Also, it causes red spots on the skin that can rise up to 5 mm. It can be treated with salt for freshwater species. Additionally, hydrogen peroxide, formalin, and ivermectin are remedies for lice.

Flukes are flatworms about 1 mm long infesting gills and skin. Symptoms are fast gill movement and release of mucus from the gills. Also, fish will have quick respiration and flopping fins. To remove the parasite, fish needs to be put in 10 mg/L of potassium permanganate in a separate tank.

Leeches are external parasites mainly introduced from the wild. The disease creates small red or white lesions, and heavy infestations lead to anemia. The parasite should not be introduced to raw plants and snails, and fish need to be washed in salt solutions.

Nematoda is a threadworm infesting all over the body but is visible when they concentrate at the anus. Infestation occurs with the introduction of wild or pond fish. Symptoms are progressive loss of weight, lethargy, void bellies, and accumulation of parasites around the anus. Colonization of viscera with 0.6–7 mm worms occurs in the intestine. Treatment is medicated feed with fenbendazole or levamisole orally.

The main issue is to balance the environmental factors, species biology and physiology, as well as plant needs to prevent the spread of disease and sustainably grow the business. The complexity forces stakeholders to prefer RAS by removing the plant variables and concentrating mainly on two other factors, which are fish and the environment.

5.6 RAS for Reducing the Possible Health Complexity

In RAS, the important water quality factors are solids, refractory organics, surface active compounds, metals, and NO_3- (Colt 2006). Although RAS does not have media beds directly utilizing the circulated organic materials, challenges are faced when operating these systems that need to be managed and monitored closely. Circulation of the water in systems plays an important role in transporting particles such as the uneaten feeds, feces, bacteria, fungi, and parasites. Figure 5.5 shows the water flow in an RAS (Good et al. 2009).

Biofilter plays a key role in the filtration of RAS. The goal of biofilter is to process NH_4+ through oxidation to NO_2- and then NO_3-. The responsible bacteria for this process will develop stable and effective biofilm that provides steady-state nitrification process at a given stock density and loading of NH_4+ into the system. According to Timmons et al. (2001), to oxidize each gram of NH_4+, 4.57 and 7.14 g of alkalinity is required. It is recommended that since both mentioned processes are oxidative, oxygen concentration be about 4 and 2 mg/L (Forteath 1993). However, when biofilters become hypoxic, the reverse process of ammonification can occur.

There are various types of biofilters in RAS depending on the type and design of the system. The main types are moving beds, rotating biological contactor (RBC), trickling fixed bed, and submerged fixed bed. In moving beds, biofilters are submerged but water currents are maintained in the media in constant motion. RBC is a system where biological media is moved within a rotating device through a sump of water. In this case, hypoxia is not a limiting factor, but RBCs have high operating costs. In submerged fixed bed types, biofilters have fixed and static surfaces where nitrification takes place and are submerged in a water flow. Finally,

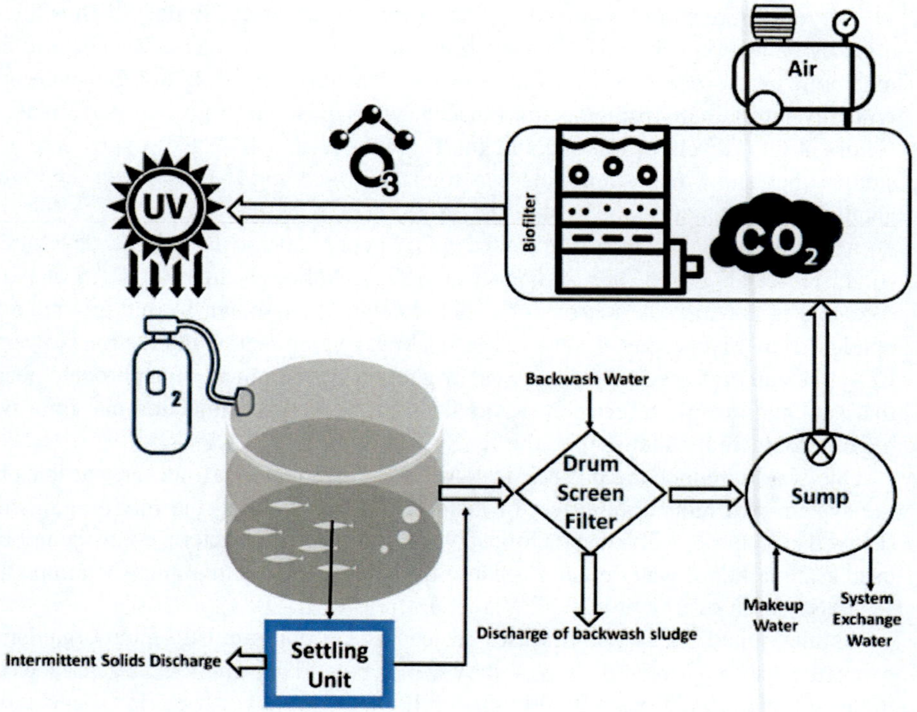

Fig. 5.5 Water flow in an RAS (Good et al. 2009)

trickling fixed beds have a fixed surface for nitrification but the water flows over the surface by gravity.

Commercial fish diets result in the production of 25–50% solid wastes. This will cause formation of particulates made from fish feces and waste feeds that range in size between 1 and 100 μm. These particles need to be removed immediately since they will begin dissolving into colloids and dissolved organic materials (Timmons et al. 2001). The presence of these materials will cause reduction in water quality and increase biofouling. Double-draining systems that consist of a small but highly concentrated flow will take out most of the settled particles. Gravity filter or screen filtration can be used for filtering these materials (Lekang 2008). An example of gravity filter is a swirl separator to remove denser particles and screen filtration such as drum, belt, and matrix filtration. According to Mead (2009), the sand matrix in sand filters is not for accumulating organic compounds that might channel filtration efficiency.

The smaller particulates such as bacteria and viruses are rarely removed by screen filtration and may rely upon the use of foam fractionation. Use of ozone (O_3) can oxidize the organic materials resulting in disinfection and water cleaning step (Good et al. 2009). O_3 is highly hazardous to fish and can create toxic bromo-halide chemicals when introduced to saltwater, according to Timmons et al. (2001).

The remaining particles in RAS are removed by a water polishing, which is often done by ozonation combined with foam fractionator (Barrut et al. 2013). In aerobic metabolic processes, O_2 is used and CO_2 is a waste product. In RAS, CO_2 levels are typically lower than 15 mg/L, and brackish water post-smolt growth performance occurs at CO_2 levels of less than 12 mg/L (Mota et al. 2019). However, there are studies that show that salmonids can tolerate 20–25 mg/L CO_2 levels in high-alkalinity freshwater (Good et al. 2010, 2018).

As a result of chronically elevated CO_2, nephrocalcinosis will be developed (Fivelstad et al. 1999, 2003; Smart et al. 1979). Although this puts RAS at high risk, this is not the sole reason for nephrocalcinosis as shown in multiple studies conducted by Fivelstad et al. (2007, 2015). However, a potential interaction between CO_2, pH, and high levels of calcium can trigger the development of nephrocalcinosis that can significantly affect fish (Myklebust 2017). Oxygen injection can improve the environmental conditions of the RAS (Timmons et al. 2001).

One way to reduce the bacteria in RAS is UV irradiation. However, particles in the system will reduce the effect of disinfection by shielding the microorganisms (Liltved et al. 2008; Liltved and Cripps 1999). Additionally, bacteria count cannot be used as a marker of water quality because of the microbial community (Munro et al. 1994; Salvesen et al. 1999, 2000; Verner-Jeffreys et al. 2003).

As mentioned, many fish diseases are caused by opportunistic microorganisms when the fish are stressed. It was shown that potential pathogenic bacteria were induced under conditions with a high supply of substrate per bacteria (Blancheton et al. 2013). Slow-growing and more beneficial bacterial community will be enhanced through a stable supply of substrate per bacteria (Attramadal et al. 2012a, b; Salvesen et al. 1999; Skjermo et al. 1997; Vadstein et al. 1993). The dosage of UV is important for the efficacy of disinfection. As shown in Table 5.1, some pathogens such as infectious pancreatic necrosis virus (IPNV) are highly resistant to UV, requiring effective dosages of over 150 mJ/cm^2 to kill them; other organisms such as *Neoparamoeba perurans* can be rendered at low dosage (Lepperød 2017; Powell and Wennberg 2016; Wennberg and Powell 2015).

In RAS, the rate of recirculation ranges between 95% and 99% and the necessity to add new water is very low, which can enter from various sources such as nearby rivers and lakes. However, adding new water from nearby waterbodies will increase the chances of infection in the RAS. For example, *Candidatus Branchiomonas cysticola* and salmon gill pox virus (SGPV) can be introduced easily into RAS facilities and are readily horizontally transmitted among salmon parr and smolts (Wiik-Nielsen et al. 2017).

Additionally, depending on seasonal variation in the water flow, water quality will be impacted. For example, upstream may release metallic ions or cause anthropogenic contamination from pesticides. Ground water can also carry pathogens and contaminations from bedrock and usually is saturated with gases. Metal ions such as Al^{3+}, Cu^{2+}, Zn^{2+}, and Mn^{2+} are found in underground water sources.

Table 5.1 UV dose required to kill 99.9% of the pathogenic fish viruses, bacteria, fungi, and parasites (Yanong 2003; Skall and Olesen 2011; Wennberg and Powell 2015; Timmons et al. 2001)

Pathogen	UV dose (μWs/cm)
Viruses	
Infectious hematopoietic necrosis virus (IHNV)	20–30,000
Infectious pancreatic necrosis virus (IPNV)	150,000
Infectious salmon anemia virus (ISAV)	4000–10,000
Viral hemorrhagic septicemia virus (VHSV)	800–3100
Nodavirus	10,000–264,000
Spring viremia of carp virus (SVCV)	1000
Bacteria and fungi	
Aeromonas salmonicida	3600
Renibacterium salmoninarum	11,000
Vibrio anguillarum	3500
Yersinia ruckeri	2700–13,400
Saprolegnia hyphae	10,000
Saprolegnia zoospores	396,000
Parasites	
Ceratomyxa shasta	30,000
Ichthyobodo necator	318,000
Ichthyophthirius multifiliis	100,000
Myxosoma cerebralis	4000–48,000
Trichodina sp.	35,000
Trichodina nigra	159,000
Paramoeba perurans	2000–25,000

5.6.1 RAS: Salinity, Diseases, and Environmental Impacts

RAS using seawater is common around the world. This is because of lack of sufficient water for expanded post-smolt production and advantage of steady and optimal temperatures from growth of the species throughout the year. Seawater RAS is not so different from freshwater RAS, but the nitrification process will be impacted by the salinity level (Chen et al. 2006). Additionally, Moran (2010) showed that the CO_2 efficiency is lower in these systems.

Seawater contains pathogenic microbes resulting in the potential need for biosecurity. Some use double barriers to protect pathogens such as UV combined with ultrafiltration membrane. In these conditions, there is an additional step that reduces the nitrate to eliminate nitrogen gas (Gutierrez-Wing and Malone 2006; Labelle et al. 2005; Sauthier et al. 1998; Van Rijn et al. 2006).

The heterotrophic denitrification drives electrons and protons for organic available materials in the seawater. Seawater contains sulfate (SO_4^{2-}), which in absence or low levels of O_2 and NO_3- will provide accepted electron by the heterotrophic denitrifiers. ORP can provide information about the production of sulfide in the system (Sauthier et al. 1998). The sulfide risk is not only restricted to systems with denitrification but also RAS with seawater and freshwater with sludge blankets in pipes or biofilms with anaerobic conditions.

Another issue is the mixed water zones forming when mixing seawater and freshwater. According to Teien et al. (2006b), adding seawater with salinity range of 1–10% can change the form of bound aluminum and it can become highly toxic.

Use of O_3 in seawater can lead to the formation of toxic bromine compounds. O_3 used in aquaculture systems can inactivate fish pathogens and destroy organic material (Goncalves and Gagnon 2011; Summerfelt 2003). According to Droste (1997), if O_3 is added before removing the particles, it could create coagulating effect of the particles. Seawater contains bromide (Br^-), so during ozonation process, bromine (Br_2) is described as total residual oxidant (TRO). Br_2 is an oxidizing agent; forms oxyacids such as hypobromous (HOBr), bromous ($HBrO_2$), and bromic ($HBrO_3$) (Liltved et al. 2006); and is toxic to fish (Fisher et al. 1999).

In RAS, lighting protocols are important and photoperiod regime as well as light quality need to be considered. The photoperiod regime can influence the maturation in most of the fish such as Atlantic salmon. In other words, RAS can provide and sustain better conditions for guaranteeing fish welfare. Palstra and Planas (2013) showed that moderate swimming exercise for salmonids can be beneficial. Additionally, rainbow trout and Atlantic salmon have better growth performance associated with prolonged swimming exercise (Houlihan and Laurent 1987; Castro et al. 2011; Waldrop et al. 2018; Jobling et al. 1993).

Moderate sustained swimming exercise has been shown to improve fin quality (Jørgensen and Jobling 1993) and decrease aggression (Postlethwaite and McDonald 1995), as well as increase disease resistance and survival (Castro et al. 2011). However, D.O. consumption will increase significantly with increased exercise (Lauff and Wood 1996); hence, in RAS, external oxygenation is required. In RAS, for culturing salmonids with sustained swimming, 80–100% saturation oxygen is required for maximum growth (Spence et al. 1996). Low oxygen saturation levels cause reduced growth and susceptibility to disease (Cameron 1971; Fischer 1963; Herrmann et al. 1962; Itazawa 1970).

Modern RAS contains circular culture where fish tend to be evenly distributed (Larmoyeux et al. 1973). Feeds are evenly carried by the flows and spread throughout the water columns for relatively easy delivery to the entire fish population (Burrows and Chenoweth 1970; Makinen et al. 1988; Parker and Barnes 2015). The upper swimming limit for salmonids is estimated in various research, and it was twice the body lengths per second (Castro et al. 2011; Davison 1997; Jobling et al. 1993; Jørgensen and Jobling 1993).

With utilization of pure oxygen in RAS, higher stocking densities can be permitted, which is fish biomass over cultured tank's water volume (North et al. 2006). There are more comprehensive methods to quantify system's biomass, which have direct impact on fish welfare (Ellis et al. 2001; Turnbull et al. 2008). Welfare concerns regarding inappropriately high stocking densities include increased susceptibility to infectious diseases (Bebak-Williams et al. 2002) and fin damage (North et al. 2006; Turnbull et al. 2005). However, low densities will also have negative impact on fish welfare (Adams et al. 1998). Grand and Dill (1999) noted that in low-density RAS, fish showed aggressive and territorial behaviors compared to schooling behavior.

RAS facility can minimize the pathogens' entry into the system. Obligate pathogens are the ones that require a host cell to replicate. Fish pathogens known as opportunistic pathogens can reproduce in the aquatic environment and are often only linked to clinical disease when the host or the environment favors their proliferation and/or virulence. According to Starliper and Schill (2011), several species of the *Flavobacterium* genus are important opportunistic fish pathogens that are considered ubiquitous in the aquatic environment. *Flavobacterium branchiophilum* is the causative agent of bacterial gill disease (BGD) than can attach and proliferate on gill tissue of a broad range of species leading to compromises in gas exchange, osmoregulation, and removal of nitrogenous wastes with 95% mortality rate in the tank. Inappropriately high rearing densities, inadequate exchange rates, and poor sanitary conditions are also some causative factors (Starliper and Schill 2011).

Flavobacterium columnare is the cause of columnaris disease. Columnaris is usually not spontaneous, but rather requires physical and/or environmental insults to instigate clinical disease with mortalities exceeding 70% (Starliper and Schill 2011). *F. columnare* can be very damaging to tissue, causing necrosis (Decostere et al. 2002) and subsequent bacteremia (Bader et al. 2006). Elevated water temperatures and high rearing densities have been shown to increase pathogen transmission and columnaris-specific mortality (Suomalainen et al. 2005), likely through density-related increases in skin abrasions allowing *F. columnare* to gain entrance and cause systemic infections (Bader et al. 2003, 2006).

Cold-water disease is related to *Flavobacterium psychrophilum*, and the disease is generally observed in colder water temperatures, most commonly between 4 and 10 °C (Starliper and Schill 2011). Mortality can reach 75% with morbidity in affected populations ranging between 1% and 50% (Post 1987). Risk factors include physical trauma, pH fluctuation, malnutrition, and dissolved toxins (Shotts and Starliper 1999), in addition to high rearing densities (Iguchi et al. 2003). Unlike *Flavobacterium branchiophilum*, *Flavobacterium columnare* and *Flavobacterium psychrophilum* can infect healthy fish (Madetoja et al. 2002). The pathogen can be isolated from both male and female sexual products (Ekman et al. 1999) and is known to be vertically transmitted from broodfish to progeny (Cipriano 2005). Early systemic infections can be successfully treated with antimicrobials (Bruun et al. 2003; Michel et al. 2003), but these lose efficacy as the disease progresses (Shotts and Starliper 1999).

From the *Saprolegnia* genus, oomycetes are causative agents of saprolegniasis in freshwater environments (Jiang et al. 2013). The fungus affects both warm- and cold-water species (Van West 2006). Torto-Alalibo et al. (2005) mentioned that in recent years and following the ban on malachite green, there has been a resurgence of this fungus in salmon farming, and it is estimated that 10% of all hatched farmed Atlantic salmon die from this fungus-caused disease (Bruno et al. 2011). The disease is first observed at the egg stage affecting the fish population reaching the market size. The agents are difficult to eradicate since spore stages are highly resistant to disinfection processes. Formalin is often used in the aquaculture industry to counter saprolegniasis; however, with human health concerns focused on the use of this therapeutant, alternative approaches, such as the use of low-dose, biofilter-friendly peracetic acid, are currently being investigated. In RAS system and by considering the treatment, hydrogen peroxide mixed with salt can treat mild outbreaks.

References

Adams CE, Huntingford FA, Turnbull JF, Beattie C (1998) Alternative competitive strategies and the cost of food acquisition in juvenile Atlantic salmon. Aquaculture 167:17–26

Attramadal KJK, Oie G, Storseth TR, Alver MO, Vadstein O, Olsen Y (2012a) The effects of moderate ozonation or high intensity UV-irradiation on the microbial environment in RAS for marine larvae. Aquaculture 330:121–129

Attramadal KJK, Salvesen I, Xue RY, Oie G, Storseth TR, Vadstein O, Olsen Y (2012b) Recirculation as a possible microbial control strategy in the production of marine larvae. Aquac Eng 46: 27–39

Bader JA, Shoemaker CA, Klesius PH (2003) Rapid detection of columnaris disease in channel catfish (Ictalurus punctatus) with a new species-specific 16-S rRNA gene-based PCR primer for Flavobacterium columnare. J Microbiol Methods 52:209–220

Bader JA, Moore SA, Nusbaum KE (2006) The effect of cutaneous injury on a reproducible immersion challenge model for Flavobacterium columnare infection in channel catfish (Ictalurus punctatus). Aquaculture 253:1–9

Barrut B, Blancheton J-P, Callier M, Champagne J-Y, Grasmick A (2013) Foam fractionation efficiency of a vacuum airlift—application to particulate matter removal in recirculating systems. Aquac Eng 54:16–21

Bebak-Williams J, McAllister PE, Smith G, Boston R (2002) Effect of fish density and number of infectious fish on the survival of rainbow trout fry, Oncorhynchus mykiss (Walbaum), during epidemics of infectious pancreatic necrosis. J Fish Dis 25:715–726

Blancheton JP, Attramadal KJK, Michaud L, D'Orbcastel ER, Vadstein O (2013) Sight into bacterial population in aquaculture systems and its implication. Aquac Eng 53:30–39

Bruno DW, Van West P, Beakes GW (2011) Saprolegnia and other oomycetes. In: Woo PTK, Bruno DW (eds) Fish diseases and disorders, Viral, bacterial and fungal infections, vol 3, 2nd edn. CABI International, Wallingford, Oxfordshire, pp 669–720

Bruun MS, Madsen L, Dalsgaard I (2003) Efficacy of oxytetracycline treatment in rainbow trout experimentally infected with Flavobacterium psychrophilum strains having different in vitro antibiotic susceptibilities. Aquaculture 215:11–20

Burrows R, Chenoweth H (1970) The rectangular circulating rearing pond. Progress Fish Cult 32: 67–80

Cameron JN (1971) Oxygen disassociation characteristics of the blood of rainbow trout, Salmo gairdneri. Comp Biochem Physiol 38:699–704

Castro V, Grisdale-Helland B, Helland SJ, Kristensen T, Jørgensen SM, Helgerud J, Claireaux G, Farrell AP, Krasnov A, Takle H (2011) Aerobic training stimulates growth and promotes disease resistance in Atlantic salmon (Salmo salar). Comp Biochem Physiol A Mol Integr Physiol 160: 278–290

Chen S, Ling J, Blancheton J-P (2006) Nitrification kinetics of biofilm as affected by water quality factors. Aquac Eng 34:179–197

Cipriano RC (2005) Intraovum infection caused by Flavobacterium psychrophilum among eggs from captive Atlantic salmon broodfish. J Aquat Anim Health 17:275–283

Colt J (2006) Water quality requirements for reuse systems. Aquac Eng 34:143–156

Copping LG (2004) The manual of biocontrol agents, 3rd edn. BCPC Publications, Alton

Davison W (1997) The effects of exercise training on teleost fish, a review of recent literature. Comp Biochem Physiol 117:67–75

Decostere A, Ducatelle R, Haesebrouck F (2002) Flavobacterium columnare (Flexibacter columnaris) associated with severe gill necrosis in koi carp (Cyrinus carpio L.). Vet Rec 150:694–695

Droste RL (1997) Theory and practice of water and wastewater treatment. Wiley, New York

Ekman E, Borjeson H, Johansson N (1999) Flavobacterium psychrophilum in Baltic salmon Salmo salar brood fish and their offspring. Dis Aquat Org 37:159–163

Ellis BW, Bradley FM (1996) The organic gardener's handbook of natural insect and disease control. Rodale Press Inc., Emmaus

Ellis T, Scott S, Bromage N, North B, Porter M (2001) What is stocking density? Trout News 32:35–37

Fischer RJ (1963) Influence of oxygen concentration and of its diurnal fluctuations on the growth of juvenile Coho Salmon. Masters of Science thesis. Oregon State University, Corvallis, OR, 48 pp

Fisher DJ, Burton DT, Yonkos LT, Turley SD, Ziegler GP (1999) The relative acute toxicity of continuous and intermittent exposures of chlorine and bromine to aquatic organisms in the presence and absence of ammonia. Water Res 33:760–768

Fivelstad S, Olsen AB, Kloften H, Ski H, Stefansson S (1999) Effects of carbon dioxide on Atlantic salmon (Salmo salar L.) smolts at constant pH in bicarbonate rich freshwater. Aquaculture 178: 171–187

Fivelstad S, Olsen AB, Asgard T, Baeverfjord G, Rasmussen T, Vindheim T, Stefansson S (2003) Long-term sublethal effects of carbon dioxide on Atlantic salmon smolts (Salmo salar L.): ion regulation, haematology, element composition, nephrocalcinosis and growth parameters. Aquaculture 215:301–319

Fivelstad S, Waagbø R, Stefansson S, Olsen AB (2007) Impacts of elevated water carbon dioxide partial pressure at two temperatures on Atlantic salmon (Salmo salar L.) parr growth and haematology. Aquaculture 269:241–249

Fivelstad S, Kvamme K, Handeland S, Fivelstad M, Olsen AB, Hosfeld CD (2015) Growth and physiological models for Atlantic salmon (Salmo salar L.) parr exposed to elevated carbon dioxide concentrations at high temperature. Aquaculture 436:90–94

Forteath N (1993) The biological filter: structure and function. In: Hart P, O'Sullivan D (eds) Recirculating systems: design, construction and management. Aquaculture Sourcebook, Turtle Press Pty Ltd., Tasmania, pp 55–63

Goncalves AA, Gagnon GA (2011) Ozone application in recirculating aquaculture system: an overview. Ozone Sci Eng 33:345–367

Good C, Davidson J, Welsh C, Brazil B, Snekvik K, Summerfelt S (2009) The impact of water exchange rate on the health and performance of rainbow trout Oncorhynchus mykiss in water recirculation aquaculture systems. Aquaculture 294:80–85

Good C, Davidson J, Welsh C, Snekvik K, Summerfelt S (2010) The effects of carbon dioxide on performance and histopathology of rainbow trout Oncorhynchus mykiss in water recirculation aquaculture systems. Aquac Eng 42:51–56

Good C, Davidson J, Terjesen BF, Takle H, Kolarevic J, Bæverfjord G, Summerfelt S (2018) The effects of long-term 20 mg/L carbon dioxide exposure on the health and performance of Atlantic salmon Salmo salar postsmolts in water recirculation aquaculture systems. Aquac Eng 81:1–9

Grand TC, Dill LM (1999) The effect of group size on the foraging behaviour of juvenile coho salmon: reduction of predation risk or increased competition? Anim Behav 58:443–451

Gutierrez-Wing MT, Malone RF (2006) Biological filters in aquaculture: trends and research directions for freshwater and marine applications. Aquac Eng 34:163–171

Herrmann RD, Warren CE, Doudoroff P (1962) Influence of oxygen concentration on the growth of juvenile coho salmon. Trans Am Fish Soc 91:155–167

Houlihan D, Laurent P (1987) Effects of exercise training on the performance, growth, and protein turnover of rainbow trout (Salmo gairdneri). Can J Fish Aquat Sci 44:1614–1621

IFOAM (2012) The IFOAM norms for organic production and processing. IFOAM, Bonn, 132 pp

Iguchi K, Ogawa K, Nagae M, Ito F (2003) The influence of rearing density on stress response and disease susceptibility of ayu (Plecoglossus altivelis). Aquaculture 220:515–523

Itazawa Y (1970) An estimation of the minimum level of dissolved oxygen in water required for the normal life of fish. Bull Jpn Soc Sci Fish 37:273–276

Jiang RHY, de Bruijn I, Haas BJ, Belmonte R, Lobach L, Christie J, van den Ackerveken G, Bottin A, Bulone V, Diaz-Moreno SM, Dumas B, Fan L, Gaulin E, Govers F, Grenville-Briggs LJ, Horner NR, Levin JZ, Mammella M, Meijer HJG, Morris P, Nusbaum C, Oome S, Phillips AJ, van Rooyan D, Rzeszutek E, Saraiva M, Secombes CJ, Seidl MF, Snel B, Stassen JHM, Sykes S, Tripathy S, van den Berg H, Vega-Arreguin JC, Wawra S, Young SK, Zeng Q, Dieguez-Uribeondo J, Russ C, Tyler BM, Van West P (2013) Distinctive expansion of potential virulence genes in the genome of the oomycete fish pathogen Saprolegnia parasitica. PLoS Genet 9:e1003272

Jobling M, Baardvik BM, Christiansen JS, Jørgensen EH (1993) The effects of prolonged exercise training on growth performance and production parameters in fish. Aquac Int 1:95–111

Jørgensen EH, Jobling M (1993) The effects of exercise on growth, food utilisation and osmoregulatory capacity of juvenile Atlantic salmon, Salmo salar. Aquaculture 116:233–246

Labelle MA, Juteau P, Jolicoeur M, Villemur R, Parent S, Comeau Y (2005) Seawater denitrification in a closed mesocosm by a submerged moving bed biofilm reactor. Water Res 39:3409–3417

Larmoyeux JD, Piper RG, Chenoweth HH (1973) Evaluation of circular tanks for salmonid production. Progress Fish Cult 35:122–131

Lauff RF, Wood CM (1996) Effects of training on respiratory gas exchange, nitrogenous waste excretion, and fuel usage during aerobic swimming in juvenile rainbow trout (Oncorhynchus mykiss). Can J Fish Aquat Sci 54:566–571

Lekang O-I (2008) Aquaculture engineering. Blackwell Publishing Ltd., Oxford

Lepperød A (2017) Effects of UV irradiation of Neoparamoeba perurans and its infectivity in Ballan wrasse (Labrus bergylta). MSc thesis. University of Bergen

Liltved H, Cripps SJ (1999) Removal of particle-associated bacteria by prefiltration and ultraviolet irradiation. Aquac Res 30:445–450

Liltved H, Vogelsang C, Modahl I, Dannevig BH (2006) High resistance of fish pathogenic viruses to UV irradiation and ozonated seawater. Aquac Eng 34:72–82

Liltved H, Bomo A-M, Handeland SO, Kristensen T (2008) UV inactivation of Moritella viscosa and other fish pathogens—inactivation, photoreactivation and by-product formation. In: Proceedings of the seventh international conference on recirculating aquaculture, 25–27 Jul 2008, Roanoke, VA

Madetoja J, Dalsgaard I, Wiklund T (2002) Occurrence of Flavobacterium psychrophilum in fish-farming environments. Dis Aquat Org 52:109–118

Makinen T, Lindgren S, Eskelinen P (1988) Sieving as an effluent treatment method for aquaculture. Aquac Eng 7:367–377

Mead S (2009) Numerical modelling of horizontal flow in sand filters. MSc thesis. University of Southern Queensland

Michel C, Kerouault B, Martin C (2003) Chloramphenicol and florfenicol susceptibility of fish-pathogenic bacteria isolated in France: comparison of minimum inhibitory concentration, using recommended provisory standards of fish bacteria. J Appl Microbiol 95:1008–1015

Moran D (2010) Carbon dioxide degassing in fresh and saline water. I: degassing performance of a cascade column. Aquac Eng 43:29–36

Mota VC, Nilsen TO, Gerwins J, Gallo M, Ytteborg E, Baeverfjord G, Kolarevic J, Summerfelt ST, Terjesen BF (2019) The effects of carbon dioxide on growth performance, welfare, and health of Atlantic salmon postsmolt (Salmo salar) in recirculating aquaculture systems. Aquaculture 498:578–586

Munro PD, Barbour A, Birkbeck TH (1994) Comparison of the gut bacterial-flora of start-feeding larval turbot reared under different conditions. J Appl Bacteriol 77:560–566

Myklebust M (2017) Nephrocalcinosis in farmed rainbow trout (Oncorhynchus mykiss). MSc thesis. University of Bergen

North BP, Turnbull JF, Ellis T, Porter MJ, Migaud H, Bron J, Bromage NR (2006) The impact of stocking density on the welfare of rainbow trout (Oncorhynchus mykiss). Aquaculture 255:466–479

Olkowski W, Dietrick E, Olkowski H, Quarles W (2003) Commercially available biological control agents. IPM Practitioner 25:1–9

Palstra AP, Planas JV (eds) (2013) Swimming physiology of fish. Springer, New York, p 429

Parker TM, Barnes ME (2015) Effects of different water velocities on the hatchery rearing performance and recovery from transportation of rainbow trout fed two different rations. Trans Am Fish Soc 144:882–890

Post G (1987) Textbook of fish health. T.F.H. Publications Inc., Neptune City, NJ, 288 pp

Postlethwaite EK, McDonald DG (1995) Mechanisms of Na+ and Cl− regulation in freshwater-adapted rainbow trout (Oncorhynchus mykiss) during exercise and stress. J Exp Biol 198:295–304

Powell MD, Wennberg AC (2016) AGD control and disinfection in cleanerfish part 2: challenge trial. Niva Report 7075-2016. Norway, Oslo

Salvesen I, Skjermo J, Vadstein O (1999) Growth of turbot (Scophthalmus maximus L.) during first feeding in relation to the proportion of r/k-strategists in the bacterial community of the rearing water. Aquaculture 175:337–350

Salvesen I, Reitan KI, Skjermo J, Oie G (2000) Microbial environments in marine larviculture: impacts of algal growth rates on the bacterial load in six microalgae. Aquac Int 8:275–287

Sauthier N, Grasmick A, Blancheton JP (1998) Biological denitrification applied to a marine closed aquaculture system. Water Res 32:1932–1938

Shotts EB Jr, Starliper CE (1999) Flavobacterial diseases. In: Woo PTK, Bruno DW (eds) Fish diseases and disorders, Viral, bacterial and fungal infections, vol 3. CABI Publishing, New York, pp 559–576

Shour MH (2000) Pesticides from nature. Iowa State University Extension. Available at: www.extension.iastate.edu/newsrel/2000/aug00/aug0007.html

Skall HF, Olesen NJ (2011) Treatment of wastewater from fish slaughterhouses. DTU Vet, National Veterinary Institute (Denmark) report, Copenhagen, Denmark, p 111

Skjermo J, Salvesen I, Oie G, Olsen Y, Vadstein O (1997) Microbially matured water: a technique for selection of a non-opportunistic bacterial flora in water that may improve performance of marine larvae. Aquac Int 5:13–28

Smart GR, Knox D, Harrison JG, Ralph JA, Richards RH, Cowey CB (1979) Nephrocalcinosis in rainbow-trout (Salmo gairdneri Richardson)—the effect of exposure to elevated CO2 concentrations. J Fish Dis 2:279–289

Soil Association (2011) Material for pest and disease control in organic crops. Fact sheet. Soil Association Trade and Producer Support, Bristol

Spence BC, Lomnicky GA, Hughs RM, Novitzki RP (1996) An ecosystem approach to salmonid conservation. TR-4501-96-6057. ManTech Environmental Research Services Corp, Corvallis, OR, 23 pp

Starliper CE, Schill WB (2011) Flavobacterial diseases: columnaris disease, coldwater disease, and bacterial gill disease. In: Woo PTK, Bruno DW (eds) Fish diseases and disorders, Viral, bacterial, and fungal infections, vol 3, 2nd edn. CABI International, Wallingford, pp 606–631

Summerfelt ST (2003) Ozonation and UV irradiation—an introduction and examples of current applications. Aquac Eng 28:21–36

Suomalainen LR, Tiirola MA, Valtonen ET (2005) Influence of rearing conditions on Flavobacterium columnare infection of rainbow trout, Oncorhynchus mykiss (Walbaum). J Fish Dis 28:271–277

Teien HC, Standring WJF, Salbu B (2006b) Mobilization of river transported colloidal aluminium upon mixing with seawater and subsequent deposition in fish gills. Sci Total Environ 364:149–164

Timmons MB, Ebeling JM, Wheaton FW, Summerfelt ST, Vinci BJ (2001) Recirculating aquaculture systems. NRAC. Cayuga Aqua Ventures, Ithaca, NY

Torto-Alalibo T, Miaoying M, Gajendran K, Waugh ME, Van West P, Kamoun S (2005) Expressed sequence tags from oomycete fish pathogen Saprolegnia parasitica reveal putative virulence factors. BMC Microbiol 5:46

Turnbull JF, Bell A, Adams C, Bron J, Huntingford FA (2005) Stocking density and welfare of cage farmed Atlantic salmon: application of a multivariate analysis. Aquaculture 243:121–132

Turnbull JF, North BP, Ellis T, Adams CE, Bron J, MacIntyre CM, Huntingford FA (2008) Stocking density and the welfare of farmed salmonids. In: Branson EJ (ed) Fish welfare. Blackwell Publishing Ltd, Oxford, pp 111–120

Vadstein O, Øie G, Olsen Y, Salvesen I, Skjærmo J, Sjåk-Breæk GA (1993) Strategy to obtain microbial control during larval development of marine fish. In: Reinertsen H, Dahle LA, Jørgensen L, Tvinnereim K (eds) First international conference on fish farming technology. A.A. Balkema, Rotterdam, pp 69–75

Van Rijn J, Tal Y, Schreier HJ (2006) Denitrification in recirculating systems: theory and applications. Aquac Eng 34:364–376

Van West P (2006) Saprolegnia parasitica, an oomycete pathogen with a fishy appetite: new challenges for an old problem. Mycologist 20:99–104

Verner-Jeffreys DW, Shields RJ, Birkbeck TH (2003) Bacterial influences on Atlantic halibut Hippoglossus hippoglossus yolk-sac larval survival and start-feed response. Dis Aquat Org 56:105–113

Waldrop T, Summerfelt S, Mazik P, Good C (2018) The effects of swimming exercise and dissolved oxygen on growth performance, fin condition and precocious maturation of early-rearing Atlantic salmon Salmo salar. Aquac Res 49:808–810

Wennberg A-C, Powell MD (2015) In vitro UV disinfection of Paramoeba perurans. NIVA report 6634-202, Oslo

Wiik-Nielsen J, Gjessing M, Solheim HT, Litlabø A, Gjevre A-G, Kristoffersen AB, Powell MD, Colquhoun DJ (2017) Ca. Branchiomonas cysticola, Ca. Piscichlamydia salmonis and salmon gill pox virus transmit horizontally in Atlantic salmon held in freshwater. J Fish Dis 40:1387–1394

Yanong RPE (2003) Fish health management considerations in recirculating aquaculture systems—part 2: pathogens. University of Florida IFAS Extension. Circular 121:1–9

Further Reading

Adikesavalu H, Patra A, Banerjee S, Sarkar A, Abraham TJ (2015) Phenotypic and molecular characterization and pathology of Flectobacillus roseus causing flectobacillosis in captive held carp Labeo rohita (Ham.) fingerlings. Aquaculture 439:60–65

Agrios GN (2004) Plant pathology, 5th edn. Elsevier Academic Press, Burlington

Aksnes A, Gjerde B, Roald SO (1986) Biological, chemical and organoleptic changes during maturation of farmed Atlantic salmon, Salmo salar. Aquaculture 53:7–20

ATTRA National Sustainable Agriculture Information Service (n.d.) Pest management. Available at: https://attra.ncat.org/pest.html

Badiola M, Albaum B, Curtin R, Gartzia I, Mendiola D (2017) Land-based on-growing of Atlantic cod (Gadus morhua) using recirculating aquaculture systems: a case study from the Basque region (Northern Spain). Aquaculture 468:428–441

Bailey DS, Rakocy JE, Cole WM, Shultz KA (1997) Economic analysis of a commercial-scale aquaponic system for the production of tilapia and lettuce. In: Tilapia aquaculture: proceedings of the fourth international symposium on tilapia in aquaculture, Orlando, Florida

Barson NJ, Aykanat T, Hindar K, Baranski M, Bolstad GH, Fiske P, Jacq C, Jensen AJ, Johnston SE, Karlsson S, Kent M, Moen T, Niemelä E, Nome T, Næsje TF, Orell P, Romakkaniemi A,

Sægrov H, Urdal K, Erkinaro J, Lien S, Primmer CR (2015) Sex-dependent dominance at a single locus maintains variation in age at maturity in salmon. Nature 528:405–408

Bennett WF (1993) Nutrient deficiencies and toxicities in crop plants. American Phytopathological Society, St. Paul

Bernstein S (2011) Aquaponic gardening: a step-by-step guide to raising vegetables and fish together. New Society Publishers, Gabriola Island

Berry W (2010) Symptoms of deficiency in essential minerals. In: Lincoln plant physiology, 5th edn. UCLA. Available at: http://5e.plantphys.net/article.php?ch=t&id=289

Bjerknes V, Duston J, Knox DJE, Harmon PR (1992) Importance of body size for acclimation of underyearling Atlantic salmon parr (*Salmo salar* L.) to seawater. Aquaculture 104:357–366

Blancheton JP (2000) Developments in recirculating systems for Mediterranean fish species. Aquac Eng 22:17–31

Blancheton JP, Bosc P, Hussenot JME, D'Orbcastel ER, Romain D (2009) The 'new' European fish culture systems: recirculating systems, offshore cages, integrated systems. Cahiers Agric 18: 227–234

Bondad-Reantaso MG, McGladdery SE, East I, Subasinghe RP (eds) (2001) Asia diagnostic guide to aquatic animal diseases. FAO Fisheries Technical Paper. No. 402, Supplement 2. FAO, Rome

Bondari K, Sheppard DC (1981) Soldier fly larvae as feed in commercial fish production. Aquaculture 24:103–109

Bromage N, Porter M, Randall C (2001) The environmental regulation of maturation in farmed finfish with special reference to the role of photoperiod and melatonin. Aquaculture 197:63–98

Buchan KAH, Martin-Robichaud DJ, Benfey TJ (2005) Measurement of dissolved ozone in sea water: a comparison of methods. Aquac Eng 33:225–231

Bullock GL, Summerfelt ST, Noble AC, Weber AL, Durant MD, Hankins JA (1997) Ozonation of a recirculating rainbow trout culture system: I. effects on bacterial gill disease and heterotrophic bacteria. Aquaculture 158:43–55

Campbell H, Marshall R (2002) Utilitarianism's bad breath? A re-evaluation of the public interest justification for planning. Plan Theory 1(2):163–187

Carrera J, Vincent T, Lafuente J (2004) Effect if influent COD/N ratio on biological nitrogen removal (BNR) from high strength ammonia industrial waste water. Process Biochem 39:2035–2041

Chan CY, Tran N, Dao DC, Sulser TB, Phillips MJ et al (2019) Prospects and challenges of fish for food security in Africa. Glob Food Sec 20:17–25

Ciplet D (2014) Contesting climate injustice: transnational advocacy network struggles for rights in UN climate politics. Glob Environ Politics 14(4):75–96

Colorado State University Extension (n.d.) Insect publication. Available at: www.ext.colostate.edu/pubs/pubs.html#insects

Cooper A (1979) The ABC of NFT. Nutrient film technique. The world's first method of crop production without a solid rooting medium. Intl Specialized Book Service Inc., Portland

Cornell University (n.d.) Plant disease diagnostic clinic. Available at: http://plantclinic.cornell.edu/factsheets.html

Cranshaw WS (2008) Bacillus thuringiensis. Factsheet 5.556. Colorado State University Extension. Available at: www.ext.colostate.edu/pubs/insect/05556.html

Danaher JJ, Pantanella E, Rakocy JE, Shultz RC, Bailey DS (2011) Dewatering and composting aquaculture waste as a growing medium in the nursery production of tomato plants. Acta Hort 891:223

Davidson J, May T, Good C, Waldrop T, Kenney B, Terjesen BF, Summerfelt S (2016) Production of market-size North American strain Atlantic salmon Salmo salar in a land-based recirculation aquaculture system using freshwater. Aquac Eng 74:1–16

De Ionno PN, Wines GL, Jones PL, Collins RO (2006) A bioeconomic evaluation of a commercial scale recirculating finfish growout system—an Australian perspective. Aquaculture 259:315–327

Delong DP, Losordo TM (2012) How to start a biofilter. SRAC Publication No. 4502. Southern Regional Aquaculture Center, Stoneville, MS

Diver S (2007) Aquaponics-integration of hydroponics with aquaculture. ATTRA—National Sustainable Agriculture Information Service

Duncan N, Mitchell D, Bromage N (1999) Post-smolt growth and maturation of out-of-season 0+ Atlantic salmon (Salmo salar) reared under different photoperiods. Aquaculture 177:61–71

Duston J, Bromage N (1986) Photoperiod mechanisms and rhythms of reproduction in the female rainbow trout. Fish Physiol Biochem 2:35–51

Duston J, Knox JDE (1992) Acclimation of Atlantic salmon (Salmo salar) parr to seawater in autumn: stimulatory effect of a long photoperiod. Aquaculture 103:341–358

Ebeling JM (2015) Water quality. LLC, New Orleans, LA. Available from: www.ag.arizona.edu

Eriksson L-O, Lundqvist H (1982) Circannual rhythms and photoperiod regulation of growth and smolting in Baltic salmon (Salmo salar L.). Aquaculture 28:113–121

Fasakin EA, Balogun AM, Fasuru BE (1999) Use of duckweed, Spirodela polyrhiza L. Schleiden, as a protein feedstuff in practical diets for tilapia, Oreochromis niloticus L. Aquac Res 30(5): 313–318

Fjelldal PG, Hansen T, Huang T-S (2011) Continuous light and elevated temperature can trigger maturation both during and immediately after smoltification in male Atlantic salmon (Salmo salar). Aquaculture 321:93–100

Folmar LC, Dickoff WW (1980) The parr-smolt transformation (smoltification) and seawater adaptation in salmonids. A review of selected literature. Aquaculture 21:1–37

Food and Agriculture Organization (FAO) (1983) FAO composition of feedstuff. In: Fish feeds and feeding in developing countries. ADCP/REP/83/18. FAO, Rome

Food and Agriculture Organization (FAO) (2012) The state of world fisheries and aquaculture. FAO, Rome

Food and Agriculture Organization (FAO) (2013) Edible insects: future prospects for food and feed security. FAO Forestry Paper 171. FAO, Rome

Galappaththi EK, Kodithuwakku SS, Galappaththi IM (2016) Can environment management integrate into supply chain management? Information sharing via shrimp aquaculture cooperatives in northwestern Sri Lanka. Mar Policy 68:187–194

Gloger KC, Rakocy JE, Cotner JB, Bailey DS, Cole WM, Shultz KA (1995) Waste treatment capacity of raft hydroponics in a closed recirculating fish culture system. World Aquaculture Society, Book of Abstracts

Goldtooth T (2014) REDD and forests case verdict. International Rights of Nature Tribunal. Available from: http://therightsofnature.org/redd-forests-verdict/

Good C, Davidson J (2016) A review of factors influencing maturation of Atlantic salmon Salmo salar with focus on water recirculation aquaculture system environments. J World Aquac Soc 47:605–632

Good C, Davidson J, Earley RL, Lee E, Summerfelt S (2014) The impact of water exchange rate and biofiltration on water-borne hormones in recirculation aquaculture systems containing sexually maturing Atlantic salmon Salmo salar. Aquac Res Dev 5:7

Good C, Davidson J, Wiens GD, Welch TJ, Summerfelt S (2015) Flavobacterium branchiophilum and F. succinicans associated with bacterial gill disease in rainbow trout Oncorhynchus mykiss (Walbaum) in water recirculation aquaculture systems. J Fish Dis 38:409–415

Good C, Marancik DP, Welch TJ, May T, Davidson J, Summerfelt S (2016a) Systemic granuloma observed in Atlantic salmon Salmo salar raised to market size in a freshwater recirculation aquaculture system. Aquac Res 47:3679–3683

Good C, Weber GM, May T, Davidson J, Summerfelt S (2016b) Reduced photoperiod (18 hours light vs 24 hours light) during first-year rearing associated with increased early male maturation in Atlantic salmon Salmo salar cultured in a freshwater recirculation aquaculture system. Aquac Res 47:3023–3027

Good C, Davidson J, Earley RL, Styga J, Summerfelt S (2017) The effects of ozonation on select waterborne steroid hormones in recirculation aquaculture systems containing sexually mature Atlantic salmon. Salmo salar. Aquac Eng 79:9–16

Gorle JMR, Terjesen BF, Mota VC, Summerfelt S (2018) Water velocity in commercial RAS culture tanks for Atlantic salmon smolt production. Aquac Eng 81:89–100

Henze M, Harremoës P (1990) Chemical biological nutrient removal: the hypro concept. In: Hann HH, Klute R (eds) Chemical, water and wastewater treatment. Springer Verlag, Switzerland, pp 499–510

Herbinger CM, Friars GW (1991) Correlation between condition factor and total lipid content in Atlantic salmon, Salmo salar L., parr. Aquac Res 22:527–529

Herrera Ugalde ME, Sucre Romero L (2019) Resultados del Proceso de Consulta: Sistematización de Cumplimiento del CLPI. Ministerio de Ambiente y Energía

Hughey TW (2005) Barrel-ponics (a.k.a. aquaponics in a barrel). Available at: www.aces.edu/dept/fisheries/education/documents/barrel-ponics.pdf

Hutchings JA, Jones MEB (1998) Life history variation and growth rate thresholds for maturity in Atlantic salmon Salmo salar. Can J Fish Aquat Sci 55(Suppl. 1):22–47

International Labor Organization (ILO) (1989) Indigenous and tribal peoples convention 169. Available from: http://www.ilo.org/dyn/normlex/en/f?p=NORMLEXPUB:12100:0::NO::P12100_ILO_CODE:C169

Kadri S, Mitchell DF, Metcalfe NB, Huntingford FA, Thorpe JE (1996) Differential patterns of feeding and resource accumulation in maturing and immature Atlantic salmon, Salmo salar. Aquaculture 142:245–257

Karakatsouli N, Papoutsoglou SE, Pizzonia G, Tsatsos G, Tsopelakos A, Chadio S, Kalogiannis D, Dalla C, Polissidis A, Papadopoulou-Daifoti Z (2007) Effects of light spectrum on growth and physiological status of gilthead seabream Sparus aurata and rainbow trout Oncorhynchus mykiss reared under recirculating system conditions. Aquac Eng 36:302–309

Karakatsouli N, Papoutsoglou SE, Panopoulos G, Papoutsoglou ES, Chadio S, Kalogiannis D (2008) Effects of light spectrum on growth and stress response of rainbow trout Oncorhynchus mykiss reared under recirculating system conditions. Aquac Eng 38:36–42

Kulabhusan PK, Rajwade JM, Sugumar V, Taju G, Hameed AS et al (2017) Field-usable lateral flow immunoassay for the rapid detection of white spot syndrome virus (WSSV). PLoS One 12(1):e0169012

Leclercq E, Taylor JF, Sprague M, Migaud H (2011) The potential of alternative lighting-systems to suppress pre-harvest sexual maturation of 1+ Atlantic salmon (Salmo salar) post-smolts reared in commercial sea cages. Aquac Eng 44:35–47

Lennard WA, Leonard BV (2006) A comparison of three different hydroponic subsystems (gravel bed, floating and nutrient film technique) in an aquaponic test system. Aquac Int 14(6):539–550

Liu Y, Rosten TW, Henriksen K, Hognes ES, Summerfelt S, Vinci B (2016) Comparative economic performance and carbon footprint of two farming models for producing Atlantic salmon (Salmo salar): land-based closed containment system in freshwater and open net-pen in seawater. Aquac Eng 71:1–12

Mangel M, Satterthwaite WH (2008) Combining proximate and ultimate approaches to understand life history variation in salmonids with application to fisheries, conservation, and aquaculture. Bull Mar Sci 83:107–130

Martins CIM, Ochola D, Ende SSW, Eding EH, Verreth JAJ (2009) Is growth retardation present in Nile tilapia Oreochromis niloticus cultured in low water exchange recirculating aquaculture systems? Aquaculture 298:43–50

Martins CIM, Eding EH, Verdegem MCJ, Heinsbroek LTN, Schneider O, Blancheton JP, D'Orbcastel ER, Verreth JAJ (2010) New developments in recirculating aquaculture systems in Europe: a perspective on environmental sustainability. Aquac Eng 43:83–93

McClure CA, Hammell KL, Moore M, Dohoo IR, Burnley H (2007) Risk factors for early sexual maturation in Atlantic salmon in seawater farms in New Brunswick and Nova Scotia, Canada. Aquaculture 272:370–379

Melo MC, Andersson E, Fjelldal PG, Bogerd J, Franca LR, Taranger GL, Schulz RW (2014) Salinity and photoperiod modulate pubertal development in Atlantic salmon (Salmo salar). J Endocrinol 220:319–332

Migaud H, Taylor JF, Taranger GL, Davie A, Cerdá-Reverter JM, Carrillo M, Hansen T, Bromage NR (2006) A comparative ex vivo and in vivo study of day and night perception in teleost species using the melatonin rhythm. J Pineal Res 41:42–52

Migaud H, Cowan M, Taylor J, Ferguson HW (2007) The effect of spectral composition and light intensity on melatonin, stress and retinal damage in post-smolt Atlantic salmon, Salmo salar. Aquaculture 270:390–404

Migaud H, Davie A, Taylor JF (2010) Current knowledge on the photoneuroendocrine regulation of reproduction in temperate fish species. J Fish Biol 76:27–68

Mota V, Martins CIM, Eding E, Canario AVM, Verath JAH (2014) Steroid accumulations in rearing water of commercial recirculating aquaculture systems. Aquac Eng 62:9–16

Mota V, Martins CIM, Eding E, Canario AVM, Verath JAH (2017) Water cortisol and testosterone in Nile tilapia (Oreochromis niloticus) recirculating aquaculture systems. Aquaculture 468:255–261

Nacional I, de Estadística y Censo (INEC) (2011) Censo Nacional de Población y VI de Vivienda: Territorios Indígenas. Instituto Nacional de Estadística y Censos, Costa Rica

New MB (1987) Feed and feeding of fish and shrimp. ADCP/REP/87/26. FAO, Rome

Noga EJ (1996) Fish disease, diagnosis and treatment. Mosby Year-Book Inc., St. Louis, 367 pp

NRC (1993) Nutrient requirement of fish. National Academy Press, Washington, DC

Oppedal F, Taranger GL, Juell J-E, Fosseidengen JE, Hansen T (1997) Light intensity affects growth and sexual maturation of Atlantic salmon (Salmo salar) postsmolts in sea cages. Aquat Living Resour 10:351–357

Oppedal F, Taranger GL, Juell J-E, Hansen T (1999) Growth, osmoregulation and sexual maturation of under yearling Atlantic salmon smolt Salmo salar L. exposed to different intensities of continuous light in sea cages. Aquac Res 30:491–499

Pal KK, McSpadden Gardener B (2006) Biological control of plant pathogens. Plant Health Instr. https://doi.org/10.1094/PHI-A-2006-1117-02

Pantanella E (2012) Integrated marine aquaculture-agriculture: sea farming out of the sea. Glob Aquac Advocate 15(1):70–72

Pantanella E, Cardarelli M, Colla G, Rea E, Marcucci A (2011) Aquaponics vs hydroponics: production and quality of lettuce crop. Acta Hort 927:887–893

Pantanella E, Cardarelli M, Colla G (2012) Yields and nutrient uptake from three aquaponic sub-systems (floating, NFT and substrate) under two different protein diets. In: Proceedings. AQUA2012. Global aquaculture securing our future, Prague, Czech Republic

Pessota CA, Åtland Å, Liltved H, Lobos MG, Kristensen T (2014) Water treatment with crushed marble or sodium silicate mitigates combined copper and aluminium toxicity for the early life stages of Atlantic salmon (Salmo salar L.). Aquac Eng 60:77–83

Peterson RH, Harmon PR (2005) Changes in condition factor and gonadosomatic index in maturing and non-maturing Atlantic salmon (Salmo salar L.) in Bay of Fundy sea cages, and the effectiveness of photoperiod manipulation in reducing early maturation. Aquac Res 36:882–889

Porter MJR, Duncan NJ, Mitchell D, Bromage NR (1999) The use of cage lighting to reduce plasma melatonin in Atlantic salmon (Salmo salar) and its effects on the inhibition of grilsing. Aquaculture 176:237–244

Rakocy JE (2007a) Aquaponics, integrating fish and plant culture. In: Simmons TB, Ebeling JM (eds) Recirculating aquaculture. Cayuga Aqua Ventures, Ithaca, pp 767–826

Rakocy JE (2007b) Ten guidelines for aquaponic systems. Aquapon J 46:14–17

Rakocy JE, Shultz RC, Bailey DS, Thoman ES (2004) Aquaponic production of tilapia and basil: comparing a batch and staggered cropping system. Acta Hort 648:63–69

Rakocy JE, Masser MP, Losordo TM (2006) Recirculating aquaculture tank production systems: aquaponics-integrating fish and plant culture. SRAC publication 454, pp 1–16

Randall CF, Bromage NR, Duston J, Symes J (1998) Photoperiod induced phase-shifts of the endogenous clock controlling reproduction in the rainbow trout: a circannual phase-response curve. J Reprod Fertil 112:399–405

Raviv M, Lieth JH (2008) Soil-less culture: theory and practice, 1st edn. Elsevier Publishing, London

Resh HM (2004) Hydroponic food production. A definitive guidebook for the advanced home gardener and the commercial hydroponic grower, 6th edn. Newconcept Press, Mahwah

Richter N, Siddhuraju P, Becker K (2003) Evaluation of nutritional quality of moringa (Moringa oleifera Lam.) leaves as an alternative protein source for Nile tilapia (Oreochromis niloticus L.). Aquaculture 217(1):599–611

Rodríguez CM (2019) Maximizing resilience, stability, and the permanence of nature-based solutions: opportunities and barriers to increase ambition. Presented at PreCOP25, San José, Costa Rica

Rondon SI, Cantliffe DJ, Price J (2001) Augmentative biological control of insects: possibilities for vegetable greenhouse producers. FACTS proceedings, pp 15–16

Rowe DK, Thorpe JE, Shanks AM (1991) The role of fat stores in the maturation in male Atlantic salmon (Salmo salar) parr. Can J Fish Aquat Sci 48:405–413

Saunders RL (1986) The scientific and management implications of age and size at sexual maturity in Atlantic salmon (Salmo salar). In: Meerburg DJ (ed) Salmonid age at maturity, Canadian journal of fisheries and aquatic sciences, vol 89. Department of Fisheries and Oceans, Ottawa, pp 3–6

Saunders RL, Henderson EB (1978) Changes in gill ATPase activity and smolt status of Atlantic salmon (Salmo salar). J Fish Res Board Can 35:1542–1546

Saunders RL, Harmon PR, Knox DE (1994) Smolt development and subsequent sexual maturity in previously mature male Atlantic salmon (Salmo salar). Aquaculture 121:79–93

Savidov N (2005) Evaluation and development of aquaponics production and product market capabilities in Alberta. Phase II. Final report—project #2004-67905621

Schmidt PM, Peterson MJ (2009) Biodiversity conservation and indigenous land management in the era of self-determination. Conserv Biol 23(6):1458

Seawright DE, Stickney RR, Walker RB (1998) Nutrient dynamics in integrated aquaculture-hydroponic systems. Aquaculture 160:215–237

Sheppard DC, Tomberlin JK, Joyce JA, Kiser BC, Sumner SM (2002) Rearing methods for the black soldier fly (Diptera: Stratiomyidae). J Med Entomol 39(4):695–698

Sheu S-Y, Chiu TF, Cho N-T, Chou J-H, Sheu D-S, Arun AB, Young C-C, Chen CA, Wang J-T, Chen W-M (2009) Flectobacillus roseus sp. nov., isolated from freshwater in Taiwan. Int J Syst Evol Microbiol 59:2546–2551

Simpson AL (1992) Role of fat stores in the maturation of male Atlantic salmon. Can J Zool 70: 1737–1742

Skilbrei OT (1989) Relationships between smolt length and growth and maturation in the sea of individually tagged Atlantic salmon (Salmo salar). Aquaculture 83:95–108

Sophie St-Hilaire IS, Cranfill K, McGuire MA, Mosley EE, Tomberlin JK, Newton L, Sealey W, Sheppard C, Irving S (2007) Fish offal recycling by the black soldier fly produces a foodstuff high in omega-3 fatty acids. J World Aquac Soc 38(2):309–313

Stefansson SO, Hansen T, Taranger GL (1993) Growth and parr-smolt transformation of Atlantic salmon (Salmo salar) under different light intensities and subsequent survival and growth in seawater. Aquac Eng 13:231–243

Stefansson SO, Björnsson BT, Ebbsesson LOE, McCormick SD (2008) Chapter 20: Smoltification. In: Finn RN, Kapoor BG (eds) Fish larval physiology. Science Publishers, Enfield, NH, pp 639–681

St-Hilaire S, Ribble C, Whitaker DJ, Kent M (1998) Prevalence of Kudoa thyrsites in sexually mature and immature pen-reared Atlantic salmon (Salmo salar) in British Columbia, Canada. Aquaculture 162:69–77

Taranger GL, Carillo M, Schulz RW, Fontaine P, Zanuy S, Felip A, Weltzien F-A, Dufour S, Karlsen O, Norberg B, Andersson E, Hansen T (2010) Control of puberty in farmed fish. Gen Comp Endocrinol 165:483–515

Taylor M, Kent ML, White WJ (2001) How activist organizations are using the internet to build relationships. Public Relat Rev 27(3):263–284

Taylor JF, Migaud H, Porter MJR, Bromage NR (2005) Photoperiod influences growth rate and insulin-like growth factor-I (IGF-I) levels in juvenile rainbow trout. Gen Comp Endocrinol 142: 169–185

Taylor JF, Davies M, Yonge N, Porter MJR, Bromage NR, Migaud H (2006) Photoperiod can be used to enhance growth rate and improve feeding efficiency in commercially farmed rainbow trout. Aquaculture 256:216–234

Teien HC, Kroglund F, Åtland Å, Rosseland BO, Salbu B (2006a) Sodium silicate as alternative to liming-reduced aluminium toxicity for Atlantic salmon (Salmo salar L.) in unstable mixing zones. Sci Total Environ 358:151–163

Texas A&M Agrilife Extension (n.d.) Texas plant disease handbook. Available at: http://plantdiseasehandbook.tamu.edu/food-crops/vegetable-crops/

Thorpe JE (1986) Age at first maturity in Atlantic salmon, Salmo salar: freshwater period influences and conflicts with smolting. In: Meerburg DJ (ed) Salmonid age at maturity, Canadian journal of fisheries and aquatic sciences, vol 89. Department of Fisheries and Oceans, Ottawa, pp 7–14

Tuhiwai Smith LT (2012) Decolonizing methodologies: research and indigenous peoples, 2nd edn. Zed Books Ltd.

Tyson RV, Simonne EH, White JM, Lamb EM (2004) Reconciling water quality parameters impacting nitrification in aquaponics: the pH levels. Proc Fla State Hort Soc 117:79–83

United Nations Secretary General (2014) Secretary-General's remarks at the opening of the world conference on indigenous peoples. Available from: https://www.un.org/sg/en/content/sg/statement/2014-09-22/secretarygenerals-remarks-opening-worldconference-indigenous

UW Madison Department of Entomology (n.d.) Insect ID. Available at: www.entomology.wisc.edu/insectid/index.php

Vera LM, Migaud H (2009) Continuous high light intensity can induce retinal degeneration in Atlantic salmon. Atlantic cod and European sea bass. Aquaculture 296:150–158

Vera LM, Davie A, Taylor JF, Migaud H (2010) Differential light intensity and spectral sensitivities of Atlantic salmon, European sea bass and Atlantic cod pineal glands ex vivo. Gen Comp Endocrinol 165:25–33

Wagner GM (1997) Azolla: a review of its biology and utilization. Bot Rev 63(1):1–26

Washington State University (WSU) (2011) Organic pest control in the vegetable garden. Community horticulture fact sheet #13. King County Extension. Available at: http://ext100.wsu.edu/king/wp-content/uploads/sites/17/2014/02/Organic-Pest-Control-in-the-Vegetable-Garden1.pdf

Wedemeyer GA, Saunders RL, Clarke WC (1980) Environmental factors affecting smoltification and early marine survival of anadromous salmonids. Mar Fish Rev 42:1–14

Poseidon-AI® Integrated Aquaculture Modules

6

Poseidon-AI aims to close the information gap between experts as well as farmers and families. The main goal is to use deep tech and trained algorithms in IAS modules to empower communities and families especially in landlocked areas and urban zones to sustainably contribute to food safety and food security of each country as well as the world. IAS can be a great solution for answering the nutritional need of families in landlocked areas. However, high infrastructure costs and costs of energy and water as well as lack of experts prevent many communities from being able to enjoy the benefits of such systems.

The Poseidon-AI® IAS modules produce domestic fish and crustacean species with a commercial value using rainwater, clay tiles, and solar energy. In the design of these modules, transportation, easy knowledge sharing, and high efficiency with low cost are considered. Poseidon-AI® IoT and algorithms act as experts for families who use these systems, assuring the production of healthy, sufficient, and nutritional food for families.

While the pandemic prevents many from working and with economies slowing down, many communities are struggling to have sufficient and nutritious food. This forces many youths to seek ways to become helping hands for their families.

Since the spread of COVID-19, lockdowns and remote working caused damage to many industries, especially agrifood industries. Under these circumstances, it is now necessary to implement multidisciplinary approaches to guarantee not only the sustainable development of nations but also secure food production for population impacted by this pandemic.

Poseidon-AI® IAS module is an innovative and efficient way for youth and their families in low-income communities to have their own source of income producing their own organic food. The closed system helps families to annually produce 80 kg of seafood and 20 kg of vegetables with the minimum required knowledge about agriculture sciences.

© The Author(s), under exclusive license to Springer Nature Singapore Pte Ltd. 2023

A. Rahimi-Midani, *Deep Technology for Sustainable Fisheries and Aquaculture*,
https://doi.org/10.1007/978-981-99-4917-5_6

The 1×1 m^2 closed system uses a low-consumption water pump working with 500 L of rainwater. The pump circulates 200 L of rainwater for moving the fish excrement to the plants allowing the production of 20 kg of vegetables.

Poseidon-AI uses SDGs to show the contribution of innovation to sustainable development. The system directly contributes to nine SDGs which are the following:

- The system helps in reduction of poverty and tackling the food security issues by producing 80 kg of seafood and 20 kg of vegetables, which provides youth with a source of income as well as creates a stable source of nutrition.
- The system helps young mothers, single moms, and girls in communities to familiarize themselves with environmental engineering and fish and plant physiology, creating a decent work with a steady income.
- The system uses rainwater and works with solar energy that contributes significantly to expansion of affordable and clean energy in local communities. Additionally, the system helps adapt to climate conditions such as fewer rainy days and warmer days. The innovation is a closed system and can maintain the water level sufficient for growing both plants and the species.
- The Poseidon-AI® IAS artisanal module is a small innovation that not only contributes to agrifood industries in microlevel but can also be fit in 1×1 m^2 area in every household. It supports urban farming initiatives in many cities and communities throughout the world.

Currently, these systems are used for vulnerable and indigenous communities in Southeast Asia and Latin America aiming to expand to other regions in the next couple of years.

6.1 Poseidon-AI® IAS in Southeast Asia

Seafood in Southeast Asia plays a big role in families' diets (Fig. 6.1). As mentioned, limitations imposed by the environment such as water quality and arable lands prevent these countries from outpacing the demands caused by the exponential population growth.

Poseidon-AI is incorporated in Singapore, aiming to support farmers, families, and communities not only in Singapore but also in all the countries around the world. The goal is to use available resources to empower communities with the help of Poseidon-AI® IAS, IoT, and algorithms.

6.1.1 Malaysia

Aquaculture contributes 8.9% of the total national agriculture GDP of Malaysia, creating an estimated 1,753,900 jobs (DOSM 2016). Hence, the sector contributes to national food security as well as globally, alleviating hunger and poverty (Allison 2011). Due to the availability of adequate food, the sector has achieved

Fig. 6.1 Singaporean chili crab is a delicious seafood dish that can be found almost in every hawker center (food court center) in Singapore

improvements in breeding programs, culture system technology, best management practices, and feed production (Béné et al. 2015; Norazmi 2017; Bentsen et al. 2017; Dickson et al. 2016; Carrier et al. 2017).

However, due to environmental concerns, disease outbreaks, declining fish stocks that drive up feeding prices, incorrect information, and unethical feeding practices from a halal standpoint, Malaysia's aquaculture industry is slowing down (Yue and Wang 2017; Fitzsimmons and LwinTun 2017; FAO 2014, 2016; Saidin et al. 2017).

Stocks in Malaysia are impacted not only by overfishing but also pollution, climate change, and foreign vessel encroachment (Begg and Waldman 1999; Little et al. 2016; Yusoff 2014; DOF 2015). Currently, the depletion of stocks is boldening the need of rapid aquaculture growth in Malaysia (Naylor et al. 2000; Ottinger et al. 2016).

Rapid spread of diseases in the aquaculture sector of Malaysia is causing great concerns among farmers, scientists, and Malaysian Government. When the water temperature in Malaysia exceeded 250 °C from June to October, cultured tilapia was afflicted by the tilapia lake virus (TiLV), which had a 90% fatality rate (Fathi et al. 2017; Eyngor et al. 2014). Figure 6.2 shows that during this period, experts

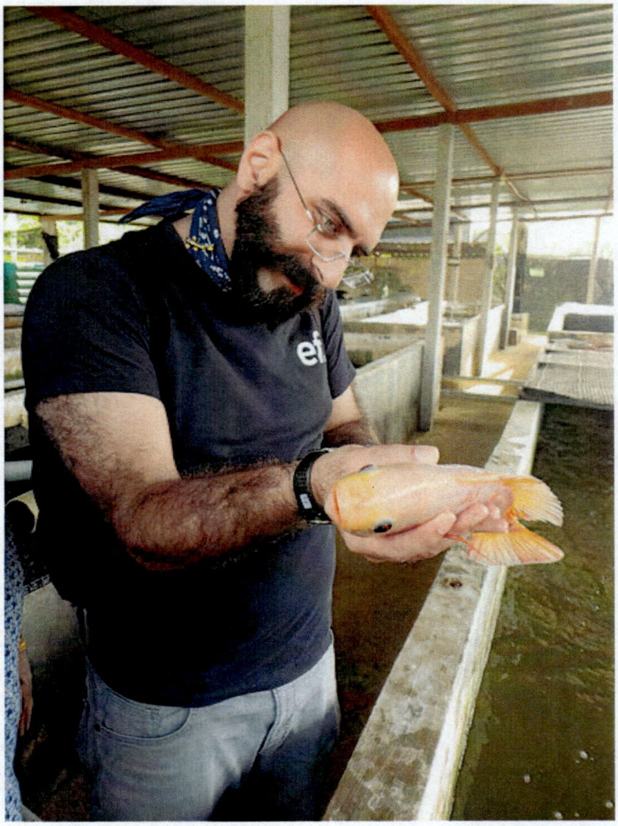

Fig. 6.2 Manual monitoring of farmed tilapia in Malaysia for possible diseases

constantly monitored the physical appearance of tilapia species as well as the water quality to reduce the risk of this disease.

Another disease with no causative pathogen that caused mass mortalities in giant tiger prawn (*Penaeus monodon*) and whiteleg shrimp (*Litopenaeus vannamei*) is early mortality syndrome (EMS) (Leaño and Mohan 2012). Additionally, white spot disease (WSD) affected many shrimp farms in Malaysia. As a result of these diseases, farmers used prohibited antibiotics causing a ban of import of Malaysian-produced shrimps into the USA. This is because 32% of Malaysian shrimp samples tested positive for nitrofuran (NF) and chloramphenicol (CAP) (USFDA 2016; Saili 2017).

Another issue was related to the publication of an article related to the nutritional value of tilapia causing a significant drop in US and European demands for Malaysian cultured tilapia (Weaver et al. 2008; Dr. Axe 2012). The article was cited by 100 other papers and websites claiming that bacon is better compared to tilapia based on omega-6 in the species.

A large majority of Malaysians are Muslims, taking cautions on the matter related to safe and halal food. That is why Malaysian Government provides certifying standards of halal food MS1500:2009 (DSM 2014). This is significant because

Fig. 6.3 Daily monitoring in farms located in south of Malaysia

reports of the use of animal byproducts, including blood, bones, and tissues derived from pig and other animal wastes, as feed, led the Muslim community in Malaysia to refrain from consuming these products, and the Department of Fisheries (DOF) guidelines state that farms using these practices are not allowed to sell their products at the local market (Norhana et al. 2012; Ahmad and Hamdan 2016; Pauzi and Man 2015; Hempel 2010).

Finally, there is a lack of communication among stakeholders active in the aquaculture sector of Malaysia, and so, it is believed that creating communication lines among farmers and use of online platforms for information sharing can help in the development of the Malaysian aquaculture sector (Shaffril et al. 2009; Kim et al. 2010; Omar et al. 2011).

Under these circumstances and to prevent stock loss and a substantial decline in yields, it is necessary to have systems that can be managed while being constantly monitored by professionals. In many areas of Malaysia, the stocks are currently being monitored manually; however, owing to a shortage of professionals and a knowledge gap between academics and farmers, Malaysia's aquaculture production experiences erratic changes every year. Figure 6.3 shows daily monitoring in aquaculture cement tanks in south part of Malaysia.

Johor Bahru (JB) is located in the south part of Malaysia, separated from Singapore by waters from Johor Strait. Many farms are located around these areas taking advantage of narrow waterbodies going in and out of the Johor Strait. Farms culturing tilapia, catfish, freshwater shrimps, and sea bass can be found along the coast and in some inland cement tank and pond cultures. There is no deep tech involved for improvement of any of these farms, and farms are very labor oriented. Traditional methods of feeding are used, and because there are no water treatment facilities nearby, water continually flows into and out of the farms and into the strait. Along with polluting the water, this has a negative impact on the environment and threatens wild species.

Under these circumstances, Poseidon-AI® semi-industrial and industrial modules can revolutionize the farming practices in this area. Each system uses a mix of 5000 L from the nearby waterbodies and rainwater. The water goes through a filter to remove possible pollution and prevent contamination from entering these systems.

The system constantly circulates 2000 L of water, and D.O. level is moderated with an air pump. The solid wastes are removed with drum screen filter, while the NO_3- is absorbed through a media bed. The media bed is for producing local consumed vegetables such as lettuce, tomato, and chili. The Poseidon-AI® IoT monitors the D.O., pH, temperature, and NH_3 level, while the cameras are used to monitor possible disease in the system. The connection in these farms is not stable, and hence, the system is not connected to the local Wi-Fi system. 4G LTE local sim card is used for transferring and visualization of data in a farmer's cellphone.

Due to concerns over the spread of disease, farmers receive disease-free fingerlings from the certified hatcheries in the area, but the production is limited, and the prices are not moderated; thus, many farmers try to build small hatcheries near their farms. However, lack of sufficient knowledge on breeding of these stocks causes many farms to fail in the production of their own brood stocks. Poseidon-AI® IoT helps these farmers to monitor their hatcheries 24/7 and maintain the water quality parameters but cannot transfer the breeding knowledge to farmers due to species' varieties and various requirements.

Furthermore, the plant seedlings are collected from the nearby greenhouses and transplanted to the system, maintaining constant consumption of NO_3- in the system. The feeding process is done manually, but Poseidon-AI® algorithms recommend the feeding amount, time of feeding, and approximate FCR based on industrial pellets provided for each system and species.

Many produced species are exported to the southern neighbor and consumed in Singaporean urban restaurant centers called hawker centers. The species are mainly mud crabs, tilapia, shrimps, and other crustaceans such as crawfish. Singapore depends on imports from Malaysia for food as well as drinking water.

6.1.2 Singapore

Singapore has the land area of 714 km^2 with limited aquaculture industry; facing various dilemmas in importing seafood, Singapore puts special food resilience efforts for sustainability of the country (AVA 2011; Ngiam and Cheong 2006). Singapore separated from Malaysia in 1965 showing tremendous development in its economy (Low 2001; Shin 2005). The development is mainly attributed to the Singaporean Government for implementing effective socioeconomic policies. In the 1970s and 1980s, the Government of Singapore concentrated its efforts and strategies to promote the private sector and invested in human resource development for providing skilled labors for government-linked companies and foreign multinational companies (Chan and Ng 2004; Dahles 2008). At first, the private sector in Singapore was Chinese family firms that according to Chan and Ng (2000) were low in productivity and intensively used the labor resources, causing decline in the economy of Singapore. Additionally, these firms with all their inefficiency displaced or subcontracted their services to the foreign multinational companies and government-linked companies (Low 1990). The concentration of Singaporean economy on multinational companies and government-linked companies continued in the 1990s (Tan and Yeung 2000).

However, government strategies focused on two main problems, which were housing and unemployment, leading to conversion of agricultural land to industrial and housing purposes. In the 1980s, the whole primary sector accounted only 1.3% of the total GDP, which was not much for fast-developing Singapore with housing problems (Hill 2013). The agriculture industry, according to Ngiam and Cheong (2006), has shrunk from 13,000 ha in the 1970s to 8000 ha in the 1980s to 1500 ha in the 2000s. The wide power of the Government of Singapore to acquire and allocate lands gave necessary means for fulfilling these public projects (De Koninck 1973; Government Publications Bureau 1966).

While putting the interests of the majority first, the land-use development eliminated numerous farming businesses that were producing (Campbell and Marshall 2002; Neo 2007). Singaporean agricultural activities declined significantly due to the government being in charge of regulating and developing the local farming and fishing industry. The Agri-Food and Veterinary Authority (AVA) changed and transformed beyond traditional role and carried out major review of the future of agriculture concentrating on food self-sufficiency and optimizing the land productivity (AVA 2013b; Ngiam and Cheong 2006). According to the AVA assessment, the 1980s saw a considerable decline in Singapore's agricultural acreage, a loss of 23,000 employment, and a significant reduction in the capacity for domestic food production and aesthetic agriculture.

To overcome these challenges, in 1985, Singapore formulated a strategy to sustain farming industry. Agrotechnology parks, according to Ngiam and Cheong (2006), were "planned for intensive agricultural systems and for an attractive and nonpolluting [sic] setting" which would fit into the urban environment; this made it possible for urban farming to exist. 1500 hectares of land was developed into six agrotechnology parks located at Lim Chu Kang, Murai, Sungei Tengah, Nee Soon, Mandai, and Loyang. In these six agrotechnology parks, the Singaporean Government invested in infrastructural provisioning and promoted private sector growth by leasing land parcels to interested parties (Ngiam and Cheong 2006). Currently, 200 farms are running in these parks to produce vegetables, livestock, eggs, milk,

Fig. 6.4 Green spaces on apartments in Singapore. (**a**) Park Royal Building located in Clarke Quay Singapore. (**b**) Singapore Airport and Jewel Changi Airport

fish feed, and breeding (AVA 2013a). Each of these plots is on an unsubsidized, totally commercial footing (Liew 1985). Figure 6.4 shows the development of green spaces on apartments due to lack of space in Singapore.

The fish farming in Singapore concentrates almost exclusively on fish cultivation from fingerlings to harvesting. The upstream and downstream activities are not carried out by these farms (Chou 1986; Ngiam and Cheong 2006). Feeding accounts for the highest cost in these farms, while the main pool of labor is derived from the owners. Typically, each farm employs less than 20 workers including the seasonal workers. The main species are tilapia, sea bass, grouper, mullets, and pomfrets (Chou 1986; Liew 1985). Farmers must decide whether to fatten their products or sell them for a profit, but the land grab has also significantly reduced Singapore's aquaculture industry. Due to a lack of available space, production has decreased, prompting Singapore to import seafood from Hong Kong, Malaysia, Taiwan, Thailand, and Vietnam (Tey et al. 2009).

The lack of cultivable space has forced many farmers to move their culturing sites off to the seawater. According to the AVA (2011), 95 out of a total of 111 aquaculture farms are in four sites along Johor Straits, namely Lim Chu Kang, Ponggol, Serangoon/Loyang, and Pulau Ubin. The main purpose of these shifts to Johor Strait is water quality standards and prevention of possible collisions of these farms with sea vessels (Chou and Lee 1997). As one of the busiest shipping ports in the world, Singapore's marine farming is constrained by available sea space (Chou and Lee 1997). To maximize the amount of acreage available for local farming, the Singaporean Government pushes farmers to boost productivity levels.

The offshore farming requires establishment of large industrial farming systems using deep net cages (Phyne et al. 2006). The farming systems are far from the coastal activities and use more of the vertical water column. The first state-led program began in 1997 for sea bass in the site off Saint John's Island. According to Ngiam and Cheong (2006), the project feasibility and sustainability have been negatively impacted by the lack of consistent supply of good-quality fish fry.

Another problem faced by Singaporean fish farms, apart from technical constraints and space, is the competitively priced seafood from nearby countries, especially Malaysia with lower labor cost and a cheaper currency. As a solution, many farmers have left the business or moved their businesses to neighboring countries such as J.B. in Malaysia and the Riau Islands in Indonesia, sending their products back to Singapore.

Singaporean aquaculture sector has located themselves in relation with wholesalers and retailers in Singapore, but the arrangement between these parties is unclear. There are studies that showed that the integration of the aquaculture sector in other value chain aspects will upgrade the farmers' activities, but it is not clear that similar conclusion can be made in case of Singapore (Phyne 2010; Phyne et al. 2006).

For fish farmers in Singapore, the pressure of space is always there. The farmers in land always face the trade-offs between keeping their business and selling their lands, which is considerably more valuable. In the case of offshore farmers, the pressure is not as much as the land-based farmers. However, the narrow sea lanes of Singapore can provide more value compared to their fish production. Farmers cannot expand their production as freely as necessary, and their establishments are always kept away from the navigational routes as a precautionary measure. These limitations have a considerable negative impact on Singapore's efforts to develop a food resilience plan and discourage farmers from operating there, compared to other nearby nations that are less complicated and more supportive of aquaculture farming within their borders.

For the mentioned reasons, Singaporean fish farms which are not supported by the government mainly concentrate on exotic species that have bigger market and higher value in both local and international markets. Horseshoe crab (*Limulidae*) or some ornamental species for aquariums such as the Asian arowana (*Scleropages formosus*) (Fig. 6.5) are cultured in Singapore for export to other countries, especially China.

Figure 6.6 shows horseshoe crab cultured in Singapore for usage in the medical sector. The horseshoe crab has blue color blood, which is due to the presence of copper in their bloods' hemocyanin. Due to the high level of amebocytes in their blood, scientists use their blood to make limulus amebocyte lysate (LAL), which is used for bacterial endotoxins.

Poseidon-AI uses its unique approach to help Singapore in sustainable farming despite the challenges set by space limitation and the environment. There are 238 farms located in Singapore producing various types of seafood as well as ornamental species. With Poseidon-AI approach, Singaporean farms can sustainably develop their businesses as well as increase their profits by efficiently reducing their feeding costs. Poseidon-AI® IoT and algorithms are used for sea bass farmers in offshore culture. The production of sea bass in Singapore is limited, and feeding cost is nearly 70% of their total costs. Additionally, in recent years, algae bloom caused high mortality in these farms. There was limited time to alert the farmers for taking

Fig. 6.5 The Asian arowana (*Scleropages formosus*) is considered as wealth among Chinese communities in Singapore

the species out from the cages and into the safety zones. Constant and live monitoring of water conditions allows farmers to prevent high mortality in their farms caused by algae bloom. Reduction in excess feed, which is caused by overfeeding, will limit the development of pathogens around these cages. The problem of certified and disease-free fingerlings is solved with the help of AVA's support, but high mortality due to unexpected environmental changes especially in offshore culture remains the big problem.

Under these circumstances, Poseidon-AI® IAS is a great solution to guarantee the seafood production in Singapore and help in sustainable development of the aquaculture sector. As already noted, the Government of Singapore has focused on two development initiatives, namely housing and employment, and Poseidon-AI® IAS can assist the Government of Singapore in attaining both objectives.

Singapore does not have a pension program, and with population getting older, many Singaporeans need to work after retiring. Housing and Development Board (HDB) was funded in the 1960s making sure that every Singaporean will have access to affordable housing. Many Singaporeans living in these buildings have access to nearby hawker centers, schools, parks, and hospitals. Rooftops of these buildings can be a perfect place for Poseidon-AI® IAS for producing organic fish and vegetables with the help of rainwater and no soil.

Poseidon-AI® IoT and algorithms used in Poseidon-AI® IAS allow communities living in HDBs to have their own green area on the rooftop while producing their own organic food. The algorithms and IoT can help the elderly communities to control and monitor the condition of each system without the necessity to climb to the rooftops. Depending on the species, Poseidon-AI® algorithms help creating the

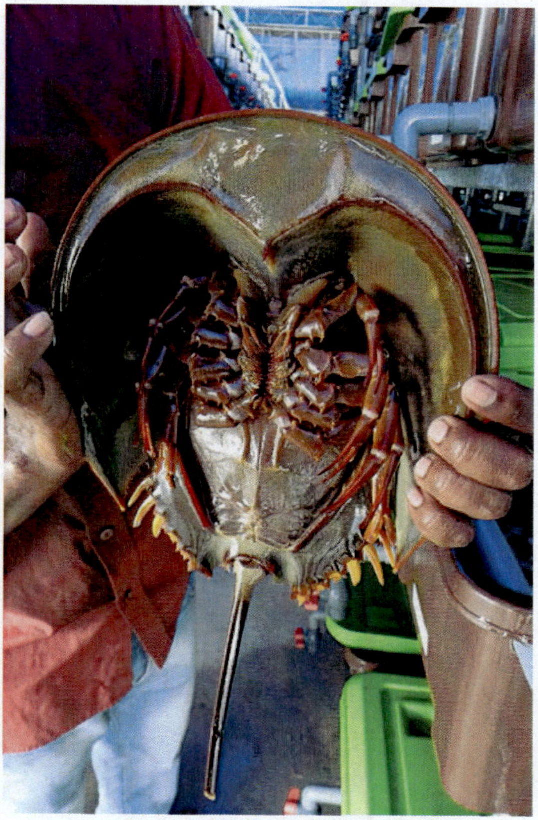

Fig. 6.6 Horseshoe crab (*Limulidae*) cultured in Singapore for medical applications

most suitable condition for the species cultured in Singapore but inside the Poseidon-AI® IAS. Due to the weight limit for rooftops of these species, only a certain number of these systems can be installed in each HBD.

The algorithms ensure that the environmental conditions are optimized throughout the life cycle of the cultured species while the systems operate continuously without the need for human interaction. The cameras monitor the plant and fish growth, while owners can have live visualization of the system's condition on their cellphones.

Poseidon-AI collaborates with different universities, government entities, and farms in Singapore to implement the innovative technologies for sustainable development of the country. The goal is to create a supply chain system with the help of various Singaporean entities that can provide fingerlings and plant seedlings and produce high-standard seafood that can be delivered to the nearby hawker center for consumption (Fig. 6.7).

With high raining amount and sufficient sunlight, Poseidon-AI® IAS helps communities in Singapore to produce tonnes of seafood and vegetables with minimum costs. Additionally, due to limited transportation, and use of solar energy, the carbon footprints of these systems are very low. The system not only helps in

Fig. 6.7 Presenting Poseidon-AI at the National University of Singapore (From left: Professor Yu Hao (Head of Biological Sciences); Professor Tan Eng Chye (President of the National University of Singapore); Dr. Amaj Rahimi-Midani (Founder and CEO of Poseidon-AI); Professor Brian Koh)

economic empowerment of communities especially elderlies but also helps in sustainable development of the aquaculture sector of Singapore.

6.2 Poseidon-AI® IAS in Latin America and Caribbean: The Case of Costa Rica

A large number of rivers that run swiftly from the mountains on both sides of Costa Rica, as well as the Pacific Ocean and the Caribbean Sea, encircle the country. Coastal and oceanographic conditions in the Pacific and the Caribbean are very different, and these differences are clearly illustrated in marine fisheries' catch levels. The Costa Rican Pacific coast is 1016 km in length. It includes numerous bays, three large gulfs, a large continental shelf, and a very large exclusive economic zone (EEZ) which, according to the National Geographic Institute data, measures 589,682.99 km^2 (due to the fact that Costa Rica exercises full sovereignty over Coco Island). The Caribbean coast is short and fairly straight, extending for 212 km, and with a very narrow continental shelf. Costa Rica's EEZ in the Caribbean Sea measures about 24,000 km^2.

The Pacific coastal area is characterized by large stretches of mangroves which, over time, have been afforded a certain degree of protection as they are the breeding and larval development sites of many marine and inland water species (FAO 2004).

Because it helps poor fishermen, the small-scale artisanal coastal fishing is particularly important from a social and economic standpoint. The most important component of this fishery is to be found in the Gulf of Nicoya. Sport fishing for rainbow trout, a practice imported from the USA, takes place in upland areas on both sides and at an elevation of 1300 m above sea level (FAO 2004).

In the absence of high sea fishing fleets, fishery productivity in the Caribbean in general is lower than in the Pacific. Fisheries in this region are not very significant in national terms but are very important for Costa Rica's Caribbean region (FAO 2004).

A major feature of the Caribbean coastal region, especially north of Puerto Limón, is the presence of coastal lagoons. Those at higher levels contain freshwater, while those at lower levels are subject to brackish incursions, especially near the mouths of large rivers such as the Matina, Pacuare, Parismina, Tortuguero, and Colorado, where sport fishing is a major activity. The coastal lagoons are interconnected by several man-made channels, which allow flat-bottomed boats to sail from Moín, near Puerto Limón, to Barra del Colorado, the most northerly village on the Caribbean coast. The whole of this area is characterized by high rainfall for most of the year, with short, relatively dry periods. Large areas of the Caribbean coastal zone have been designated as national parks, protecting land as well as marine areas, where small-scale fishing is prohibited or severely restricted. Costa Rica does not have a large-scale commercial fishery in the Caribbean, and, under specific legislation, the first 12 miles from the coast (territorial waters) are reserved solely for small-scale artisanal fishing.

Total capture production of Costa Rica stood at 19,508 tonnes in 2014. According to the FAO (2016), from 1990 to 2014, the highest capture species in Costa Rica were sharks, tuna, shrimps, prawns, clupeoids, swordfish, and shortbill spearfish. The Atlantic coast of Costa Rica has traditionally been far less important for fisheries, with only about 10% of registered fishing vessels (Alpízar 2006) and representing approximately 2–3% of reported landings (Gutierrez-Rodriguez 1990; Guzmán-Mora 2009). Appraisal studies carried out by the FAO in the late 1960s determined that it would be difficult to increase fisheries production in the Caribbean waters, mostly because of low availability of resources due to oceanographic conditions (FAO 1970).

Small pelagic species are fished at an industrial scale and consist of sardines (mainly *Opisthonema medirastre*, *O. bulleri*, and *O. libertate*) and anchovies (*Engraulidae*). Prior to the 1950s, the Pacific anchoveta (*Cetengraulis mysticetus*) was an important baitfish for tuna until its abrupt decline in 1953. Secondary bait species were taken, such as thread herring (*O. libertate*) and smaller anchovies (e.g., *Anchovia macrolepidota*). During this period, US tuna vessels were also collecting baitfish in these regional waters, contributing to the decline of the stocks. Overfishing due to national demand led to a decline in sardine catches in the late 1980s, and moratoriums were put in place to safeguard the endangered stocks, which now make up around 7% of total landings (Vega-Corrales 2010). Catches of large pelagic species have increased during the last decade, now making up around 50% of reported landings. These are dominated by a few families such as Carangidae, including jacks (*Caranx* spp.), moonfish (*Selene* spp.), and amberjacks (*Seriola* spp.); Scombridae, including bonitos (*Sarda* spp.) and skipjack tuna (*Katsuwonus pelamis*); Coryphaenidae (dolphinfish, *Coryphaena hippurus*); and Sphyraenidae (*barracudas*, *Sphyraena* spp.). In addition to commercial fishing, all of these species are actively targeted by recreational fisheries. Pelagic sharks are also an important target group for this sector, contributing around 15% of reported landings. The principal taxa caught are requiem sharks (*Carcharhinidae*), mainly silky shark (*Carcharhinus falciformis*) and hammerheads (*Sphyrnidae*).

Demersal species found in Costa Rica are diverse, as is typically seen in many tropical ecosystems. In a Pacific shrimp trawl survey performed in 1984, 221 fish species were caught as bycatch, only a few of which were of perceived commercial values (Campos 1986). Important taxa for the commercial fishery include drums or croakers (*Sciaenidae*, particularly *Micropogon altipinnis* and *Cynoscion* spp.), snappers (*Lutjanidae*), groupers (*Epinephelus* spp.), and grunts (*Haemulidae*). Elasmobranch catch includes various rays (Rajidae and Torpedinidae, particularly *Raja equatorialis* and *Torpedo tremens*), as well as small demersal sharks (e.g., *Alopias superciliosus*). In the southern Pacific coast, artisanal catches are heavily dominated by snappers (*Lutjanus* spp.) (Guzmán-Mora 2009).

Around 41 shrimp vessels were in operation along the Pacific coast of Costa Rica during the 2000s (Álvarez and Ross 2010). Target species in shallow coastal waters (5–40 m depth) include white shrimp (*Litopenaeus occidentalis*, *L. stylirostris*, *L. vannamei*), conchudo (*Rimapenaeus byrdi*), and tití shrimp (*Xiphopenaeus kroyeri*), and in deeper waters (35–120 m) pink (*Farfantepenaeus brevirostris*) and brown shrimp (*F. californiensis*). Deepwater shrimp fisheries (120–1000 m depth) focus mainly on three species, *Heterocarpus affinis*, *H. vicarius* (camello), and *Solenocera agassizii* (fidel).

Shrimp fisheries in Costa Rica have been characterized by a progressive move to deeper waters as stocks become overexploited and depleted (Álvarez and Ross

2010). Shrimp landings from near-shore waters have significantly declined, such that only tití shrimp are still commercially viable. In the case of deepwater shrimp, landings of around 220 t/year of each of the three species were recorded in the mid-2000s. Since then, *H. affinis* catch has dropped dramatically, such that there are no landings on record since 2006. On the other hand, landings of *H. vicarius* and *S. agassizii* are relatively stable or slightly increasing (Wehrtmann and Nielsen-Muñoz 2009).

Shrimp trawl fisheries have the most bycatch of any of the Costa Rican fisheries sectors (Gutierrez-Rodriguez 1990; Kelleher 2005). Given the fact that bycatch in deepwater shrimp trawlers is entirely discarded, a particularly worrisome statistic is the amount of bycatch relative to shrimp. In 2008, for example, yearly catch for this fishery was of about 5% target shrimps (almost exclusively *S. agassizii*), 55% stomatopods, and 40% fish (Wehrtmann and Nielsen-Muñoz 2009).

Aquaculture started to gain popularity in Costa Rica in 1963 with the introduction of the tilapia species by the Ministry of Agriculture and Husbandry. By 1974, the Costa Rican Department of Aquaculture was established to promote the expansion of aquaculture production and facilitate the construction of aquaculture stations throughout Costa Rica (FAO 2010). From that point forward, aquaculture output grew gradually until the 1990s. New products (like trout and shrimp) gained popularity over time, and the methods used to grow them were standardized. In the early 1990s, aquaculture gained popularity, possibly because of technological advancements like better feed quality (Ramírez 2007). It was necessary to create a new institution to organize the growing practices. The Costa Rican Government created the Instituto Costarricense de Pesca y Acuicultura (INCOPESCA) in 1994 to meet this need. Their mission is to "promote the development of fishing and aquaculture by regulating, protecting, and managing marine resources and aquaculture products," according to a translation of the original Spanish. In order to contribute to the development of the national economy, this regulation encourages sustainable practices for the management and protection of marine and aquatic resources (INCOPESCA 2010). The creation of this institution may have contributed to the unexpected boom in the middle of the 1990s.

The lockdowns brought on by the pandemic had serious socioeconomic repercussions for many Costa Ricans, just like they did for everyone else in the world. Since many in these communities worked in the service industry and lockdowns made many lose their jobs, they were particularly hard-hit. Many of these workers lack the educational background necessary to find new employment and financial security provided by social security. In light of these circumstances, Poseidon-AI® IAS sought to assist Costa Rica's indigenous and vulnerable communities by developing a source of income that required the least amount of technical expertise. The Poseidon-AI® IAS artisanal modules were made to run on 500 L of rainwater, solar power, and water monitoring devices with the least amount of energy possible.

In low-income areas, women and girls frequently work from home and marry at noticeably young ages, which prevents them from completing their primary education. By doing so, they are typically prevented from acquiring the necessary professional experience and knowledge to enter the workforce as skilled workers.

In addition, women and girls in these communities face a variety of difficulties when attempting to meet the needs of their families because of climate change and global warming. Climate change affects the agricultural sector's production yields on the one hand, which leads to instability in the food supply chains and higher market prices; on the other hand, it affects the community's water quality, making it more difficult for women and girls to maintain a healthy and safe lifestyle for themselves and their families.

In these communities, Poseidon-AI uses a multidisciplinary approach to improve a variety of aspects of the lives of women and girls. In addition to producing fish, shrimp, and vegetables using rainwater, Poseidon-AI® IAS also supports the participation of women and girls.

The Poseidon-AI® system uses 500 L of rainwater, so in areas with high water costs and inadequate water piping and storage facilities, women and girls can easily collect the rainwater for use in these systems.

Without extensive training, the women and girls can contribute to the production of enough protein for their families. The system can transmit fundamental information about physiology, biology, feeding, growth rate, and feeding amount of fish and shrimp. Using Poseidon-AI® IoT and algorithms, the system, which annually produces 80 kg of fish and shrimp as well as 20 kg of vegetables, acts as the go-to source of knowledge for women and girls who use the system and ensures the production of wholesome, ample, and nutrient-rich food for families.

Poseidon-AI® not only provides families with wholesome and nutritious food, but also empowers women and girls to support their families financially. Women and girls can make pure financial profit by producing 80 kg of fish and shrimp each year, as well as 20 kg of vegetables, while using solar energy and rainwater at costs that are close to zero. The women and girls can generate income for their families by selling all of the agrifood products they produce with little to no production experience.

The multidisciplinary approach will create the platform for women and girls, to be able to use these modules for producing more fish and vegetables helping their families both financially and producing sufficient food. Additionally, the capacity building made for water management, market development, and seafood processing, and clean energy, will help women and girls to sustainably grow their businesses. As a result, women and girls will be able to have their own source for producing fish fingerlings, vegetable seeds, and knowledge in aquaculture and agriculture, taking advantage of their lands and excess water.

As mentioned, the approach was designed considering the limitations and barriers of all these vulnerable communities, creating a platform with the systems and capacity building for the women and girls to be able to manage their resources, learn to process and sell their products in markets, and generate revenue. Additionally, they can have basic education on fish and shrimp physiology, biology, feeding habits, and growth.

Fig. 6.8 Impacts of Poseidon-AI® IAS modules. The five pillars are nutrition, environment, agriculture, aquaculture, and local market

Figure 6.8 illustrates the Poseidon-AI® IAS module's five distinct effects on communities, particularly on women and girls.

- Nutrition: Poseidon-AI® IAS improves the nutritional intake of women and girls with diversification of their diets and continuous follow-up.
- Environment: Use of solar panels, rainwater, and locally used materials such as clay tiles to produce healthy and nutritious food for women and girls with zero energy and water costs.
- Agriculture: Creating small greenhouse and build capacities for women and girls, giving them the ability to produce plant seedlings for Poseidon-AI® IAS modules.
- Aquaculture: Creating hatchery and nursey along with capacity building for women and girls, giving them the ability to produce fingerlings for Poseidon-AI® IAS modules.
- Local market: Giving new options for women and girls to provide their agricultural and seafood products, both raw and processed, to the local market and gain revenue.

6.2.1 Urban Farming with Poseidon-AI® IAS Modules

Everyone was impacted when the restrictions were put in place and the pandemic began. While there was a significant disruption in food chains globally, millions were forced to stay at home for months. On the other hand, every VC stopped funding any kind of idea. Even though there were and continue to be growing social needs (UNSDGs can be a good reference for addressing these needs), funds from countries were frozen out of fear of a potential global economic crisis.

Under these circumstances, the only solution that comes to mind is to shift from uni-disciplinary approach to multidisciplinary ones. However, no single person, organization, and/or entity has enough capacity for implementing multidisciplinary approaches on its own, and hence, it is required to create alliances that can guarantee the success of such projects.

Additionally, with increasing requests for funds, donors are facing difficult questions of what to fund and what community must be prioritized. Usually, donors try to find communities in remote areas with impassable roads where the impact of pandemic is felt much harder. However, this will shift the attention from vulnerable communities near the big cities and capitals who are hidden under the luminosity of these cities with increasing crimes, inequality, and poverty.

Knowing and understanding these barriers, Poseidon-AI® IAS artisanal modules were first presented in Santa Ana, San José, Costa Rica, for vulnerable communities living close to the capital but suffering as much as the ones living in remote areas from the impact of pandemic. Furthermore, due to rapid urban development in most areas and lack of space for activities such as gardening that connect the community with nature, Poseidon-AI® IAS modules created a platform for families to do their own gardening, listen to the movement of the water, and observe the growth of fish.

The main vision was to achieve multiple goals in line with SDGs while aiming to attract the support of government and international organizations. Below are the goals set with relation to UNSDGs:

The primary goal is to produce sufficient and nutritious food (SDGs 1 and 2) for families, so the module was presented with the angle of food security and safety to Ministerio de Agricultura y Ganadería (MAG) and the Municipalidad de Santa Ana. Municipalidad de Santa Ana has a department called "Seguridad Alimentario" with a group of experts in nutrition and organic agriculture production that agreed to support.

Since the module uses an efficient pump that requires electricity, the system is designed in a way that can use solar energy. This allows families to produce their own food without needing to pay for any energy cost (SDGs 7 and 13).

One of the main goals of implementing the Poseidon-AI approach is to empower women, enable families to have stable incomes, while contributing to the economic growth of Costa Rica (SDGs 5, 8, and 10).

One of the great characteristics of Costa Rica is tonnes of rainwaters that every year pour down to its fertile soils. Taking advantage of this fact, the modules use rainwater to create a suitable environment for fish and shrimps without adding any cost for families (SDGs 6 and 14). Also, the system uses clay tiles instead of soil,

Fig. 6.9 Map of Santa Ana's districts

making it convenient to transport and lighter to set up the system on rooftops, small areas in the houses, preventing soil and water contamination (SDGs 9, 12, and 15).

Finally, these systems contribute significantly to SDG 11, which is sustainable cities and communities.

Santa Ana, the ninth canton in the province of San José, is 17 km^2 from the national capital and is known to its residents as the "Valley of the Sun," the birthplace of the onion and the pottery. With a total land area of 61.4 km^2, it is split into six districts: Centro, Salitral, Pozos, Uruca, Piedades, and Brazil (Fig. 6.9). According to data from INEC, Santa Ana's population was expected to reach 59,701 people in 2019. Pozos, which has a total population of 19,795 people, is the district with the highest density, while Brazil, which is the district that is located the furthest from the city center and has the lowest density, has a total population of 3099 people. The reality of food and aquaponic production in Santa Ana is not dissimilar from the reality of the nation, which has low production volumes, few areas overall, and high-impact degradative effects more closely linked to pond culture aquaculture.

Regarding the climate, it is described as of the dry humid tropical type in the canton of Santa Ana. Because of the high level of cloud cover present even during the dry season, the southwest region experiences more precipitation than the northern region. The canton is drained by the Uruca River and its tributaries. These include the Oro River, the Navajas, Pilas, Canca, La Cruz, San Marcos, and Muerte ravines, as well as the Corrogres River, which is linked by the Lajas and Rodriguez streams. The Rio Grande de Tárcoles basin's Pacific slope and the canton of Santa Ana's fluvial system are both part of this basin. The majority of these waterbodies are severely polluted and cannot be used for aquaponics, aquaculture, or providing drinking water to people in Santa Ana due to rapid urban development and excessive fertilization used for onion culture.

Around 100 in. (2.54 m) of rain falls on average each year in Costa Rica. Santa Ana experiences between 2 and 3 m of precipitation on a yearly average. Despite having a respectable amount of annual rainfall, about 350,000 people in Costa Rica still struggle due to a lack of access to clean drinking water, according to UNDP Costa Rica. With the effects of climate change, this situation has been getting worse.

To encourage the growth and enhancement of the residents' quality of life, the Association "Desarrollo Humano Manos Amigas la Promesa" was founded in April 2008. Most of the people living in La Promesa (the Promise) are vulnerable and work in the service sector. The Cantonal Human Development Atlas 2021 of Costa Rica reports that La Promesa, a low-income neighborhood hidden away to the north of Highway 27, was awarded the highest cantonal Human Development Index (HDI) in 2019 for this generally prosperous municipality. The promise is made up of 90 families from various age groups in the district of Brazil where the Association's HQ is located.

In this area, deployment of Poseidon-AI® IAS artisanal modules aimed to address various climatic and social aspects. In relation to climatic aspects, one of the most important is CO_2 emission. The average amount of CO_2 emission intensity (ECO_2) for beef is 67.8 kg ECO_2/kg product, while Poseidon-AI approach can help in significantly reducing CO_2 emissions.

The approach aimed to shift community consumption behavior from beef and pork to consumption of fish and shrimp with a lower carbon footprint. According to the FAO, the average emission associated with tilapia production was 1.58 kg of ECO_2 per kg of live fish. This amount increases to 1.81 kg of ECO_2 per kg of live fish when including aquatic feed footprint.

Since the Poseidon-AI system uses an energy-efficient solar-powered pump, the carbon footprint of the system related to energy consumption is exceptionally low (approximately 0.11 kg of ECO_2 per kg of live fish).

However, the system is not directly involved in the aquafeed production, making it difficult to estimate the amount of CO_2 emission for this system and related to aquafeed production. The estimated carbon emission initiated for tilapia production from this system is approximately 0.34 kg of ECO_2 per kg of live fish, which is 1.47 kg of ECO_2 per kg of live fish lower than the FAO estimate for tilapia production. On the other hand, the average emission associated with lettuce for primary production was 0.18 kg of ECO_2 per kg of lettuce, 0.33 kg of ECO_2 per kg of lettuce for wholesalers, and 0.43 kg of ECO_2 per kg of lettuce for retailers.

The Poseidon-AI® modules are installed in all homes and do not need any type of fertilizer; thus, families can have their own source of production without having to visit supermarkets, wholesalers, and/or retailers that place them in the primary production stage, reducing approximately 0.25 kg of ECO_2 per kg of carbon footprint lettuce.

Therefore, the total amount of carbon emission for the system is estimated to be 0.52 kg of ECO_2 per kg of live weight of fish and lettuce, which is 67.28 kg of ECO_2 per kg of product less than the production of meat.

Additionally, Poseidon-AI® IAS modules take advantage of rainwater by collecting and using 500 L of rainwater to create a comfortable environment for 20 tilapia and 10 shrimp, producing approximately 20 kg of seafood every 4 months.

The installation of Poseidon-AI® IAS modules faced problems in La Promesa starting with stabilizing and firming the ground where the modules were going to be

Fig. 6.10 Community living in La Promesa helping in making the area ready for the installation of Poseidon-AI® IAS artisanal module

installed. For this reason and before installation, the leveling and firming the land took place in the common area (Fig. 6.10). This location was chosen based on several criteria, including security, harsh sunlight, and shade. The community came together to remove all the stones, including the large stone shown in Fig. 6.10, and assisted in leveling the ground by putting cement in the designated areas. Since every resident of La Promesa has access to the common area and it is challenging to identify the factors causing the modules to become unstable, security concerns were another problem encountered when installing such systems. As a result, security cameras and fences were installed around the first four modules inside the shared space. The cameras were primarily used to keep an eye on the entry of animals like cats and monkeys.

As mentioned, the waterbodies are polluted, and so, Poseidon-AI® IAS modules could not be filled with water from the nearby waterbodies. On the other hand, the drinking water in Santa Ana is filled with chlorine, which is toxic for fish, and so, for using it, the water needed to be set aside for 24–48 h. Before installation, the water characteristics of La Promesa were tested to ensure the usage of best available water in this area (Fig. 6.11). The average water temperature was 26.05 °C, while the average pH was 6.71. The average water TDS level was 131 mg/L and the average conductivity 260.28 μS/cm. As it is shown, the conductivity level is three times higher than the amount suitable for freshwater species and TDS level is close to showing hardness in the water.

Municipality provided the plant seedlings and fingerlings to the community in La Promesa. This is because there are very few licensed hatcheries in Costa Rica which can provide and guarantee disease-free and healthy fingerlings; hence, Santa Ana City Hall accepted to seek disease-free and healthy fingerling and provide them to the community. Additionally, plant seedlings were also purchased and provided by the City Hall of Santa Ana from certified greenhouses shown in Fig. 6.12. The

Fig. 6.11 Manual water monitoring before the implementation of Poseidon-AI® IAS modules in La Promesa

Fig. 6.12 Certified greenhouses providing healthy seedlings to the community

seedlings were various types of lettuce, tomato, green onions, parsley, cucumber, eggplants, and pumpkins.

A solar panel can provide power for the highly effective water pump used by the Poseidon-AI® IAS module. The modules at La Promesa were joined to a single, sizable 450 W solar panel and a 150 W, 12 V battery. The structure was constructed with the community assistance, as was the installation of the solar panel in the common area. Figure 6.13 depicts the construction of the building and the installation of the first solar panel at La Promesa.

Poseidon-AI® IAS modules with solar panels were also delivered to other homes, but each system only had one 250 W solar panel and one 100 W, 12 V battery.

Similar to how water near urban areas carries contaminants from various activities, these soils cannot be used to fill modules. In addition to the water circulation, the main concern was the possibility of heavy metals and other toxic contaminants in the soil. Because of this, the Poseidon-AI® IAS module uses clay

Fig. 6.13 Setting up the first solar panel in La Promesa

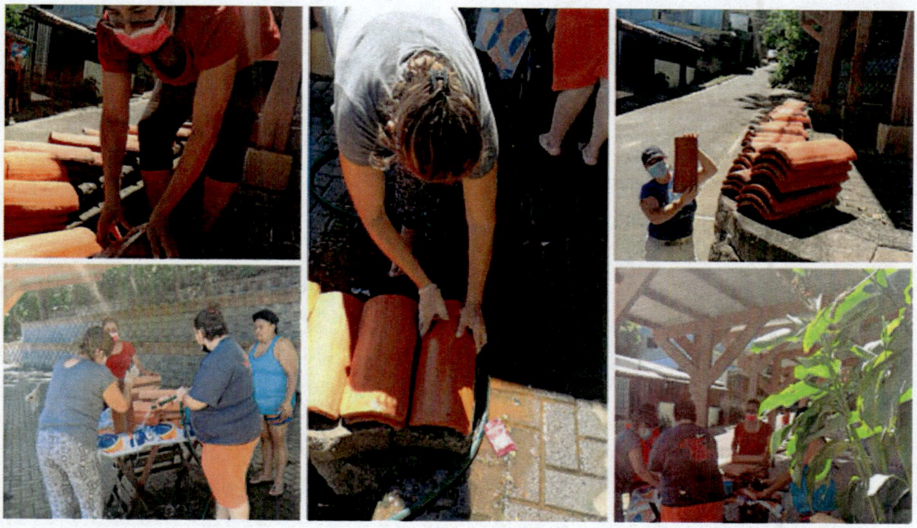

Fig. 6.14 Community in La Promesa working together in washing and cleaning the clay tiles

tiles rather than any kind of soil. Fine soil with hydrous aluminum phyllosilicates is called clay.

Historically, Santa Ana is known for its unique pottery methods using clay. The habitants in this part of the country used to collect fine soil from the rivers and shape it with the firing method. Due to hardness and not absorbing water, most of the rooftops in Costa Rica use clay tiles. However, after 20 years, the clay tiles will be changed and substituted with the new ones.

These tiles are valuable products that can be used in aquaponic systems. Additionally, due to their sizes, it is easier to carry and transport to different areas compared to stones, volcanic rocks, and clay balls. The used tiles need to be cleaned and washed before the installation, and the community was involved in every step from washing to setting up the media bed with these clay tiles. Figure 6.14 shows the

Fig. 6.15 Community in La Promesa learning and installing different aspects of Poseidon-AI® IAS module

community working together in washing and cleaning the clay tiles before installation of the Poseidon-AI® IAS modules.

For every Poseidon-AI® artisanal IAS module, there are 10 broken and 50 whole clay tiles. These clay tiles are arranged in layers on top of one another, with the top layer facing upward to create a platform for the plants, much like clay pipes. The quantity of water flowing through the system determines how many of these clay tiles are needed to prevent the plants from drowning. The community received training in the theory and practice of installing these tiles, the science of water circulation, and the development of microorganisms (Fig. 6.15).

The system holds 50 pots for 50 plants, which are set in rows above each clay tile. The pots have small holes in the bottom and are filled with broken tiles. The broken clay tiles create a suitable environment for the microorganisms, which are crucial for healthy growth of the plants. The plants are monitored and measured before transplanting into the modules to prevent external and possible diseases from being introduced into the systems. Figure 6.16 shows the transplanting process carried out by members of the community in La Promesa.

Two hours from Santa Ana, in a place called San Carlos, are where the fingerlings are bought. This is because there are not many accredited hatcheries in the entire country of Costa Rica, making it challenging to find fingerlings that are disease-free and healthy. San Carlos is a city surrounded by rivers and mountains, which makes it the perfect location for tilapia farming, especially in the spring and summer. But not all of these hatcheries are licensed and have the right tools for tilapia farming. The use of river waters for shrimp and fish farming and their subsequent recirculation into the rivers are the main issue in this region. This raises the possibility that various diseases will spread to all of the aquaculture farms in San Carlos. To avoid the spread of a potential fungal infection, the fish are transported from San Carlos to Santa Ana

Fig. 6.16 Plants are measured before providing to the community and being transplanted into Poseidon-AI® IAS modules

Fig. 6.17 Monitoring and measurement of the species before introducing into Poseidon-AI® IAS modules

in water containing methylene blue solution. Furthermore, all fish are watched over, weighed, and examined for any parasites, fungi, or bacteria after the Santa Ana City Hall purchases the species. A maximum of 30 tilapias or 15 freshwater shrimp can be kept in each Poseidon-AI® artisanal IAS module (each tilapia requires about 20 L of rainwater). This procedure for these modules in La Promesa is shown in Fig. 6.17.

It is advised to only use male tilapia in these modules because they grow bigger and faster than females. Each fingerling weighs 25 g on average and measures about 10 cm in length (Fig. 6.17).

For the duration of each module and for a total of 4 months, commercial pellets are offered to ensure that the species grows effectively. The calculation of feeds is based on the species' appetite in relation to the environmental condition because the water temperature drops below 25 °C at night and during the rainy season.

6.2.1.1 Socioeconomic Aspects of Poseidon-AI® IAS Modules in La Promesa

When the pandemic started, most of the families in La Promesa lost their jobs, and without any life savings, life was showing its harder side. Poseidon-AI presented the modules to the City Hall of Santa Ana to create a new source of income for these families while helping in sustainable development of the country.

It was clearly explained to the officials in the City Hall of Santa Ana that Poseidon-AI is a private company and cannot act like a nongovernmental organization (NGO); hence, the technology can only be provided if enough funding be provided for purchasing Poseidon-AI® artisanal IAS modules.

Like other local government entities, the City Hall of Santa Ana did not have enough funding, and so, it was decided to apply for funding from international organizations working in Costa Rica. The goal was to empower the communities into producing their own seafood and vegetables instead of giving government-bought food packages.

The United Nations Development Programmes (UNDP) in partnership with Global Environment Facility (GEF) in Costa Rica agreed to fund providing these modules for this community in La Promesa with the support of the City Hall of Santa Ana in providing them with plant seedlings, fingerlings, and capacity buildings in themes such as agrifood, nutrition, added value, supply chain, and marketing.[1]

Poseidon-AI® IAS artisanal modules in La Promesa helped families to produce 3 tonnes of fish and shrimps and nearly 1 tonne of vegetables during the 24-month period. Additionally, 20 tonnes of rainwater were saved and used within these systems. Approximately 200 tonnes of CO_2 were prevented from entering the atmosphere thanks to the solar panels and efficient production systems. Finally, 30 women were directly involved in sustainable production, marketing, and selling process of seafood and vegetables (Fig. 6.18).

Every Sunday, there is a farmers' market held in the center of Santa Ana. Farmers bring their fresh products such as fruits, vegetables, eggs, and dairy products to sell to the public. According to personal experience, the farmers charge a dollar (~500 colones) for a kilo of lettuce, 3 USD for a kilo of tomato, 2 USD for a kilogram of cucumber, and a dollar for a basil roll. Additionally, a kilogram of freshwater shrimp costs 50 USD, and a kilogram of tilapia costs 20 USD. As mentioned, the Poseidon-AI® IAS artisanal module can produce 20 kg of vegetables and 80 kg of shrimp and

[1] https://pnudcr.exposure.co/manna-from-heaven.

Fig. 6.18 Seafood and vegetables produced by Poseidon-AI® artisanal IAS module

tilapia. So, the community in La Promesa can generate 50,000 USD in revenues within the 24 months of implementation.

6.2.1.2 Environmental Aspects of Poseidon-AI® IAS Modules in La Promesa

With cultural differences, language barriers, rapid environmental changes, and now global pandemic, the solution is to implement multidisciplinary approaches that can tackle multiple needs at once. As shown, Poseidon-AI approach used the capacity of international and national organizations not only to improve the living condition of a community in Costa Rica but also to take adaptation and mitigation measures toward these rapid environmental changes.

Poseidon-AI® IAS artisanal module uses 500 L of rainwater. The community collects rainwater with tanks and stores it for later usage in the module. As mentioned, 200 L of water is circulated every hour, and this creates higher evaporation rate in the system. For this reason, every 2–3 weeks, there is a necessity to add 100 L of water to each model. Without adding enough water to the system, the fish welfare will be compromised, and thus, it was recommended to add 100 L every 2–3 weeks.

Usually in Poseidon-AI® IAS modules, the fish welfare is managed by Poseidon-AI® algorithms located in the AWS. However, due to the geographical location of community, 4G connectivity is not available. Furthermore, many families in this community do not have access to Wi-Fi internet. Hence, fish species' living conditions, growth, feeding, and maturity time were managed by each owner of the Poseidon-AI® IAS module. For this reason, more than 15 h of capacity-building classes (in Spanish) were conducted by Poseidon-AI water and aquaculture experts (Fig. 6.19).

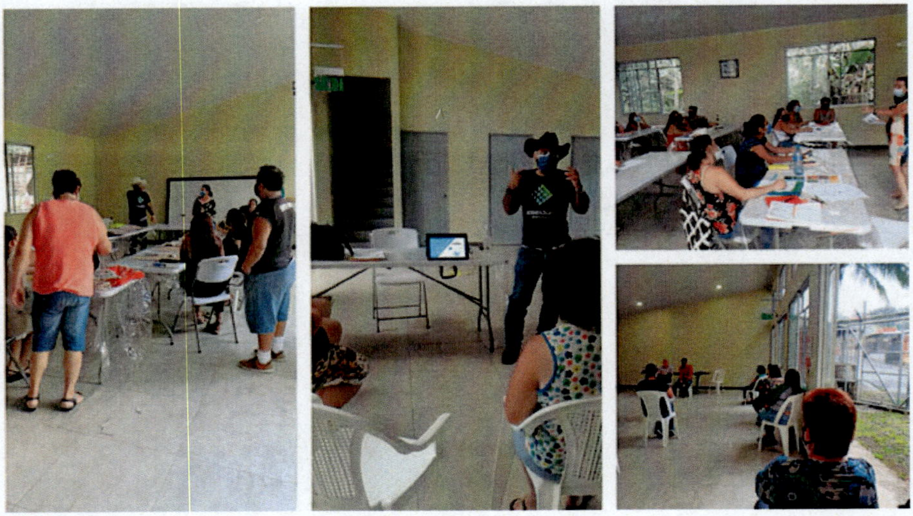

Fig. 6.19 Capacity-building classes conducted by Poseidon-AI's experts

As mentioned, the community faced financial difficulties, and so, shifting energy costs, no matter how small, to the community was not feasible. For this reason, solar panels were installed in every house with Poseidon-AI® IAS module. The amount of energy consumption was calculated for every module, and so, during the rainy season, each module can work for 8 days without any sunlight. Removing the unnecessary transportation and use of solar panels as well as lack of fertilizer usage, in 24 months, the Poseidon-AI approach saved up to 200 tonnes of CO_2 emissions from entering the atmosphere. This is a great example of "thinking global and acting local."

6.2.1.3 Political Aspects of Poseidon-AI® IAS Modules in La Promesa

As mentioned, due to the pandemic, the community in La Promesa was facing economic difficulties and Poseidon-AI approaches were applied to overcome these challenges. During this period, Costa Rican presidential election was going on, making it challenging to implement these approaches neutrally. For this reason, Poseidon-AI signed an agreement with the community to not use Poseidon-AI and its approaches in favor of any political party. Poseidon-AI's Intellectual Property (IP) was submitted in Costa Rica to prevent any complication during and after the implementation of Poseidon-AI's approach.

The presidential election was between Mr. José María Figueres from the National Liberation Party and Mr. Rodrigo Chaves Robles from the Social Democratic Progress Party. National Liberation Party has many years of presence in different layers of Costa Rican politics, while Social Democratic Progress Party is relatively new. Hence, it is only fair to not involve public beneficial works in favor of any political party, especially during the sensitive times such as the presidential election.

Fig. 6.20 Healthy and organic final products produced by Poseidon-AI® IAS modules in La Promesa

However, due to the longer history of National Liberation Party, most of the high-ranking officials in the local governments such as mayors are from this party. During the period of implementation of Poseidon-AI approach and seeing the positive outcomes both for the community and the city (Fig. 6.20), the City Hall of Santa Ana decided to involve National Liberation Party's officials in the implementation phase and showcase the outcome as an achievement of National Liberation Party.

The mayor, running vice president, delegates running for Costa Rican congress, and youth leaders, all from National Liberation Party, visited and presented the achievements in favor of their political party, creating serious complications in the path of implementation phase (Fig. 6.21).

Poseidon-AI's lawyers officially requested the National Liberation Party to stop involving Poseidon-AI® IAS modules and approaches into their political campaign and requested removal of all photos, videos, and social media posts. Additionally, they informed the UNDP officials, reminding the neutral position of Poseidon-AI in any political procedure within Costa Rica and around the world. Unfortunately, UNDP officials did not take this reminder in a positive manner and responded very harshly. This created the question about the neutrality of the international organizations in Costa Rica and forced Poseidon-AI to stop the implementation process until after the election. In a close race, Mr. Rodrigo Chaves Robles from the Social Democratic Progress Party won the presidential election, and a new chapter in Costa Rican political scenery started. However, Poseidon-AI kept its neutrality and continued in implementing its approach for many other communities in Costa Rica.

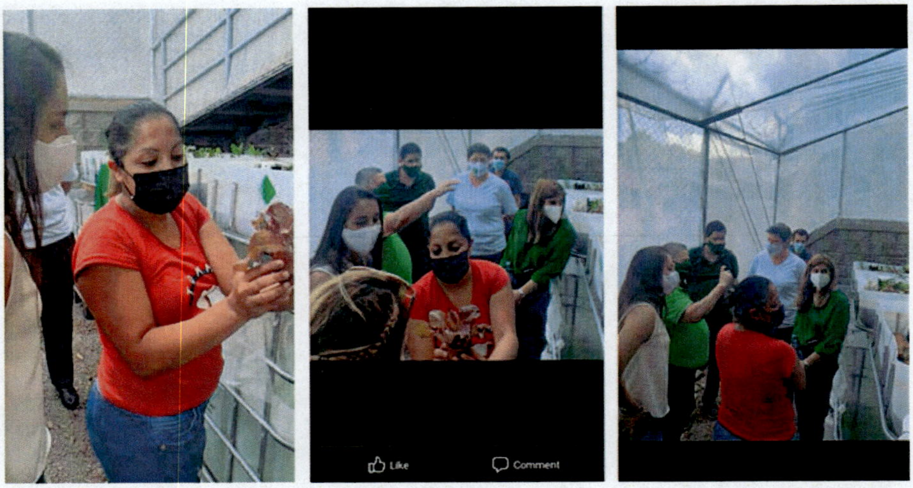

Fig. 6.21 Officials from National Liberation Party unlawfully involving Poseidon-AI approaches in their campaigns

6.2.2 Poseidon-AI® IAS for Indigenous Community in Costa Rica

Costa Rica is an example as a leader in the recognition of indigenous people's rights as well as the global environmental sustainability. According to Article 3 of the United Nations Declaration on the Rights of Indigenous Peoples (UNDRIP): "By virtue of that right they freely determine their political status and freely pursue their economic, social and cultural development," outlining the self-determination right (UNDRIP 2008). However, this convention is not legally binding, but Costa Rica is legally required to act in other Indigenous and Tribal Peoples Convention signed with the International Labour Organization (ILO)[2] (1989).

Working with indigenous community requires mutuality, reciprocity, and deep connection (Kovach 2009; Tuhiwai Smith 2012). In other words, developing personal relationships with indigenous communities plays a fundamental role in implementing any sort of approach.

Costa Rica has legally established its indigenous territories in 1977 by the Indigenous Law.[3] 3344 km^2 for 24 indigenous territories with 8 different indigenous groups were outlined by the secretariat of the convention on biological diversity.[4] The largest group are Bribri and Cabécar (Molina Murillo et al. 2014).

Indigenous communities in Costa Rica gained the voting right in 1994 and received national IDs in 1992.[5] However, the failure of Costa Rica to enforce

[2] http://www.ilo.org/dyn/normlex/en/f?p=NORMLEXPUB:12100:0::NO::P12100_ILO_CODE:C169.

[3] http://www.wipo.int/edocs/lexdocs/laws/es/cr/cr057es.pdf.

[4] https://www.forestpeoples.org/sites/fpp/files/publication/2015/07/cerd-report-finaleng.pdf.

[5] http://www.tse.go.cr/pdf/normativa/leyinscripcionycedulacionindigena.pdf.

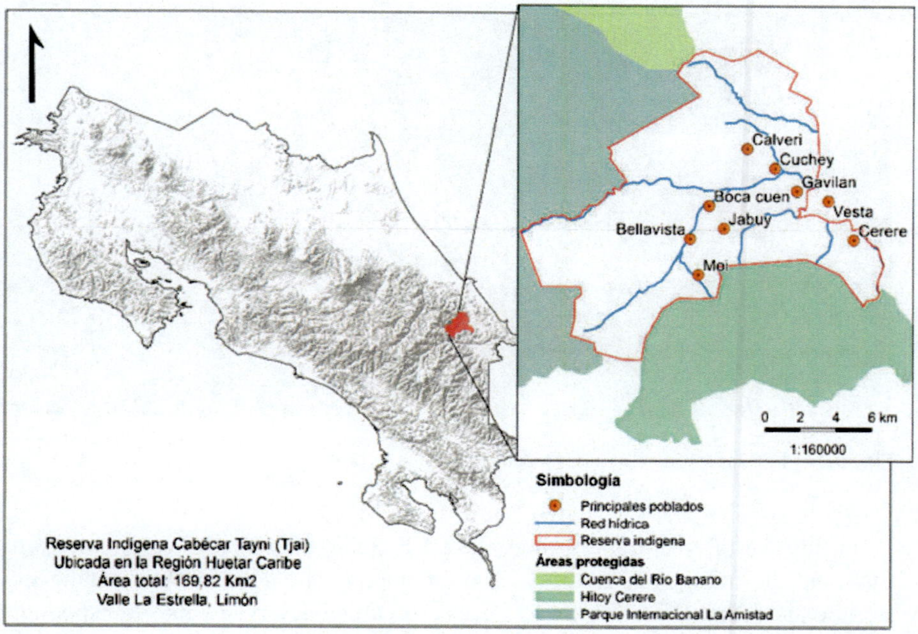

Fig. 6.22 Cabécar Tjai indigenous territory, Limón, Costa Rica

laws to protect indigenous people forced UN Secretary General Mr. Ban Ki-Moon[6] and Inter-American Court of Human Rights (IACHR) to intervene in 2015 (Arguedas Ortiz 2015).

Poseidon-AI® IAS artisanal module helps the indigenous communities, especially women and girls, in Costa Rica to address food security concerns as well as issues caused by poverty in these territories. The modules provide women and girls not only a source of income but also a stable source of nutrition. Furthermore, it helps young teenage mothers, single mothers, and girls in indigenous communities to be familiarized with environmental engineering, fish, and plant physiology, as well as creates decent work with a stable income.

Depending on the territory, most of the communities do not have any source of energy, and thus, solar energy used in Poseidon-AI® IAS artisanal module can contribute significantly to the expansion of clean and affordable energy, adapting to climatic conditions.

6.2.2.1 Poseidon-AI® IAS Modules for Indigenous Community Tayní

Poseidon-AI® IAS modules were presented to the province of Limón, specifically in the indigenous territory called Cabécar de Tayní (Xiphopenaeus kroyeri No. 5904-G 1976), located in the Valle La Estrella district of the Limón (Fig. 6.22).

[6] https://www.un.org/sg/en/content/sg/statement/2014-09-22/secretarygenerals-remarks-opening-world-conference-indigenous.

Fig. 6.23 Entrance to the Tayní-Cabécar indigenous territory

In the Tjai (Tayní)-Cabécar in Valle La Estrella, there are 19 communities that make up the indigenous territory. The entrance to the Tayní-Cabécar indigenous territory is depicted in Fig. 6.23. Access to essential services like transportation, healthcare, and education has proven difficult for the locals. People only consume food produced and raised within the territory, so agrifood consumption greatly depends on the cultured goods produced within these territories.

The majority of the producers in this region are micro-producers, and 32.7% of those producers are women. These women face significant obstacles when trying to implement sustainable businesses that make the best use of the resources (land, water, and climate) that are available on their farms and protect the food security of these communities.

The main sources of protein in the staple diets of indigenous families are sausages, canned tuna and sardine fish, and chicken and pork meats (Fig. 6.24). Due to changes in the traditional food structure and restrictions on the variety, accessibility, and availability of food sources, these regions currently struggle with malnutrition, overweight, and obesity. These areas have less access to vegetables and are becoming less conducive to activities like fishing because there are no fish in the rivers.

For the women and girls who make up the Alakalawa ishaka tami group, Poseidon-AI® IAS artisanal modules were installed. Owners of the land in Cabécar, also known as Alakalawa ishaka tami, are a group of women who work in food production using sustainable agroecological systems that, above all else, ensure the safety of their families' food supply. Additionally, by incorporating the long-preserved songs, tales, customs, traditions, and cuisine of the indigenous people, these women create spaces where they can advocate for and save their culture.

This group's mission is to pass on their ancestral knowledge and environmentally friendly way of life to the next generation so that they can appreciate the importance of women in maintaining culture and providing for the nutritional needs of their

Fig. 6.24 Examples of the types of foods grown and consumed in the indigenous territory of Tayní-Cabécar

families. Women are in charge of clearing the land, planting, and harvesting food, as was already mentioned.

In addition, this group challenges stereotypes by speaking out as legitimate landowners in a situation where there is inequality between men and women, primarily as a result of machismo. Although there are many differences between men and women in the region, women have been able to advance and realize their dreams.

Additionally, since women are the heads of households, they are required to produce food sustainably, cut back on waste, and make the most of their resources in order to protect the health of their families. This is consistent with their worldview and approach to caring for the environment.

6.2.2.2 The Environmental Condition in Tayní-Cabécar

Each household's environmental condition was assessed and recorded prior to the installation of the Poseidon-AI® IAS artisanal modules. Since there is no potable water available in the community, the majority of people use water from the nearby river for drinking and personal hygiene. The locals also wash their cars, mostly motorbikes, and occasionally swim in the river during the warm summer months.

Banana and plantain farms, which use industrial-grade pesticides and fertilizers, are located all around the native territory. The majority of these extra pesticides and fertilizers end up in the same river that flows through this region. It is crucial to avoid using water from the nearby river because of this and instead focus on gathering rainwater.

Fig. 6.25 Monitoring the environmental variables in four different locations in the indigenous territory

The region also lacks an energy source that would allow families to have electricity in their homes. As a result, the solar panels in Poseidon-AI® IAS modules can be used to supply these families with electricity.

For the purpose of choosing locations to install the solar panels and testing the water quality, four areas in the area were chosen (Fig. 6.25). In these four locations, the daytime average water temperature was 30.5 °C, and the average pH was 7.05. The average TDS and EC were 131.75 mg/L and 268.75 μS/cm, respectively, because there are no nearby water treatment facilities. During the rainy season, the TDS of the water is higher, and the water turns milky white in color. In addition, the pH shifts in the rainy season due to increased soil contamination in the water and faster water flow.

Since many of the homes in the region are constructed of wood, palm, and banana leaves, solar panel installation cannot be done on their roofs. Additionally, the wires were unable to be placed close to the ground or high in the trees due to the presence of farm animals. For this reason, structures had to be constructed before solar panels could be installed.

As previously mentioned, there are no nearby accredited and/or regulated hatcheries that could supply fingerlings to the native territory. To use the species in Poseidon-AI® IAS modules, it was necessary to transport them from San Carlos. Additionally, there are no greenhouses for producing plant seedlings, and hence, the plants needed to be transported. It takes 4 h from San Carlos to the indigenous territory and nearly 7 h from San Jose; hence, the fish species need to be transported with enough oxygen and low stress conditions.

460 fingerlings were transported from San Carlos to the indigenous territory for growing in Poseidon-AI® IAS modules as well as distributing among the community

Fig. 6.26 Transportation and distribution of 460 fingerlings for the indigenous community in Tayní-Cabécar

Fig. 6.27 Plants transported from San Jose to the indigenous community in Tayní-Cabécar

members (Fig. 6.26). Previously, the FAO has built a cement tank in various areas in this indigenous territory, but due to lack of good-quality water, most of these tanks have high mortality rates. For this reason, it was recommended that before taking these fingerlings, sufficient amount of rainwater be collected in these tanks.

Nearly 1000 plant seedlings were delivered to the indigenous territory to be transplanted into Poseidon-AI® IAS artisanal modules (Fig. 6.27). The plants included basil, tomato, lettuce, and parsley. Some of these plants such as tomato

Fig. 6.28 Transport of clay tiles to the indigenous territory

were first time introduced for growing in the territory due to lack of knowledge among the indigenous community.

Nearly 400 clay (Fig. 6.28) tiles were transported to the community for usage in the media beds of Poseidon-AI® IAS modules. Additionally, 400 W solar panels with 100 W and 12 V batteries were transported to the indigenous territory, to bring electricity for this community as well as provide energy for the pumps used in Poseidon-AI® IAS modules.

6.2.2.3 Production of Seafood and Vegetables in the Indigenous Territory

The houses in the indigenous territory are not close to one another, and so, Poseidon-AI® modules needed to be carried to each house. Indigenous community helped in carrying the tanks, solar panels, batteries, plants, and fish species.

As mentioned, it was first time that some of the plant species were introduced for culturing, and so, some of the women planted them inside a pot filled with soils gathered from nearby area. When it was time to transplant the plants into the system, the contaminated soil entered the circulated water in the system, and so, gathered

Fig. 6.29 Installation of solar panels in the indigenous territory in Tayní-Cabécar

rainwater in some of the modules needed to be changed. Poseidon-AI® IAS modules use clay tiles to prevent outside pollution and contaminant from entering the system.

The solar panels were set up in an open space near the homes, but far from the trees (Fig. 6.29). The PVC pipes, which were positioned inside the holes dug by the indigenous community, were used to transport the wires. As seen in Fig. 6.29, the solar panels power the neighborhood and, for the first time, enable residents to have lights in their homes.

Before transferring the fish and plant species to all the Poseidon-AI® IAS modules, environmental variables were tested and recorded. The average recorded temperature was 28.9 °C with an average pH of 7.7. EC and TDS were, respectively, 300.4 μS/cm and 146.4 mg/L.

The territory does not have access to the internet and has zero mobile connectivity. Community members need to walk down from the territory to make calls or connect to the internet. For this reason, Poseidon-AI® algorithms and IoT device could not be used in Poseidon-AI® IAS modules; hence, 8 h of capacity-building workshops were conducted for the women and girls in this community (Fig. 6.30).

The feeding amount under environmental condition in the territory was estimated and presented to the households. The community uses the final products for their family's consumption and to answer to their nutritional needs. Figure 6.31 shows the first tomato produced by Poseidon-AI® IAS module in the indigenous territory. It is an important milestone since these tomatoes were produced with clay tiles and by indigenous women in their first attempt.

The first cycle, which was for the first 6 months, was faced by many complexities. This was due to communication issues raised during the capacity-building procedure. Most of the women and girls in indigenous territory do not speak Spanish, and thus, capacity building in their local language requires a translator. Unfortunately,

Fig. 6.30 Capacity-building workshops conducted in indigenous community

Fig. 6.31 First tomato produced by Poseidon-AI® IAS module in the indigenous territory

there are very few translators in the whole indigenous territory, and with various entities going in and out, these translators are occupied whole year around. Under these circumstances, capacity-building workshops were short but effective proved by the outcome of each Poseidon-AI® IAS module.

The community has lights in their houses thanks to the solar panels installed in the community. The community in the first year produced 4 tonnes of fish and 100 kg of vegetables. Additionally, more than 20 tonnes of rainwater were gathered and used in the modules. As a result, more than 15 women and girls were trained and

Fig. 6.32 Map of indigenous territory of Guaymí in Coto Brus

familiarized with fish growth, feeding habits, soilless culture, and different water management themes.[7]

6.2.2.4 Poseidon-AI® IAS Modules for Indigenous Community of Guaymí

The Guaymí indigenous territory, in Coto Brus, is made up of ten communities. The area includes the communities of Brusmalis and La Casona, in the Cabécar Guaymí indigenous territory (Fig. 6.32). In the territories, in general, the inhabitants have problems accessing basic services, such as transportation, health, education, and commerce, among others. Also, people only eat food grown and raised in their homes. Agriculture depends a lot on the diets kept by farmers.

The economy in these areas is made up of micro-producers; some coffee and bananas are produced; however, they find many limitations to generate income and, above all, feed their families; hence, they mostly produce food for self-consumption. The protein contribution to the basic diet of indigenous families comes mainly from the consumption of chicken meat, pork meat, canned food (tuna/sardine), and sausages.

La Casona, an indigenous community in the South of Costa Rica, is situated 40 min from San Vito, a small town. The majority of the community's economic pursuits are related to handicrafts, cultural activities, tourism, service sector, and coffee, cacao, and banana farms. As to be expected, the pandemic made a large number of community members unemployed, and within 2 years, poverty, hunger, and malnutrition spread rapidly. Changes in the traditional food structure and restrictions on the variety, accessibility, and availability of foods are to blame for the malnutrition, overweight, and obesity issues that are prevalent in these regions. In these areas, it is getting harder to engage in activities like fishing because there are no fish in the rivers and few vegetables are available.

[7] https://delfino.cr/2021/10/iniciativa-permite-a-territorios-indigenas-producir-acuicultura-organica-a-partir-de-agua-de-lluvia.

The Poseidon-AI® IAS module uses a new technique to tackle the food security problem with sustainable fish and vegetable production in the community focused on children and youth. Furthermore, these modules help students, teachers, and parents in the community to become familiar with environmental engineering, fish, and plant physiology, as well as create a decent job with a stable income. Furthermore, use rainwater and clay tiles to replace the land for growing vegetables. The innovation is in maintaining a sufficient water level for the cultivation of both plants and fish, and the efficient use of fish excretion as fertilizer for vegetables without the need for chemicals or fertilizers. Finally, the use pf Poseidon-AI® IAS module will strengthen the community access to foods with high nutritional value at a low cost, thus allowing a variety of nutritious food to be incorporated into the community's food patterns while promoting a circular economy and sustainable food microsystems.

The main beneficiaries using Poseidon-AI® IAS modules are 670 children and young students from 3 education centers, namely La Casona High School, Brüs Malis primary school, and school of Ngöbegüe in this indigenous territory. The module works as a tool to support the availability of high-protein food sources to meet the nutritional needs of indigenous youth while involving parents and teachers from different educational centers.

6.2.2.5 Environmental Condition in the Indigenous Community of La Casona

The indigenous community is tucked away in a mountainous region with a variety of microclimates. The community's water supply comes from a nearby river or rainwater collection systems. Since there is no tap water, there is no chlorine in the water; however, rainwater is collected from rooftops, and this can occasionally contaminate the water. With an average pH of 7 and TDS levels under 100 mg/L, the water quality in the native territory is generally good (Fig. 6.33).

The soil condition in the territory is similar to other areas in Costa Rica with human activities and farming having contaminated it. To reduce the risk of introducing contaminants into the Poseidon-AI® IAS modules, 1600 clay tiles were transported to the community (Fig. 6.34).

San Vito, a small town 40 min away from where the community is located, provided many materials necessary for Poseidon-AI® IAS modules such as the plant seedlings. This allowed providing various plants for the community as well as for the modules. More than 1300 plants were transplanted into Poseidon-AI® IAS modules (Fig. 6.35).

In this territory, there is a single aquaculture farm for culturing tilapia. They use the water from nearby rivers into the earth ponds without any water filtration and/or water treatment facilities. Most of the water from these ponds is returned into the river, creating high risk of the spread of disease, water pollution, and endangering wild species.

As mentioned, Poseidon-AI® IAS modules use male species which can grow faster and larger in size. In the indigenous territory, the male and female tilapia are not separated, and so, manually the healthy male species were selected from the mentioned aquaculture farm. Figure 6.36 shows the process of selection and

Fig. 6.33 Microclimates in La Casona and rainwater collection tanks

Fig. 6.34 Transportation of clay tiles into the indigenous territory of La Casona

monitoring of male tilapia in the aquaculture farm in La Casona. Each male tilapia was weighted and checked for fungi or parasites. The average weight of these species were 150 g with the length of 20 cm with no observed parasites and fungi on their skins, gills, and fins.

Fig. 6.35 Seedlings transported in the community in La Casona

Fig. 6.36 Selection of male tilapia in La Casona

6.2.2.6 Production of Seafood with Poseidon-AI® IAS Modules in La Casona

Thanks to the people-friendly and eco-friendly policies taken by the Government of Costa Rica, the students in these schools can have breakfast and lunch at their schools. La Casona High School has 325 students, Brüs Malis primary school has 113, and school of Ngöbegüe has 225 students; thus, 663 students consume breakfast and lunch at their schools. With schools' staffs and some parents, this number reaches to 800 people. Majority of the materials for cooking these two food courses are bought from surrounding cities, especially San Vito.

During the pandemic and when the schools were closed, the students needed to eat in their houses, but with many losing their jobs, one can only guess how difficult it was for families to feed themselves. For this reason and to prevent such issues from happening in the future, Poseidon-AI® IAS modules were presented to receive funding from Sustainable Fund of Banco Nacional de Costa Rica (BN). The modules were accepted and installed in these schools based on the number of students in each school to provide healthy and nutritious food and train families, students, and teachers to have their own source of protein, just in case of another unexpected event such as the pandemic.

Fig. 6.37 Construction of greenhouses in three schools to protect Poseidon-AI® modules and products from wild animals, dogs, insects, and pests

Depending on the number of students in each school, 10% of the total pupils along with their teachers and parents were selected to learn about the system, physiology, and biology of fish and plants. In the school of Ngöbegüe 25, Brüs Malis primary school 15, and La Casona High School 35 students, teachers, and parents were trained. Due to the location of these schools and being surrounded by mountains, the internet connectivity is unstable, and so, Poseidon-AI® algorithms could not be used in these modules. For this reason, a total of more than 15 h of capacity-building classes were conducted in all these schools.

Additionally, due to the presence of wild animals and street dogs, greenhouses were built to keep these modules and prevent insects, wild animals, and dogs from harming the fish and the vegetables (Fig. 6.37). As shown in Fig. 6.37, constant rain created a muddy and harsh environment to build these greenhouses and problems in balancing the modules. Electric power was taken from the buildings located near to the greenhouses, and with heavy rains and high humidity, the wires were isolated and protected by PVCs.

Students and their parents as well as their teachers were trained on how to organize the clay tiles in each Poseidon-AI® module as well as transplant the plants into the modules. The broken clay tiles were used for planting the vegetables and plants as shown in Fig. 6.38.

The fish were transported from the aquaculture farm within the indigenous territory in plastic bags and were moved to the tanks filled with rainwaters. The students, parents, and teachers were trained on the themes of water quality and circulation, feeding, and fish growth (Fig. 6.39). The aquaculture farmers also were trained about distinguishing male and female tilapia, FCR, water quality management, and the importance of water filtration (Fig. 6.40).

Fig. 6.38 Capacity building for students, parents, and teachers in La Casona for the use of clay tiles in Poseidon-AI® IAS modules

Fig. 6.39 Capacity building for students, parents, and teachers on themes of water quality and circulation, feeding, and fish growth

A kilo of fish is produced and sold by the community at around 10 USD, and so Poseidon-AI® IAS modules in a year can save up to 20,000 USD in food cost. Furthermore, the community can produce nearly half a tonne of vegetables in a year. Figure 6.41 shows the products of Poseidon-AI® IAS modules in the first 40 days.[8]

[8] https://ticotimes.net/2022/10/11/teamwork-feeds-800-indigenous-costa-rica-students.

Fig. 6.40 Training local aquaculture farmers about fish sex, water quality management, and the importance of water filtration

Fig. 6.41 Product of Poseidon-AI® IAS modules in the first 40 days

References

Ahmad S, Hamdan NM (2016) Penemuan ikan tilapia diberi makan babi. Sinar Online. Retrieved from http://www.sinarharian.com.my/

Allison EH (2011) Aquaculture, fisheries, poverty and food security. The WorldFish Center, Penang

Alpízar M (2006) Participation and fisheries management in Costa Rica: from theory to practice. Mar Policy 30:641–650

Álvarez J, Ross E (2010) La pesca de arrastre en Costa Rica. MarViva, San José, 55 p

Arguedas Ortiz D (2015) IACHR tackles violence against native peoples in Costa Rica. Inter Press Service News Agency. Available from: http://www.ipsnews.net/2015/05/iachr-addressesviolence-against-native-peoples-incosta-rica/

AVA (2011) Media release: AVA ups food supply resilience efforts: helping local fish farms improve productivity. Agri-Food and Veterinary Authority, Singapore

AVA (2013a) Agrotechnology Parks. Retrieved from http://www.ava.gov.sg/AgricultureFisheriesSector/FarmingInSingapore/AgroTechParks/

AVA (2013b) History. Retrieved from http://www.ava.gov.sg/AboutAVA/History/

Begg GA, Waldman JR (1999) An holistic approach to fish stock identification. Fish Res 43(1): 35–44

Béné C, Barange M, Subasinghe R, Pinstrup-Andersen P, Merino G et al (2015) Feeding 9 billion by 2050—putting fish back on the menu. Food Security 7(2):261–274

Bentsen HB, Gjerde B, Eknath AE, de Vera MSP, Velasco RR et al (2017) Genetic improvement of farmed tilapias: response to five generations of selection for increased body weight at harvest in Oreochromis niloticus and the further impact of the project. Aquaculture 468:206–217

Campbell H, Marshall R (2002) Utilitarianism's bad breath? A re-evaluation of the public interest justification for planning. Plan Theory 1(2):163–187

Campos J (1986) Fauna de acompañamiento del camarón en el Pacífico de Costa Rica. Rev Biol Trop 34(2):185–197

Carrier TJ, Eddy SD, Redmond S (2017) The SeaCoRe system for large scale kelp aquaculture. Aquac Int 25(1):355–366

Chan KB, Ng BK (2000) Myths and misperceptions of ethnic Chinese capitalism. In: Chan KB (ed) Chinese business networks: state, economy and culture. Prentice Hall & Nordic Institute of Asian Studies, Singapore, pp 285–302

Chan KB, Ng BK (2004) Singapore. In: Gomez ET, Hsiao MHH (eds) Chinese business in Southeast Asia: contesting cultural explanations, researching entrepreneurship. Curzon, Surrey, pp 38–61

Chou CGH (1986) Kelongs in transition: an aspect of the fishing industry in Singapore

Chou R, Lee HB (1997) Commercial Marine Fish Farming in Singapore. Aquac Res 28:767–776

Costa Rica (1976) Decree No. 5904-G

Dahles H (2008) Entrepreneurship and the legacies of a developmental state: Singapore enterprises venturing across national borders. J Dev Entrep 13(4):485–508

De Koninck R (1973) Farmers of a City state: the Chinese smallholders of Singapore. Universite Laval, Quebec

Department of Fisheries (DOF) (2015) National plan of action for the management of fishing capacity in Malaysia (plan 2). DOF, Putrajaya

Dickson M, Nasr-Allah A, Kenawy D, Kruijssen F (2016) Increasing fish farm profitability through aquaculture best management practice training in Egypt. Aquaculture 465:172–178

DOSM (2016) Selected agricultural indicators, Malaysia

Dr. Axe (2012) Eating Tilapia is worse than eating bacon. Retrieved from https://draxe.com/

DSM (2014) MS 1500: 2009—Halal food. Department of Standard Malaysia

Eyngor M, Zamostiano R, Tsofack JEK, Berkowitz A, Bercovier H et al (2014) Identification of a novel RNA virus lethal to tilapia. J Clin Microbiol 52(12):4137–4146

FAO (2010) National Aquaculture Legislation Overview Costa Rica, Rome, Italy

Fathi M, Dickson C, Dickson M, Leschen W, Baily J et al (2017) Identification of Tilapia Lake Virus in Egypt in Nile tilapia affected by 'summer mortality' syndrome. Aquaculture 473:430–432

Fitzsimmons K, LwinTun K (2017) Supply and demand in tilapia markets. Paper presented at the Asian-Pacific aquaculture 2017 "transforming for market needs", Malaysia

Food and Agriculture Organization (FAO) (2014) FAO feed resources database. In: FAO aquaculture feed and fertilizer resources information system. FAO, Rome

Food and Agriculture Organization (FAO) (2016) The state of world fisheries and aquaculture. Contributing to food security and nutrition for all. FAO, Rome

Food and Agriculture Organization of the United Nations (FAO) (1970) Proyecto pesquero para la gradual conversión de la pesca de subsistencia en actividad orientada hacia el mercado [Fishing project for gradual conversion of subsistence fishing in market-oriented activity]. Circular 70/1. Food and Agriculture Organization of the United Nations (FAO), Livingston

Food and Agriculture Organization of the United Nations (FAO) (2004) Fishery country profile—Costa Rica, Rome, Italy, Report no. FID/CP/CRI

Government Publications Bureau (1966) Land acquisition act. Singapore Act 1966, no. 41. Government Publications Bureau, Singapore

Gutierrez-Rodriguez DM (1990) Diagnostico de la actividad pesquera en Costa Rica. Food and Agriculture Organization of the United Nations (FAO), San Jose, 365 p

Guzmán-Mora AG (2009) Diagnóstico de la composición de capturas de peces en la pesca artesanal de Golfo Dulce, Costa Rica [Diagnosis of the composition of fish catches in artisanal fisheries of Golfo Dulce, Costa Rica]. In: Acción Social y Docencia del Programa Institucional Osa-Golfo Dulce, 4–6 Nov 2009

Hempel E (2010) Value chain analysis in the fisheries sector in Africa. AU/NEPAD Programme

Hill RD (2013) Agriculture in the Malaysian region. NUS Press, Singapore

Instituto Costarricense de Pesca y Acuicultura INCOPESCA (2010). Retrieved from http://www.incopesca.go.cr

Kelleher K (2005) Discards in the world's marine fisheries: an update. FAO Fisheries Technical Paper 470, Food and Agriculture Organization of the United Nations (FAO). xix+131 p

Kim S, Park J-H, Wertz EK (2010) Expectation gaps between stakeholders and web-based corporate public relations efforts: focusing on fortune 500 corporate web sites. Public Relat Rev 36(3):215–221

Kovach M (2009) Indigenous methodologies: characteristics, conversations and contexts. University of Toronto Press Incorporated, Toronto

Leaño EM, Mohan C (2012) Early mortality syndrome threatens Asia's shrimp farms. Glob Aquac Advoc 2012(7/8):38–39

Liew PJ (1985) Agricultural change in Singapore with special reference to aquaculture. BA (Hons) Dissertation. Geography. National University of Singapore, Singapore

Little DC, Newton R, Beveridge M (2016) Aquaculture: a rapidly growing and significant source of sustainable food? Status, transitions and potential. Proc Nutr Soc 75(3):274–286

Low L (1990) State entrepreneurship. In: Lee TY, Low L (eds) Local entrepreneurship in Singapore: private and state. Institute of Policy Studies, Singapore, pp 150–175

Low L (2001) The Singapore developmental state in the new economy and polity. Pac Rev 14(3): 411–441

Molina Murillo SA, Pérez Castillo JP, Herrera Ugalde ME (2014) Assessment of environmental payments on indigenous territories: the case of Cabecar-Talamanca, Costa Rica. Ecosyst Services 8:35

Naylor RL, Goldburg RJ, Primavera JH, Kautsky N (2000) Effect of aquaculture on world fish supplies. Nature 405(6790):1017

Neo H (2007) Challenging the developmental state: nature conservation in Singapore. Asia Pac Viewp 48(2):186–199

Ngiam TT, Cheong L (2006) Singapore, Laguna: Southeast Asian Regional Center for Graduate Study and Research in Agriculture (SEARCA)

Norazmi LNH (2017) It's application in aquaculture research MFS seminar 2017: new paradigms for aquaculture in Malaysia

Norhana MW, Dykes G, Padilah B, Hazizi AA, Masazurah A (2012) Determination of quarantine period of African catfish (Clarias gariepinus) fed with pig offal to assure compliance with halal standards. Food Chem 135(3):1268–1272

Omar SZ, Hassan MA, Shaffril HAM, Bolong J, Drsquo JL (2011) Information and communication technology for fisheries industry development in Malaysia. Afr J Agric Res 6(17):4166–4176

Ottinger M, Clauss K, Kuenzer C (2016) Aquaculture: relevance, distribution, impacts and spatial assessments—a review. Ocean Coastal Manag 119:244–266

Pauzi N, Man S (2015) Jurnal Fiqh, No. 12(57–58)

Phyne J (2010) A comparative political economy of rural capitalism: Salmon Aquaculture in Norway, Chile and Ireland. Acta Sociologica 53(2):160–180

Phyne J, Hovgaard G, Hansen G (2006) Norwegian Salmon goes to market: the case of the Austevoll seafood cluster. J Rural Stud 22(2):190–204

Ramírez R (2007) Mejoramiento de los Mercados Internos de Productos Pesqueros (No. 3111). INCOPESCA, San José. Retrieved from http://www.incopesca.go.cr/Mercadeo/Varios/MERCADOS%20INTERNOS%20PROYECTO%20FAO.pdf

Saidin N, Rahman FA, Abdullah N (2017) Animal feed: halal perspective. Paper presented at the int. conf. on humanities, social sciences and education (HSSE'17), London

Saili NAB (2017) Interview of aquaculture status in Malaysia: import alert and DOF reporting systems/Interviewer: Fathi S. Department of Fisheries, Putrajaya

Shaffril M, Hassan MS, Hassan MA, D'Silva JL (2009) The use of mobile phone among farmers for agriculture development. Eur J Sci Res 36(1):41–48

Shin J-S (2005) Substituting and complementing models of economic development in east Asia. Glob Econ Rev Perspect East Asian Econ Ind 34(1):99–118

Tan C-Z, Yeung HW-C (2000) The internationalization of Singaporean firms into China: entry modes and investment strategies. In: Yeung HW-C, Olds K (eds) Globalization of Chinese business firms. St. Martin's Press, New York, pp 220–243

Tey YS, Suryani D, Emmy FA, Illisriyani I (2009) Food consumption and expenditures in Singapore: implications to Malaysia's agricultural exports. Int Food Res J 16:119–126

Tuhiwai Smith LT (2012) Decolonizing methodologies: research and indigenous peoples, 2nd edn. Zed Books Ltd

United Nations Declaration on the Rights of Indigenous Peoples (UNDRIP) (2008) United Nations General Assembly 107th Plenary Meeting, Sep. 13, 2007. Available from: http://www.un.org/esa/socdev/unpfii/documents/DRIPS_en.pdf

United States Food and Drug Agency (USFDA) (2016) FDA, Import Alert 16–131 (DWPE). USFDA

Vega-Corrales LA (2010) Evaluación poblacional y pautas de ordenamiento pesquero del complejo Opisthonema (Pisces: Clupeidae), Golfo de Nicoya, Costa Rica [Population assessment and fisheries management guidelines Opisthonema complex (Pisces: Clupeidae), Gulf of Nicoya, Costa Rica]. MSc thesis, Universidad Nacional de Costa Rica, Heredia, Costa Rica

Weaver KL, Ivester P, Chilton JA, Wilson MD, Pandey P et al (2008) The content of favorable and unfavorable polyunsaturated fatty acids found in commonly eaten fish. J Am Diet Assoc 108(7): 1178–1185

Wehrtmann I, Nielsen-Muñoz V (2009) The deepwater fishery along the Pacific coast of Costa Rica, Central America. Latin Am J Aquat Resour 37(3):543–554

Yue G, Wang L (2017) Current status of genome sequencing and its applications in aquaculture. Aquaculture 468:337–347

Yusoff A (2014) Status of resource management and aquaculture in Malaysia. Paper presented at the int. workshop on resource enhancement and sustainable aquaculture practices in Southeast Asia, Philippines

Further Reading

Asamblea Legislativa de la República de Costa Rica (2010) Ley sobre Desarrollo Autónomo de los Pueblos Indígenas, Expediente N°14352. http://proyectos.conare.ac.cr/asamblea/14352%203M137.pdf

AVA (2012) Statistics. Retrieved from http://www.ava.gov.sg/Publications/Statistics/

Backyard Aquaponics (2011) The IBC of aquaponics. Edition 1.0. Backyard Aquaponics, Success Western, Australia

Baker R (2014) Impulsando la Participación de los Pueblos Indígenas en REDD+: La Inclusión Temprana y la Consulta en Costa Rica

Costa Rica (2017) Consulta Indígena Cumple con la Norma y se Desarrolla en Total Respeto a Derechos de Pueblos Indígenas. www.presidencia.go.cr/comunicados/2017/05/consulta-indigena-cumple-con-lanorma-y-se-desarrolla-en-total-respetoa-derechos-de-pueblos-indigenas/

Declaración Territorio Bribri Libre de REDD+, Katsi (2016). http://www.feconcr.org/doc/Territorio%20bribri%20libre%20de%20redd%2B.pdf

Decostere A, Ducatelle R, Haesebrouck F (2002) Flavobacterium columnare (Flexibacter columnaris) associated with severe gill necrosis in koi carp (Cyprinus carpio L.). Vet Record 150:694–695

Decreto Ejecutivo N° 40932-MP-MJP (2018) Mecanismo General de Consulta a Pueblos Indígenas. Sistema Costarricense de Información Jurídica. Available from: https://www.consultaindigena.go.cr/mecanismo/

EatThis.com (2022) Report Eat This, Not That! Retrieved from http://www.eatthis.com/

Florian Rivero EM, Sucre L, Díaz Briones A (eds) (2014) Cambio Climático y Bosques: Promoviendo la Participación del Pueblo Bribri y Cabecar. Centro Agronómico de Investigación y Enseñanza (CATIE), Turrialba

Forest Carbon Partnership Facility (2016) Costa Rica's readiness preparation proposal readiness fund of the FCPF. World Bank—grant reporting and monitoring (GRM) report. Available from: https://www.forestcarbonpartnership.org/sites/fcp/files/2015/June/GRM_Costa%20Rica_2015.PDF

Forest Peoples Programme (2015) Report on the grave and persistent violation of indigenous peoples' rights in Costa Rica. United Nations committee on the elimination of racial discrimination, 87th session. Available from: https://www.forestpeoples.org/sites/fpp/files/publication/2015/07/cerd-report-finaleng.pdf

Hasan MR, Chakrabarti R (2009) Use of algae and aquatic macrophytes as feed in small-scale aquaculture: a review. FAO Fisheries and Aquaculture Technical Paper No. 531. FAO, Rome

Houlihan D, Laurent P (1987) Effects of exercise training on the performance, growth, and protein turnover of rainbow trout (Salmo gairdneri). Can J Fish Aquat Sci 44:1614–1621

Kogan M (1998) Integrated pest management: historical perspectives and contemporary developments. Annu Rev Entomol 43(1):243–270

Kothari A, Corrigan C, Jonas H, Neumann A, Shrumm H (2012) Recognising and supporting territories and areas conserved by indigenous peoples and local communities: global overview and national case studies. Secretariat of the Convention on Biological Diversity, Technical Series no. 64. p 160. Available from: https://www.cbd.int/doc/publications/cbd-ts-64-en.pdf

La Nación (2012) Conai Incumple con Defensa de los Derechos de Indígenas. Available from: https://www.nacion.com/el-pais/servicios/conai-incumplecon-defensa-de-los-derechos-de-indigenas/4HK3IRSZBFBEPJMQ3FJ3CZG6UI/story/

Larson AM (2011) Forest tenure reform in the age of climate change: lessons for REDD+. Glob Environ Chang 21(2):540–549

Larson AM, Brockhaus M, Sunderlin WD, Duchelle A, Babon A, Dokken T et al (2013) Land tenure and REDD+: the good, the bad and the ugly. Glob Environ Chang 23:678–689

Ley Indígena 6172 (1977) La Asamblea Legislativa de la República de Costa Rica. Available from: http://www.wipo.int/edocs/lexdocs/laws/es/cr/cr057es.pdf

Mendoza Salamanca T (2013) Indigenous participation and rights in the REDD/CCAD/GIZ program in Central America. Institut für Ökologie und Aktions-Ethnologie e.V.. Available from: https://www.infoe.de/images/stories/pdf/casestudy_redd_ccad_giz_final.pdf

Mesa Nacional Indígena de Costa Rica (2009) "Asi Vivimos los Pueblos Indígenas...": Diagnóstico Niñez y Adolescencia Indígena. Available from: https://www.unicef.org/costarica/docs/cr_pub_Asi_vivimos_los_pueblos_indigenas.pdf

Ministerio de Ambiente y Energía (MINAE) (2018) Marco de Gestión Ambiental y Social (MGAS). Available from: http://reddcr.go.cr/sites/default/files/centro-de-documentacion/mgas_versionfinalsetiembre1sc.docx

Ministerio de Ambiente y Energía (MINAE) & Fondo Nacional de Financiamiento Forestal (FONAFIFO) (2015) Diseño de un Sistema de Información País sobre las Salvaguardas de REDD: Normativa, Institucionalidad, Información e Indicadores. Available from: http://reddcr.go.cr/sites/default/files/centro-de-documentacion/propuesta_sis-redd_informe_final_-_fonafifo.pdf

Mora Rojas J (2015) Fiscalía Impulsa Acceso a la Justicia de Ciudadanos Indígenas. Seminario Universidad. Available from: https://semanariouniversidad.com/pais/fiscalia-impulsa-acceso-a-lajusticia-de-ciudadanos-indigenas/

Moreton-Robinson A (2009) Imagining the good indigenous citizen: race war and the pathology of patriarchal white sovereignty. Cult Stud Rev 15(2):61–79

Plan de Trabajo (2013) Programa regional REDD/CCAD-GIZ. Available from: http://www.reddccadgiz.org/mediadores/concepto_proceso.pdf

Rojas D (2004) Indígenas Ticos Pierden Tierras. Ambientico 133:6. Available from: http://www.ambientico.una.ac.cr/pdfs/ambientico/133.pdf

REDD+ (2017) Ministerio de Ambiente Oficializa Inicio de la Última Etapa del Proceso de Consulta de la Estrategia REDD+. Available from: http://reddcr.go.cr/es/ministro-de-ambiente-oficializa-iniciode-la-ultima-etapa-del-proceso-deconsulta-de-la-estrategia

Tribunal Supremo de Elecciones (1991) Ley de Inscripción y Cedulación Indígena de Costa Rica, Ley N° 7225. Available from: http://www.tsego.cr/pdf/normativa/leyinscripcionycedulacionindigena.pdf

7 Sustainable Development Goals, Deep Tech, and the Path Forward

Although using SDGs in texts is easy, as shown in previous chapters, any contribution to these goals faces unpredictable and unexpected challenges. Lack of experts and implementation hardships such as language barriers and financial barriers are only a few of these challenges. Under these circumstances, deep tech can be a great solution in bringing closer the knowledge gap among stakeholders, removing language barriers, and even lowering the implementation costs.

Since the aquaculture industry is expanding quickly, it is crucial to ensure its sustainability because it consumes limited resources, has a rich cultural history, and has the potential to make a significant economic contribution to the county. However, the sector's adoption of technology is not keeping up with its rate of expansion, so deep tech adoption and impact investment can speed the sector up.

It is believed that at the end of implementation phase of UNSDGs (end of 2030) and in order to guarantee the world's sustainability, more aggressive and straightforward approaches need to be applied to which deep tech has the potential to considerably contribute. This chapter uses Poseidon-AI's approaches as examples for possible contributions made to UNSDGs while trying to sketch the path forward under rapid global changes.

7.1 Real-World Implementation of the UNSDGs

There are 17 UNSDGs (Fig. 7.1) which are SDG 1: No Poverty, SDG 2: Zero Hunger, SDG 3: Good Health and Well-Being, SDG 4: Quality Education, SDG 5: Gender Equality, SDG 6: Clean Water and Sanitation, SDG 7: Affordable and Clean Energy, SDG 8: Decent Work and Economic Growth, SDG 9: Industry, Innovation and Infrastructure, SDG 10: Reduced Inequalities, SDG 11: Sustainable Cities and Communities, SDG 12: Responsible Consumption and Production, SDG 13: Climate Action, SDG 14: Life Below Water, SDG 15: Life on Land, Peace, SDG 16: Justice and Strong Institutions, and SDG 17: Partnerships for the Goals.

© The Author(s), under exclusive license to Springer Nature Singapore Pte Ltd. 2023

A. Rahimi-Midani, *Deep Technology for Sustainable Fisheries and Aquaculture*,
https://doi.org/10.1007/978-981-99-4917-5_7

Fig. 7.1 United Nations Sustainable Development Goals (https://www.un.org/development/desa/disabilities/about-us/sustainable-development-goals-sdgs-and-disability.html)

Poseidon-AI approaches advance these objectives both directly and inadvertently as the private enterprise expands. Contrary to popular opinion, which holds that private businesses can only expand by focusing on their own human resource growth, Poseidon-AI's sustainable deep tech techniques prioritize enhancing the abilities of those employed in the aquaculture industry. The UNSDGs are used to monitor and assess these consequences because it is challenging to determine the true impact of these measures.

7.1.1 Poseidon-AI's Contribution to Poverty Reduction (SDG 1)

Many people living in poverty lack education, work in low-paying service jobs, and are not supported by insurance, pensions, or social security. These are the first demographics to lose their jobs in the event of a financial crisis, including the most recent COVID-19 pandemic. For this reason, employing methods that need the least amount of knowledge as well as utilizing AI/ML, families can produce income and escape poverty. With the help of Poseidon-AI® IAS modules, families can have their own sources of income and encourage entrepreneurship in the neighborhood rather than hunting for work.

7.1.2 Food Security and Food Safety (SDG 2 and SDG 3)

The epidemic exposed the fragility of both the domestic and global food supply chains. As a result of the pandemic, it is essential to rebuild the world's supply chain networks, remove barriers, and strengthen vulnerable areas. Communities are empowered with Poseidon-AI® IAS modules because they can grow their own fruits, vegetables, and fish. In the event of another lockdown, a family of four would have enough food for at least a year thanks to each module's annual production capacity of 80 kg of fish and shrimp and 20 kg of vegetables (food security). Moreover, communities can create organic, nutritious seafood by using rainwater and clay tiles. By avoiding the use of antibiotics and/or chemicals, families are able to contribute to the national and international food supply chains with products that are free of these substances.

7.1.3 Educating Women and Girls (SDG 4 and SDG 5)

With the exception of a few countries like Finland and Norway, the majority of the workforce is made up of men. Primary sectors like aquaculture are highly labor-intensive. This is partly due to the demanding and isolated working circumstances that greatly favor men in the aquaculture industry. The bulk of employees in the sector are women who work in the processing facilities. A processing plant in the Mekong River Delta with a large female workforce is shown in Fig. 7.2.

Typically, aquaculture farmers work alongside their sons as they assist them and learn about the culture techniques. On the other hand, women and girls usually stay at home supporting their husbands, fathers, and brothers by cooking, cleaning, and taking care of the kids. These women and girls often reside in cultures where marriage at a young age is the norm, and they either never attend school or just receive basic education.

Women and girls can work alongside their husbands, brothers, and fathers in fish farming with the aid of Poseidon-AI® algorithms and IoT devices. The goal is for

Fig. 7.2 A fish processing plant located in Mekong River Delta with majority female workers

women and girls to be able to interpret the findings of Poseidon-AI® algorithms and learn how to work with IoT devices to sustainably build their families' aquaculture businesses rather than having undergo a lengthy and difficult educational procedure. Additionally, Poseidon-AI® IAS modules enable women and girls to sustainably grow vegetables and seafood in their backyards, patio gardens, and rooftops with the support of trained Poseidon-AI® algorithms.

7.1.4 Clean Water and Clean Energy (SDG 6 and SDG 7)

Vulnerable communities live in areas with limited access to clean water and, in some cases such as indigenous communities, access to any sort of energy. Rapid urbanization and increasing demand forced tremendous pressure on farms to increase their production with the help of fertilizers and pesticides, contaminating the surrounded environments especially the nearby waterbodies. Hence, most of the waterbodies cannot be safe water provider for aquaculture farming. Poseidon-AI® IoT devices can support monitoring the waterbodies to prevent the usage of contaminated water both for human consumption and aquaculture farming (Fig. 7.3). The device can be located and anchored in different waterbodies, and industrial grade sensors can be used both for water and soil monitoring. The Poseidon-AI® IoT device does not require installation, and farmers, researchers, and officers working in government entities and international organizations can easily use it for monitoring all waterbodies every day and all year around.

High energy prices prevent many vulnerable communities from using any source of energy. Lack of electricity in some communities creates many obstacles in human development and implementation of many development projects. Poseidon-AI approaches use solar energy to provide cheap, clean, and more sustainable energy sources for the communities and farmers. The Poseidon-AI® IAS modules use bigger solar panels to allow communities to use solar energy for other purposes such as lights, internet access, and even water filtration (Fig. 7.4).

7.1.5 Innovation for Reducing Inequality and Economic Growth (SDG 8, SDG 9, and SDG 10)

Deep tech provides innovative solutions and contributes to a country's economic growth. Impact investing allows deep tech companies to tackle the most challenging issues faced at national and/or international level. In the case of aquaculture sector and as shown, there are firsthand challenges that can only be addressed with new and innovative technologies. Affordable IoT and intelligent software developed by Poseidon-AI allow farmers to save 20% of their feeding costs while closing the knowledge gap between farmers and experts. Rapid spread of diseases, high mortality rates, and lower yields prevent the aquaculture sector from showing its real contribution to a country's economy. Using special and cutting-edge solutions

Fig. 7.3 Poseidon-AI® IoT is used for monitoring waterbodies in Ghana as part of FAO programs

Fig. 7.4 Installation of solar panels for vulnerable communities

developed by deep tech companies can aid in achieving sustainability by lowering inequality and promoting national economic growth.

7.1.6 Sustainable Cities and Responsible Production of Seafood (SDG 11 and SDG 12)

Urban development shifted many societies from farming their own food to buying it from wholesalers, retailers, and small groceries. Through time, this made the majority of society to shift their priorities to other aspects such as housing, healthcare, and transportation. However, phenomena such as the pandemic made many to reconsider the priorities, especially in large urban areas. Rainwater recirculation systems, organic greenhouses, and laboratory-made protein sources are some of the solutions developed for answering the needs of the populations in large cities.

Singapore can be a great example for building a sustainable city in a short period of time due to external and internal constraints faced by this country. The country does not have pension and social security systems in place, and thus, many elderlies need to continue working until the end of their lives. Although affordable housing is available for every Singaporean, with high food prices, many struggle in having healthy and nutritious food on a daily basis. Under these circumstances, Poseidon-AI® IAS modules can help the elderlies' communities in Singapore to produce their own seafood and vegetables for self-consumption or selling in local markets. The Singapore rainy season supplies enough water for these modules, and Poseidon-AI® algorithms can assist the elderly in remotely managing these modules. It will take 4–24 months, depending on the species, before the community may buy or sell the seafood in local markets.

7.1.7 Climate Change, Ocean Resources, and Life on Land (SDG 13, SDG 14, and SDG 15)

Many aquatic species are cold-blooded animals, and so global warming and changes in environmental condition will impact their appetite, growth, maturity, and even mortality. For this reason, monitoring environmental condition in aquaculture farms can play an important role in healthy growth of these species. On the other hand, the species feed intake will change with fluctuation in environmental variables. This is extremely important since in every feed, there are three components (fish meal, fish oil, and crops) of which two of them are constantly taken from the oceans around the world. With stock depletion and increasing demand, sustainable use of these feeds will give the stocks that are living in the waters around the world enough time to recover from decades of rapid exploitation.

Farmers may monitor their waters and comprehend the health of the farmed species thanks to Poseidon-AI® IoT. Although many farmers are unable to adopt expensive climate change mitigation strategies, Poseidon-AI® IoT and intelligence algorithms make adaption strategies accessible to farmers. Farmers can effectively increase their farms' output while reducing feed waste by 20%.

On the other hand, with rapid urbanization, in many developing countries, many farmers change their jobs and/or shift their activities to services such as nature treatment facilities rather than using their lands for agriculture production. Under these circumstances, efficient modules that use smaller land areas but have higher yield productions can improve the life quality in many urban and rural areas. The use of clay tiles with no fertilizers and pesticides, along with rainwater used in Poseidon-AI® IAS modules, contributes significantly to improving populations' livelihood around the world.

7.1.8 Global Peace with the Help of Deep Tech (SDG 16)

The massive waves of immigrants running from wars, injustice, and/or unlivable environmental conditions such as floods have created a dilemma for advanced countries on whether to accept these immigrants or close their borders, keeping the resources for the population currently living inside these borders. The complications caused in Europe with the immigrants entering from the Middle Eastern countries and construction of border walls in the southern part of the USA for preventing immigrants from the Latin American countries are some of the examples of the twenty-first-century challenges impacting the global peace and security.

It is mainly believed that immigrants and the local population will compete for available jobs and scarce resources in the destination countries. It is difficult to integrate the new immigrants without properly educating them about the new culture, languages, and necessary techniques before entering into the new countries' job markets.

As a solution for rapid integration on these immigrants, deep tech can assist the host countries. For example, Costa Rica hosts many Nicaraguan immigrants living in different communities. These immigrants usually work in construction, or as helping hands in wealthier communities. However, pandemic, and the economic difficulties faced by Costa Rica, created similar dilemma in this country. Poseidon-AI® IAS modules in these communities were provided to some of these Nicaraguan immigrants, especially women and girls, allowing them to have a source of income while slowly integrating in the society.

7.1.9 Partnership With Various Stakeholders (SDG 17)

With the rapidly changing world, uni-disciplinary approaches cannot answer all these needs; thus, there is a global shift from uni-disciplinary to multidisciplinary approaches. However, unlike uni-disciplinary approaches, there is no single entity, organization, or person who can implement multidisciplinary approaches. For this reason, twenty-first-century challenges can only be faced by involving multiple stakeholders in academia, private and public sector, and international organizations.

Fig. 7.5 Communities continue producing their own seafood and vegetables nearly 24 months. (The images were gathered from social medias of these communities as part of evaluation and monitoring mechanisms)

Being a Singapore-based company, Poseidon-AI provides a deep tech solution that contributes to the global sustainability. Although Poseidon-AI® algorithms can provide an accurate approach for empowering farmers and communities around the world, expanding to various markets cannot be achieved without sufficient knowledge and understanding of these countries' culture, languages, aquaculture sector, and micro- and macro-economy. For this reason, various partnerships with local and international organizations were made before implementation of Poseidon-AI approaches. This allowed creating platforms both for implementation and follow-up phases, saving costs, energy, and time for various parties involved.

In most of the communities, after installation and capacity buildings conducted by Poseidon-AI, government entities such as City Hall or NGOs follow up by providing seedlings and supporting the communities by conducting classes in themes such as nutrition. Additionally, in case of complication using the Poseidon-AI® modules, these entities try to support the communities before involving Poseidon-AI team for resolving the issues. Finally, international organizations such as the UNDP, the GEF, and the FAO carry the monitoring and evaluation weights for smooth implementation of Poseidon-AI approaches in accordance with international laws, regulations, and standards. Figure 7.5 shows communities producing their own seafood and vegetables, nearly 24 months after the implementation of Poseidon-AI approaches, thanks to the partnership made with various local and international stakeholders.

7.2 Sea Level Rise, Salinity, and Food Security Problem

In dry and semiarid regions, soil salinity can reduce pastures, horticulture, and crop yield. In the subsoil, salt may form naturally or may be added by brackish irrigation streams. Land clearing, unsustainable irrigation methods, and pressure to put marginal land to use for agriculture are all contributing to an increase in salinity. Agronomic and engineering approaches have reached their limits; thus, breeding existing crops to be more salt tolerant and introducing new species for cultivation are the best options moving ahead to reduce the impact of saline land on world food production.

Plants consume most of the energy produced by photosynthesis, which is converted into *C* molecules, for general maintenance (Amthor 2000; Jacoby et al. 2011). Even in the best cases, a small amount is used directly for biomass accumulation. Energy usage can be used to characterize stress. According to the definition, stress happens when plants acquire less energy or when energy is diverted from growth to defense against stress. More fixed *C* could be transferred to grain, increasing production, by increasing the metabolic and physiological energy efficiency of plants, notably during floral growth and grain fill. When crops are stressed, less energy is required for plants to tolerate salt, resulting in higher grain loss. Several features are used by plants to resist the salt in the soil solution. The most important characteristic is osmotic adjustment; to sustain turgor, all cells must acquire enough solutes to counteract the increased osmotic pressure in the soil solution. In order to do this, plants employ two different strategies to varying degrees. They either exclude Na^+ and Cl^-, particularly from leaves, and rely on organic solutes for osmotic adjustment (a strategy known as "ion exclusion"), or accumulate enough of these ions to balance them in the soil solution while maintaining strict ionic regulation in different cell compartments (a strategy known as "tissue tolerance").

In comparison to the external solution, salt-tolerant plants typically have leaves with high Na^+ and Cl^- contents. This is especially true for halophytes and non-halophytes that are more salt tolerant, like barley, where the feature of tissue tolerance is clearly present. To keep the cytosolic and organellar concentrations below hazardous levels, such plants must compartmentalize the majority of the leaf's *Na* and *Cl* in vacuoles. They also use organic osmolytes (including K^+) to balance the osmotic pressure in these cytoplasmic compartments (Shabala 2013). According to estimations for the cytosolic concentration of Na^+ (or Cl^-), it is hazardous at about 30 mM (Munns and Tester 2008; Conn and Gilliham 2010), although chloroplasts and mitochondria appear to be able to handle 100–200 mM Na^+ and Cl^- (Flowers et al. 2015). The more delicate species typically exhibit lower Na^+ concentrations in their leaves than in the surrounding solution, indicating that their primary mode of adaptation is "ion exclusion." There is a connection between salt tolerance and Na^+ exclusion in any species where there is significant genotypic variation in Na^+ buildup in leaves. According to Munns (2005), this is true for delicate species like rice and durum wheat, but it might also be true for salt-tolerant species like barley (Chen et al. 2005).

7.2.1 Salt-Affected Soils

For efficient reclamation and management, it is essential to comprehend the variations in qualities among salt-affected soils (SASs). Salt-affected soils are categorized into saline, sodic, and saline-sodic groups based on the amount of total soluble salts (TSSs) (measured by EC), sodium adsorption ratio (SAR; the ratio of Na^{+} to Ca^{2+} and Mg^{2+} on the exchange sites of soil), exchangeable sodium percentage (ESP; the relative amount of the Na^{+} ion expressed as a percentage to the cation exchange capacity (CEC) or the sum of exchangeable bases), and soil pH. Sodic soils typically have a pH higher than 8.5, but it can go as high as 10.5. The high pH of sodic soils may be caused by more exchangeable *Na* being hydrolyzed than more strongly bound ions like *Ca* and *Mg*. The increase in soil pH is a result of the OH^{-} ion concentration rising. Soil pH is limited to 8.5 due to the limited hydrolysis of $CaCO_3$ and $MgCO_3$, but soils containing Na_2CO_3 have a pH of more than 8.5 or even 10–10.5 due to their increased solubility (Abrol et al. 1988; Brady and Weil 2005; Doula and Sarris 2016).

Salinization refers to the mechanisms that result in the production of salty soils, whereas sodification refers to the processes that result in the formation of sodic soils. The reasons for salinization and sodification are complex, and one item that influences one of these processes may also have an impact on the other ones (Doula and Sarris 2016). The amount and makeup of salts in the soil, the quantity and quality of irrigation, and the properties of the soil all have an impact on soil salinization and sodification (Endo et al. 2011). Soils that have been impacted by salt are the result of either natural processes or human activity. Primary salinization/sodification refers to processes that occur naturally, whereas secondary salinization/sodification refers to processes that are caused by humans (Ghassemi et al. 1995). Figure 7.6 summarizes the main sources of salts under natural and anthropogenic factors.

Salinity and sodicity of the soil have contributed to changes in land use/land cover characteristics through time, which are directly related to land degradation and produce numerous environmental changes (Hussain et al. 2019). High salinity affects more than 20% of all agricultural lands, which makes up more than 7% of the world's total land area (Ghassemi et al. 1995; Jamil et al. 2011; Machado and Serralheiro 2017; Sairam and Tyagi 2004; Szabolcs 1974, 1994). According to reports, the majority of the SASs are said to be concentrated in Asia, Australia, and South America's arid and semiarid regions, covering an estimated 1 billion acres (Doula and Sarris 2016). According to the data compiled from Abrol et al. (1988) and Szabolcs (1974), over 932 million ha of land is occupied by SAS worldwide.

7.2.2 Salinity Effects on Plants

According to definitions given by many sources (Alam 1999; Fahad et al. 2015a, 2017; Gull et al. 2019), the term "stress" in plants refers to an environmental restriction that prevents the morphological, physiological, and biochemical

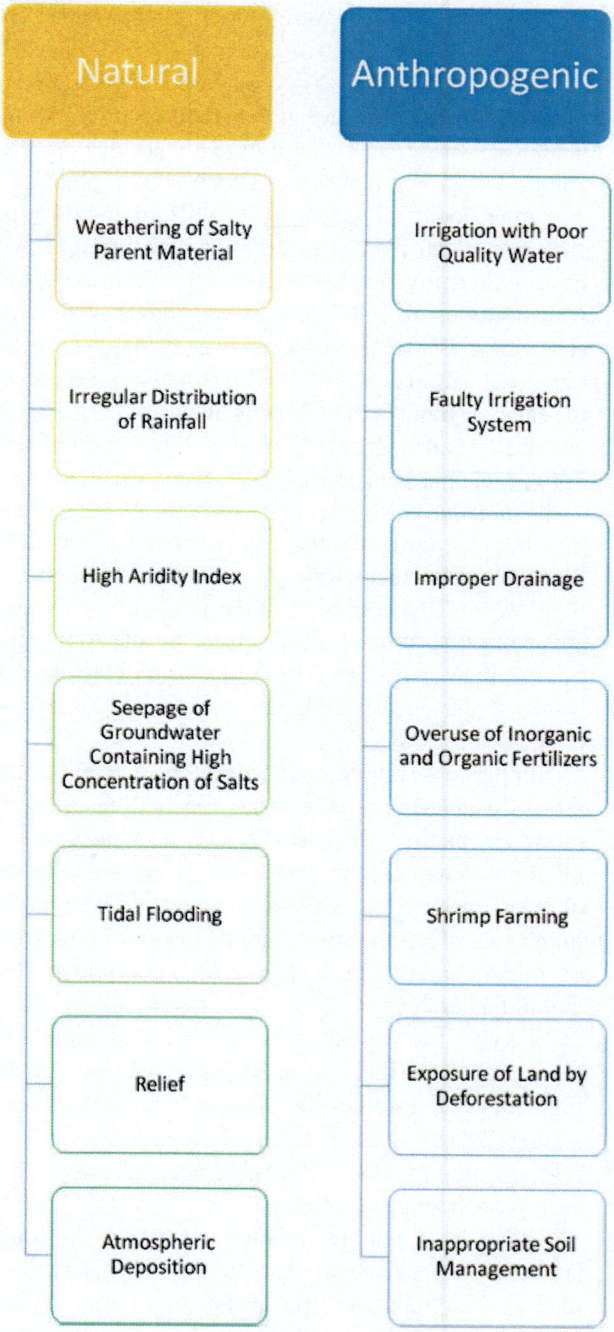

Fig. 7.6 Natural and anthropogenic sources of salts in soils

functioning of plants, negatively influencing their growth and development. Pests, diseases, weeds, and other biotic and abiotic stresses (soil salinity, radiation, water stagnation, drought, extremely high temperatures, organic and inorganic pollutants, etc.) (Gull et al. 2019; Akram et al. 2018a; Fahad et al. 2013, 2018, 2019a; Khan et al. 2017; Shah et al. 2014; Yang et al. 2017; Zia et al. 2017; Al-Karaki 2000; Talei et al. 2012; Zribi et al. 2018) may operate singly or in combination to reduce crop productivity and threaten global food security. More than 20% of all agricultural lands worldwide are covered by SASs, which are the most harmful environmental abiotic stresses on plants (Al-Karaki 2000; Talei et al. 2012; Zribi et al. 2018).

Saline, sodic, and saline-sodic soils are the three types of salt-affected soils, and each one affects plants in a different way through a variety of mechanisms. Osmotic impact, ionic toxicity, and nutritional imbalances are the main causes of salt stress in plants (Gull et al. 2019; Akram et al. 2018a; Fahad et al. 2013, 2018, 2019b; Khan et al. 2017; Shah et al. 2014; Yang et al. 2017; Zia et al. 2017; Al-Karaki 2000; Talei et al. 2012; Zribi et al. 2018). The reduction in *N* and *P* concentrations brought on by the salt constraint is thought to be the result of the antagonistic relationships between *Na* and *Cl* and NH_4^+, NO_3, and H_2PO_4 (Etesami and Noori 2019; Khan and Duke 2001; Maksimovic and Ilin 2012; Sahin et al. 2018).

By preventing different processes of *N* metabolism, including uptake, assimilation, and amino acid and protein synthesis, the salt stress severely impacts plant development (Dluzniewska et al. 2007). The uptake of *K* is negatively impacted by high *Na* ion concentrations at the root surface (Maathuis and Amtmann 1999). Na has a negative impact on *K* uptake by the root, specifically through high-affinity potassium transporters (HKTs) and nonselective cation channels (NSCCs), as a result of the chemical similarity between *Na* and *K* ions (Wakeel 2013; Maria et al. 1997).

Similar to *K*, high *Na* concentrations typically found in salty soils can negatively affect Ca and Mg absorption and transport, resulting in decreased *Ca*:*Na* and *Mg*:*Na* ratios in plants (Hadi et al. 2008). Under salt stress, the concentration of micronutrients in plants may rise, fall, or remain unchanged depending on the type of plant, tolerance of the plant to salinity, concentration of macro- and micronutrients in the soil, pH of the soil solution, adsorption phenomena on the surface complexes of mineral and organic particles, and various environmental factors (Hu and Schmidhalter 2001).

Leaching is one method for removing salts from the root zone. Other methods for managing SAS include mulching, including organic and inorganic fertilizers, regulating the groundwater table, and growing crops that can withstand salt stress (Osman 2018; Shaaban and El-Fouly 2002). Several techniques are used in salt-affected areas to remove excess soluble salts from the root zone of plants to enhance crop growth and production.

The most popular procedures are leaching, flushing, and scraping, although these are highly expensive (Jouyban 2012). The use of organic amendments is a successful method for reducing the physical, chemical, and microbiological problems connected with SAS. Due to their bonding or adhesive qualities, organic amendments aid in the flocculation of mineral particles to organic polymers, resulting in good structural stability, a requirement for maintaining an optimal soil structure (Diacono and Montemurro 2012). When organic matter is added to SAS, the aggregate stability and porosity can be improved, which leads to increased *Na* leaching and decreased EC and exchangeable sodium percentage (ESP) values (Wang et al. 2014). The inclusion of organic matter was found to significantly boost the microbiological and enzymatic activity of SAS soil in several experiments (Liang et al. 2003; Oo et al. 2015; Tejada et al. 2006).

Plant salt resistance is the ability of the plant to withstand the effects of high salts on the root zone or on the leaves without a significant adverse effect. The rules for analyzing the salinity level in plants are the following (Lunin et al. 1963):

- The actual tolerance will vary according to the growth stages at which salinization occurs.
- The salt tolerance values are estimated with the portion of the plant to be marketed.

Hence, salt tolerance is a complex, quantitative, genetic character controlled by many genes (Shannon and Noble 1990; Shannon 1996). Thus, measurement of salt tolerance is described as a complex function of yield decline across a range of salt concentrations (Maas and Hoffman 1977; van Genuchten and Hoffman 1984). Maas (1986) studied 127 crops, which include 68 herbaceous crops, 10 woody species, and 49 ornamentals to provide information concerning potential hazards of a given saline water and soil (Maas 1990).

Generally, salinity will reduce the growth rate of plants leading to smaller and fewer leaves as well as shorter stature. The primary effect of salinity is due to salt's osmotic effects (Munns and Termaat 1986; Jacoby 1994). Furthermore, roots are also affected in length and mass by becoming thicker or thinner. Depending on the species, the maturity rate will be delayed or advanced. According to Shannon et al. (1994), depending on environmental interactions such as radiation and air pollution, severity of salinity can be mediated. Ion toxicity and/or nutritional deficiencies will arise depending on the composition of saline solutions due to predominance of a specific ion or competition among anions (Bernstein et al. 1974). The ionic effects of salinity are manifested in leaf and meristem maturity rate, while the osmotic effects mainly contribute to growth rate reduction and changes in leaf color and developmental characteristics. High salinity will accumulate *Na* and/or *Cl* in leaves causing "firing" of leaves. Furthermore, nutritional deficiencies such as *Ca* deficiency are common among plants growing in high *Na*/*Cl* ratio. If pollutants, such as O_3, are present, reduction in air exchange due to osmotic stress may also reduce the volume of pollutants that enter the plant, thereby decreasing the adverse effects of salinity (Maas and Hoffman 1977).

Depending on the growth stage, different plants will have different tolerance levels. Lettuce is sensitive during the early seedling stages and during flowering (Shannon et al. 1983); sugar beet is tolerant during later growth stages but is sensitive during germination (Bernstein and Hayward 1958); and turnip is more salt tolerant at germination but is more sensitive at seedling growth stage than during yield (Francois 1984).

Salinity level will have some favorable effects on yield, quality, and disease resistance. For example, Osawa (1963) demonstrated yield increase in spinach in low or moderate salinity. In carrot with increasing salinity, the sugar content decreases, while in potato, the starch content decreases (Bernstein 1959). At low salinity levels, cabbage heads are more solid, but with increasing salinity, the heads become less compact (Osawa 1961). Studies showed that celery is both resistant and susceptible to salinity (Osawa 1963; Aloni and Pressman 1987).

In melon and tomato, moderate salinity during fruit development will change the partitioning of photosynthates and improve soluble solids (Shannon and Francois 1978; Mizrahi and Pasternak 1985; Mizrahi et al. 1988; Cornish 1992). Hence, small yield decrease due to salinity might be partially offset by the higher marketable quality of the fruit. Furthermore, salinity often affects the timing of development. According to Pasternak et al. (1979), flowering of tomato (*Lycopersicon esculentum*) is delayed by salinity while the flowering in onions occurs earlier under salt stress.

Salinity stress can also impact the yield components and growth parameters. In comparison to shoot growth, root growth is frequently less affected by salinity at moderate salinities, and in some cases, salinity may even accelerate root growth. Root growth was less responsive to salinity than the aboveground growth of turnip (Francois 1984) and carrot (Bernstein and Ayers 1953a). Salinity had a less significant impact on the yield of asparagus spears than fern production (Francois 1987), and artichoke bud growth was more negatively impacted by salinity than shoot growth (Francois 1995). In iceberg lettuce, Shannon (1980) applied selection pressure to both the vegetative growth and the head/frame ratio and discovered that both features were under strain. According to Shannon and Francois (1978), salt tolerance in muskmelons decreased in the following order: total vegetative dry weight, total vine yield, fruit yield, and marketable yield, highlighting both the importance of accounting for quality traits and the differences in measurement criteria. According to the evaluation criteria, the level of salt tolerance between and within species is therefore likely to differ. In a critique, Jones and Qualset (1984) claim that in order to identify growth stages that are particularly salt sensitive, plant growth parameters must be monitored over the course of the growth cycle.

Many environmental conditions, including the kind of soil, affect how different crops respond to salt (Levitt 1972). High concentrations of many types of dissolved salts are found in saline soils and streams, any one of which may be a severe limiting factor for plant growth. A wide range of soil types and moisture conditions are covered by saline soils, which may be sodic or acidic. Genotypes with comparable salt tolerance in one environment could respond differently in a different environment. The complexity of soils and the interactions between the environment have been identified by Rana (1985) as significant barriers to successful breeding for salt tolerance. In contrast to non-alkaline saline soils, Rana (1985) demonstrated that crops adapted to alkali soils are more tolerable.

High amounts of boron, selenium, arsenic, or other ions may be present in drainage waters or waters reused from agricultural processing or manufacturing processes, which may cause environmental risks (Francois and Clark 1979; Clark 1982). Many plant species have shown a wide range of variance in their capacity to endure, exclude, or accumulate the harmful effects of certain ions (Flowers and Yeo 1986; Shannon et al. 1994). Despite so, one of the research areas that has not been sufficiently studied is the possibility of variation between species and varieties. In-depth reviews of the genetic variability connected to plant tolerance to specific ions can be found in the works of Vose (1963), Epstein and Jefferies (1964), Läuchli (1976), Wright (1976), Jung (1978), and Christiansen and Lewis (1982).

Salinity is measured using the unit of EC of a saturated soil paste extract taken from the root zone of the plant over time and zone. USDA (1954) described soil paste extracts as soil samples that are brought up to their water saturation points. It is advantageous to measure and reference salinity using saturation extracts since, for the majority of soils, this value is directly related to the field moisture range (USDA 1954). Across a wide range of soil textures from medium to fine, the soluble salt concentration in a saturation extract is typically half as concentrated as the soil water at saturation. It is simpler to make and measure soil-to-water extracts of 1:1 or 1:5, and back calculations to EC for a particular soil can be constructed. Modern techniques require even less time and effort to determine EC by using electronic probes or electromagnetic pulses (Rhoades 1976, 1993).

7.2.2.1 Amaryllidaceae

Grass and grasslike flowering plants with only one embryonic leaf, or cotyledon, in their seeds are known as monocotyledons, or simply monocots. Amaryllidaceae family such as onion, garlic, chive, and leek as well as Liliaceae family such as asparagus fall under the monocot clade.

Leeks (*Allium ampeloprasum*), garlic (*Allium sativum*), and onions (*Allium cepa*) most likely originated in central Asia. By 3200 BC, all were grown in Egypt. The chive (*Allium schoenoprasum*) is a wild plant that grows in Europe, northern Asia, and North America. Since the sixteenth century, it has been growing in Europe. Allium species—with the exception of chive—are grown for their bulbs and occasionally the base of the flattened leaf blades. For garnish and flavor, chive is only utilized in the leaf blades. Onion, garlic, leek, and chive are typically thought to be salt sensitive based on yield loss, although only onion and garlic have strong data.

Onions are relatively excluders of both Na^+ and Cl^-, are sensitive to sulfate, and react negatively to salt. Many cultivars have been examined, but little genetic variation has been found. The tolerance level improves again at the time when the seedlings have three to five leaves, after initially being with quite low-throughout seedling growth. Salt stress causes a change in the color of the leaves, turning them from a vibrant green to a drab blue-green, and the leaf tips sometimes exhibit burn symptoms.

Five onion cultivars (Yellow Sweet Spanish, Texas Early Grano, San Joaquin, Crystal Wax, and Excel) were examined by Bernstein and Ayers (1953b) for their ability to withstand salt in field plots at the US Salinity Laboratory. Initial yield decline started at a threshold EC of 1.4 dS/m, and 50% yield reduction was at 4.1 dS/m. According to Bernstein and Ayers (1953b), salt raised the osmotic potential of the produced sap without significantly raising sucrose or reducing sugar levels at the same time. As salts were applied, the percentage of dry weight increased, which led to a rise in the amount of bulk ions. Salinity reduced the number of leaves per plant, the number of leaves per bulb, the weight of the bulb, and the growth of the roots. Growing onions in salty environments may cause them to mature a week earlier.

According to Pasternak et al. (1984), the shallow and condensed root systems of young plants may contribute to their sensitivity throughout the early growth stages. Whether rooting systems may be genetically altered to increase tolerance or even if variability exists for this trait is not known. If variability can be added, it must be done so without impairing the bulb's capacity to be sold commercially. Five onion cultivars that are frequently grown in Texas were investigated by Wannamaker and Pike (1987) for their germination and growth responses to salinity. Using $NaCl$ and $CaCl_2$ solutions, it was discovered that germination was unaffected at EC up to 20 dS/m but significantly decreased after that without any obvious cultivar differences. After 8 days, germination in all cultivars was inhibited by 50% by solutions of 30–35 dS/m.

Despite being an Asian native, garlic was first cultivated in Egypt in 2780 BC (Yamaguchi 1983). According to a 2-year study by Francois (1994), garlic's threshold salinity was 3.9 dS/m, and at 7.4 dS/m, the production was reduced by 50%. With increased salinity, both percent solids, a key factor in bulb quality, and all yield components (bulb weight and diameter, and plants per unit area) decreased. Although leaf tissues collected much greater Cl^-, Na^+, and Ca^{2+} concentrations than bulbs did, shoot dry weight was less responsive to salt changes than bulb weight.

7.2.2.2 Liliaceae

Native to the scrub habitats of southern Europe, western Asia, and northern Africa is asparagus (*Asparagus officinalis*). Ancient Egyptians, Greeks, and Romans all practiced cultivation of it, but it appears to have been abandoned during the Middle Ages, with the exception of the Arabs, until the seventeenth century, when France came to favor it as a luxurious vegetable. Asparagus is frequently discovered as a garden escape in subsaline waste areas in North America. Although asparagus is widely regarded as the vegetable crop that can withstand the most salt commercially, it thrives in sandy, well-drained soils rather than those with a thick texture. According to Francois (1987), spear output was only decreased by 2% for every unit rise in soil salinity above a threshold of 4.1 dS/m in the first year following plantation. Salinity levels in the soil water up to 9.4 dS/m in the same study had no discernible impact on germination, while higher salt levels delayed germination and reduced final percentage.

According to Francois's (1987) observations, the germination of asparagus on filter papers using mixed $NaCl$ and $CaCl_2$ (1:1 by weight) solutions occurred at about 14.3 dS/m. However, Uno et al.'s (1996) studies on filter paper showed that the germination of asparagus would be reduced by 50% at about 60 mM $NaCl$. It is possible that Ca^{2+} was a factor in the discrepancy between these trials, but more analysis is necessary. The vulnerability of asparagus to salinity during its early growth phases must also be investigated.

Studies of asparagus tissues performed in vitro revealed that organogenesis and cellular organization were closely related to tolerance, with rooted and unrooted plantlets displaying comparable levels of tolerance (Mills 1989).

7.2.2.3 Apiaceae

The carrot (*Daucus carota L.*) is prized for its underground fleshy structure, which is primarily made up of the swelling base of the taproot but is also partially derived from the hypocotyl. The species, *Daucus carota*, is indigenous to western Asia, most likely Afghanistan (Shannon and Grieve 1999). Carrots were initially employed as a medicine before becoming popular as a cuisine. Prior to the tenth century, carrots were grown in Europe. The earliest Virginian immigrants brought carrots to North America. The carrot is classified as a salt-sensitive crop (Bernstein and Ayers 1953a; Malcolm and Smith 1971). For every unit rise in salinity above the 1.0 dS/m cutoff, root yield decreases by 14% (Maas 1986). Carrot germination and seedling growth were both inhibited by soil moisture potentials of less than 0.01 MPa, although these stages of growth were unaffected by osmotic potentials as low as 0.5 MPa. At matrix potentials between 0.1 and 0.3 MPa, root development rose dramatically; however, an equivalent osmotic potential had different results.

Schmidhalter and Oertli (1991) drew the following conclusions from these observations: water stress has a more detrimental impact on carrot growth than salt stress, and germination and seedling growth are impacted by equivalent matric and osmotic stresses in distinct ways. Under non-limiting water supply circumstances, the effects of salinity, soil aeration, and nutrient level on the transpiration coefficient of carrot were assessed. This measurement measures the quantity of water transpired per unit biomass generated (Schmidhalter and Oertli 1991). Schmidhalter and Oertli (1991) found no change in the transpiration coefficient at salt concentrations up to 16 dS/m in the soil solution and hypothesized that, in the absence of harmful ion effects and nutritional imbalances, salinity had minimal impact on the transpiration coefficient.

The wild stock that gave rise to cultivated celery (*Apium graveolens L.* var. *dulce* (*Mill.*) *Pers.*) was found in marshy areas of Sweden, Algeria, Egypt, and Abyssinia. Apparently growing in brackish marshes, by tidal waterways, and close to the sea, wild stock could be categorized as a halophyte. Thus, some salt tolerance may have been kept while celery was being developed as a crop (Francois and West 1982). The juvenile petioles of the plant, which are noticeably ridged on the outside and thicker at the base, are the edible parts.

According to Lingle and Carolus (1956) and Osawa (1961), *NaCl* salinity has the ability to stimulate the growth of celery. Francois and West (1982) classified celery as moderately sensitive based on the findings of a field study. Their threshold EC was at 1.8 dS/m, and their slope was 6.2% per dS/m. With salt-stressed celery grown in a greenhouse, Sonneveld (1988) found a greater slope value (7.7% per dS/m). This may be due to a variation in cultivar response or environmental factors. Salinity values are 10 dS/m for trimmed shoots and 11 dS/m for untrimmed plants (Osawa 1961; Francois and West 1982). Celery yield increased 10% under arid conditions in response to irrigation waters with EC levels between 4.2 and 5.4 dS/m but declined 10% when EC was in the range of 6.2–8.0 dS/m (Pasternak and De Malach 1994).

Celery is prone to "blackheart," a physiological condition that affects young, quickly developing leaves in the center of the plant. The symptoms, tip burn and necrosis, may spread to the petioles and significantly reduce the amount of crop that can be sold.

Although its incidence may depend on the cation makeup of the saline medium, the role of calcium status in blackheart has not been conclusively proven.

According to Osawa (1963), too much Na^+ and Mg^{2+} in the root media restricted Ca^{2+} uptake and led to damage. In a similar vein, Sonneveld (1988) found that blackheart symptoms were significantly worse when celery was exposed to Na^+-, Mg^{2+}-, and K^+-based salinity than when Ca^{2+} was the salinizing salt. Takatori et al. (1961) discovered that spraying the plants with either $Ca(NO_3)_2$ or $Sr(NO_3)_2$ partially regulated the symptoms and also linked the condition to low substrate Ca^{2+}. In contrast, Aloni and Pressman (1987) found that although *NaCl* salinity decreased Ca^{2+} levels in young, vulnerable leaves, *Na* provided some level of protection against blackheart, and the disorder's occurrence was minimal.

The Mediterranean plant fennel (*Foeniculum vulgare* Mill.) is a biennial aromatic herb. Waste areas such as inland areas of England and Wales, and southern and central California, frequently contain both wild and dulce fennel (*var. dulce*). The main product with marketable quality is the "bulb" with an anise flavor, which is made up of the modified basal part of the leaf petioles. As a garnish, the fluttery leaves are also employed. According to Graifenberg et al. (1996), two fennel cultivars, "Monte Bianco" and "Everest," showed a sensitivity to *NaCl* salinity. For fennel bulb yield and plant fresh weight tolerance parameters (threshold and slope), EC of irrigation water and saturated extract of sandy soil were used as the units of measurement. With a slope of 17.8–18.9% and an EC value of about 3.8 dS/m, the EC of irrigation threshold for bulb formation was 1.15 dS/m. The threshold for EC saturated extract was once more 1.15 dS/m, but the slope was between 14.3 and 15.7, and EC was roughly 4.8 dS/m. Fennel bulbs gathered more *Na* and *Cl* than either the leaves or the roots, despite minor varietal differences.

Native to the eastern Mediterranean, parsnip (*Pastinaca sativa* L.) is a typical plant of waste areas and wayside vegetation, especially on calcareous soils. It has been grown for its larger, tapering taproot at least since the Roman era, but better varieties were not likely created until during the Middle Ages. Parsnip has been classified as salt sensitive, with considerable yield losses anticipated when EC surpasses 0.8 dS/m, even though there is a dearth of quantitative evidence on the crop's tolerance to salt (Malcolm and Smith 1971).

The impact of *NaCl* salinity on four types of umbelliferous plants was studied by Zidan and Elewa (1995). The germination concentration in the first 24 h was 120 mM *NaCl* in anise (*Pimpinella anisum*), 150 mM *NaCl* in coriander (*Coriandrum sativum*, popularly known as cilantro), and 200 mM *NaCl* in caraway (*Carum carvi*) and cumin (*Cuminum cyminum*). Yet, seedling growth of caraway and cumin appeared to be encouraged by *NaCl* concentrations up to 80 mM. Seedling dry weights in anise and coriander often declined in tandem with increasing salinity. Anise and coriander seedlings exhibited a salinity-dependent increase in total free amino acid and proline levels, although the level of proline increased at the expense of the other amino acids in cumin and caraway.

7.2.2.4 Araceae

A vital subsistence crop in many islands and nations in the South Pacific, Asia, and Africa is the nutrient-rich root crop taro (*Colocasia esculenta* (L.) *Schott*). Since taro is often grown vegetatively and many cultivars produce irregular flowers and seeds, tissue cultures have been suggested as a means of increasing salt tolerance (Nyman et al. 1983).

7.2.2.5 Asteraceae

Ancient Egyptians, Greeks, and Romans all grew lettuce (*Lactuca sativa*, L.). The Arabs extensively disseminated improved forms. Except for the warmest tropical lowlands, lettuce has been produced on every continent from the time of European colonization. It was discovered that lettuce is moderately salt sensitive in field plot tests at the US Salinity Laboratory in Riverside, California, with a threshold EC of 1.3 dS/m and a slope of 13% (Ayers et al. 1951). Results of a field trial in Israel, however, showed that irrigation water salinity at 4.4 dS/m had no impact on the yield and quality of iceberg lettuce (Pasternak et al. 1986). Subsequent research revealed that lettuce's ability to tolerate salt improved with age and that romaine varieties were much more salt tolerant than iceberg varieties. Lastly, it was found that, contrary to what had been observed in cultures grown in salinized solution, no apparent osmotic adaptation appeared to have taken place as a result of the irrigation water's increased salinity (Shannon et al. 1983). Why the reported data differs so widely is impossible to ascertain. It is probable that the non-salinized control treatments were under more stress and that imposed salinities were not administered as early as those in the Riverside trials because the data from Israel did not cover a high number of salinity treatments. The Israel experiment's soils were gypsiferous, and Russo (1987) notes that over-irrigation can partially mitigate the impacts of salt. There is also a lot of data to support the idea that different lettuce cultivars tolerate salt in different ways. To mitigate the effects of field variability, Shannon (1980) selected salt tolerance in the lettuce cultivar "Empire" using the four-probe electrical conductivity apparatus (Rhoades 1979). Successful choices for a large increase in plant fresh weight (frame) or a high head-to-frame ratio were made in one screening cycle. A wide variety of cultivars and plant introductions of *L. sativa* were evaluated for salt tolerance during early seedling growth in later investigations, which were carried out in greenhouse sand cultures under more controlled conditions than in the field (Shannon et al. 1983; Shannon and McCreight 1984). *L. sativa* plant introductions demonstrated higher mean average salt tolerance and a wider range of salt tolerance than standard cultivars. Some wild cousins of cultivated lettuce, including *L. serriola*, *L. vignata*, and *L. saligna*, were shown to have an even wider range of tolerance than the introductions, according to later investigations. The germination for lettuce appears to be around 8 dS/m and is very varied among cultivars, according to germination studies with *NaCl* (Odegbaro and Smith 1969;

Coons et al. 1990). According to research by Lazof and Läuchli (1991), lettuce tissues next to the apical meristem experience an increase in Na^+ and Cl^- and a corresponding decrease in Ca^{2+}, K^+, and PO_4^{2-}. Such changes in ion compositions were predicted to impact the apical meristem's ability to grow, which could indicate reduced leaf growth. According to other investigations, growth was not improved even though exogenously given Ca^{2+} increased nutritional levels of Ca^{2+} during salt stress and decreased Na buildup (Cramer and Spurr 1986a). Higher root Cl^- levels were discovered to be advantageous in maintaining root water content using two lettuce cultivars that differed in their susceptibility to salt (Cramer and Spurr 1986b).

Native to North America, the Jerusalem artichoke (*Helianthus tuberosus*) was first grown by indigenous people before the arrival of the Spanish conquistadors. Early in the 1600s, the palatable tubers were brought to Europe, where the crop became popular in places where white potatoes cannot grow due to excessive dryness or bad soil conditions. The Jerusalem artichoke has been evaluated as fairly salt tolerant with a threshold EC of 8.3 dS/m, slope of 1.2%, and a yield drop at an EC of 7.5 dS/m based on final tuber yield per plant in field testing. Salinity treatments considerably decreased plant density when tuber production was evaluated in terms of land area, which led to the crop being rated as sensitive to moderately sensitive. Thus, the slope was 9.62%, the predicted yield drop was 5.8 dS/m, and the salt tolerance threshold EC was 0.4 dS/m (Newton et al. 1991). Salinity caused the *Cl* in stems to grow, but leaf *Na* stayed low and was probably under some kind of control.

The Greeks and Romans were aware of the globe artichoke (*Cynara scolymus*) as a food plant because it is a native of the Mediterranean region. It is likely that the huge succulent forms originated during the Renaissance. Although it has not been identified as a wild plant, it might be related to the wild cardoon (*C. cardunculus*). The immature flower head of the artichoke, which consists of the fragile bract bases and the fleshy receptacle or "heart," is what is grown. Certain cuisines may use small, extremely immature whole heads.

Based on research done in a greenhouse (Graifenberg et al. 1993) and a field trial done in an irrigated desert location, the artichoke has been classified as a crop that is only moderately salt tolerant (Francois 1995). Graifenberg et al. (1993) indicated that the tolerance threshold EC was 4.9 dS/m and the slope was 10.7% based on crop performance in the greenhouse. Francois (1995) found comparable results for field-grown artichokes (threshold 6.1 dS/m, slope 11.5%), but an interior browning drastically decreased the quantity of marketable buds. With a rise in saline level, the disorder's frequency and severity increased. According to Francois et al. (1991), the root pressure-driven Ca^{2+} transfer to the shoot apex was hindered in dryland environments. Economically significant vegetable crops include *Cichorium intybus* (chicory, witloof chicory, Belgian endive, chicon, radicchio, and Italian dandelion) and *C. endivia* (endive, escarole), which are both members of the tribe Cichorieae (*Asteraceae*). Given the chance, plants from the two species would freely hybridize, creating several intermediate kinds and complicating taxonomic classification. Whereas C. *intybus* most likely hails from the Mediterranean, *C. endivia* might be a Himalayan native. Chicory was brought to North America as a garden plant in the middle of the eighteenth century. Chicory is a weed that grows in semiarid waste areas where it is likely salty; hence, it is possible that this species has kept some salt tolerance. There was no information about this tribe's salt tolerance.

7.2.2.6 Brassicaceae

Brassica is a large, diversified genus of leafy vegetables with many genomic groups that are highly compatible with one another. A polymorphic species of well-known vegetables called *B. oleracea* presumably evolved from wild sea cabbage. Around 4000 years have passed since the cultivation of this eatable plant. Vegetables come in a broad variety thanks to various changes made to the leaf or shoot system. The more extreme morphological variations emerged later than head cabbages. Although the cultivated vegetables' mature appearances can vary greatly, their root, fruit, and seed structures are remarkably similar to one another, making it impossible to tell them apart as seedlings.

Kale (*Brassica oleracea, Acephala* group) and cabbage are closely related, but kale has open leaves that grow from a simple, upright, robust stem rather than a compact head. The oldest Brassica species appears to be kale (Brouk 1975). While Malcolm and Smith (1971) speculate that the crop may be productive when watered with waters that have electrical conductivities in the range of 2.3–5.5 dS/m, there is little information available regarding the salt tolerance of kale.

Broccoli (*Brassica oleracea, Botrytis* group) typically grows with small, loose heads from buds that appear in the leaf axils of both the main stem and side shoots. In contrast to cauliflower, which is primarily made of fleshy blooms, broccoli stalks make up the majority of the edible portion since their stems are much thinner and longer than those of cauliflower (Brouk 1975). With an estimated threshold EC of 2.8 dS/m and a slope of 9.2% for each unit rise in salinity, broccoli is a crop that is only moderately sensitive to salt (Bernstein et al. 1974).

According to the oldest recorded account of cauliflower (*Brassica oleracea, Botrytis* group), which was written in a book by the Dutch botanist Dodoens in 1559, it appears to be a native of Asia Minor and was known in Europe in the sixteenth century. The racemose inflorescence, which is made up of fleshy, short, and densely clustered abortive flower stalks, has a firm head that is edible. Although there is a modest salt tolerance rating for the crop (Bernstein 1959), there is little quantitative information accessible.

Cabbage (*Brassica oleracea, Capitata* group) has been grown for at least 2000–2500 years, and the Romans brought it to Britain. On a short stem, the smooth, meaty leaves form a tight, firm head. According to studies by Bernstein and Ayers, Osawa (1965), and Bernstein et al. (1974), yield, as determined by head weight, is evaluated as being somewhat sensitive to salinity. With a slope of 9.7% per dS/m, the threshold salinity is 1.8 dS/m. Compared to nonsaline settings, cabbage heads are often more compact, and the leaves are fleshier.

The Brussels sprout (*Brassica oleracea, Gemmifera* group) is thought to have originated in the northern region of what is now Belgium in the fifteenth century and was first mentioned in writing in 1587. The lateral buds that develop on the stems in place of lateral branches are edible. According to Maas and Grattan (1998), the crop is fairly sensitive to salt because of its evolutionary connections to other Brassica species.

The base of the stem, which is swollen to create a spherical, 5–12 cm in diameter bulge resembling a turnip, is the part of the kohlrabi (*Brassica oleracea, Gongylodes* group) that can be eaten. Kohlrabi's history is hazy, although it appears that it was grown throughout Europe before the early Middle Ages. The sensitivity to salt in kohlrabi is average. Field tests have shown that irrigation fluids with EC between 4.2 and 5.4 dS/m significantly decreased yield by roughly 30% (Pasternak and De Malach 1994).

Pe-tsai, also known as Chinese cabbage (*Brassica campestris, Pekinensis* group), has been grown in China since the fifth century AD. It appears to be a native of that country. Petioles and leaves are eaten. Lettuce and Chinese cabbage were evaluated for their interactions with salinity and nitrate nutrition by Feigin et al. (1991). Chinese cabbage biomass output was not considerably decreased until soil salinity approached 3.2 dS/m. Following that, yield decreased by roughly 10% per dS/m, placing Chinese cabbage in the same category as other *Brassica* crops, such as cabbage, cauliflower, and Brussels sprouts, which are highly salt sensitive (Maas 1986). Certain cultivars may be more prone to severe tip burn condition than lettuce in response to salinity (Pasternak and De Malach 1994). Chinese cabbage is highly susceptible to the type of nitrogen provided when grown in salty environments (Feigin et al. 1991). When $NH_4{-}N$ was used, marginal tip burns on younger leaves became more common and more severe. Leaf curling was observed in all salt treatments in sand cultures salinized with *NaCl*, and leaves were dark bluish green when EC exceeded 14 dS/m (Osawa 1961). According to Paek et al. (1988), sulfate salt was more than twice as inhibitive to growth and fresh weight-to-dry weight ratios in callus cultures as *NaCl*.

Pak choi (*Brassica campestris, Chinensis* group) is grown for its green blades and plump, white leaf petioles. It is a Far Eastern native and is widely used in China, Japan, and Southeast Asia. It matters how N is delivered to plants under salt stress. Pak choi leaves were less damaged by $NO_3{-}N$ than $NH_4{-}N$, likely because it prevented the absorption of harmful quantities of *Cl* (Osawa 1955). Pak choi and three other vegetables were grown in sand cultures by Osawa (1966) and were irrigated with *NaCl* or concentrated nutritional salt solutions beginning at the cotyledon stage of growth. Pak choi's EC for yield was determined to be 17 dS/m in NaCl and a little bit higher in nutritional salts. When pak choi was irrigated with Na_2SO_4-dominated salty drainage waters, tests conducted in outdoor sand cultures at the US Salinity Laboratory suggest that the EC for pak choi was roughly 14 dS/m. At salinities ranging from 3 to 23 dS/m, yield decreased at a rate of around 4% per dS/m.

Several researchers have noted that *B. juncea* is tolerant to salt (Jain et al. 1990; Ashraf and Naqvi 1992). However, the focus of research has been on brown mustard, often known as Indian mustard (*B. juncea Czern.* and *Coss.*), a crop valued for the production of seed oil (Ashraf and McNeilly 1990; Sharma and Gill 1994). About the effects of salinity on certain leafy mustard kinds, which are significant and well-liked specialties in cuisines all over the world, there is little to no evidence, if any. The shape and color of the leaves can vary greatly depending on the variety, being either broadly round or narrow and deeply notched (bright green, purple, or brownish red). When *B. juncea* plants were cultivated in solution cultures with 50 mM *NaCl*, their shoot weight dropped to less than 50% of the non-salinized controls (Ashraf and McNeilly 1990). There is a lot of variation across cultivars. Five different *B. juncea* cultivars were irrigated with 100 mM *NaCl* solutions for 4 weeks in sand cultures, and the relative reductions in shoot growth ranged from 28% to 72% (Ashraf 1992). The ability to decrease stomata frequency in response to salt stress, stronger K^+/Na^+ selectivity, and higher leaf succulence were found to correlate with salt tolerance in *B. juncea* cultivars (Kumar 1984). Supplemental Ca^{2+} does not appear to considerably increase the salt tolerance of *B. juncea* (Schmidt et al. 1993).

Because of their quick development and relatively tiny genomes, some Brassica species have proven to be valuable model plants for research in genetics, molecular biology, and physiology. In four cultivated Brassica species planted in sand-filled pots and irrigated with *NaCl* solutions, Ashraf and McNeilly (1990) compared the vegetative development. *B. juncea* and *B. campestris* were shown to be more sensitive to *NaCl* salt than *B. napus* and *B. carinata* when the plants were collected right before flowering. He and Cramer (1992, 1993a, b) used solution culture studies to examine the effects of seawater salinity dilutions on the relative salt tolerance, growth, and ion relations of six rapid-cycling genetic strains, including *B. campestris*, *B. nigra*, *B. oleracea*, *B. juncea*, *B. napus*, and *B. carinata*. Based on shoot growth of plants picked right before flowering, *B. napus* was the species most tolerant of salt, whereas *B. carinata* was the species most sensitive to salt. This finding supported earlier investigations. The remaining four species were deemed to be just moderately sensitive to salt. He and Cramer (1993c, 1996) investigated the effects of salinity on the development and physiological characteristics of the two species, *B. napus* and *B. carinata*, which represented the two extremes in salt sensitivity.

Turnips (*Brassica rapa* L., *Rapifera* group) have been farmed for human and animal consumption for many thousands of years. They are a native crop of Russia, Siberia, and the Scandinavian nations. The enormous, lobed green leaves and fleshy roots (hypocotyl) of the biennial herbaceous plants are what people grow them for. Turnip tops can tolerate salt far better than the roots (Osawa 1961). Roots were classified as being moderately sensitive. According to Francois (1984), production of root biomass decreased by 8.9% for each unit rise in salinity above a threshold of 0.9 dS/m. Nevertheless, Malik et al. (1983) discovered that turnip roots grown in salinized soil-filled pots (between 1.1 and 2.1 dS/m) did not lose any fresh weight. According to both studies, an EC of roughly 6.5 dS/m corresponds to the drop in root development. With a threshold of 3.3 dS/m and a yield loss of 4.8% for every unit of salinity increase, turnip shoots were only moderately salt tolerant (Francois 1984). During the early stages of growth, turnips are more salt tolerant than later stages. Final germination percentage was unaffected by a salt level of 11.6 dS/m, which should have reduced root growth by 95% (Francois 1984).

Eruca sativa is most likely native to western Asia and southern Europe. It frequently grows on severely salt-affected soils in arid and semiarid locations (Deo and Lal 1982; Ashraf and Noor 1993). The seeds of arugula are a rich source of oil and protein and are used as salad greens. Two genotypes of *E. sativa* were compared to *Brassica carinata* or Ethiopian mustard for their relative salt tolerance and ion relationships (Ashraf and Noor 1993). In terms of yield and relative growth rate, the normal line and Ethiopian mustard were outperformed by the Eruca line, which was taken from a field that had been impacted by salt. The former line's salt tolerance seems to be correlated with Na exclusion, strong *K/Na* selectivity, and high Ca^{2+} uptake. In order to further compare the two populations of Eruca, Ashraf (1994) looked at the relative salt tolerance of soluble sugars, proline, free amino acids, and soluble proteins. As comparison to the non-tolerant population, the tolerant line accumulated much more sugars, proline, and amino acids in the leaves. Yet soluble protein did not differ between the genotypes. In salinized sand cultures, the drop in vegetative growth of the tolerant line happened at about 300 mM *NaCl*.

Most likely, western Asia is where radish (*Raphanus sativus L.*) first appeared. It was first grown 4500 years ago in Egypt and Assyria, and it reached China at least 2000 years ago. There are numerous varieties of radish, including the large daikon. *Radicula*, the most common kind, can be lengthy (6–7 cm) or spherical (2 cm in diameter). The swelling hypocotyl is the component that can be eaten. Radish is a crop vulnerable to salt (Osawa 1965; Malcolm and Smith 1971). When growing radish under low relative humidity (45%), Hoffman and Rawlins (1971) investigated the interaction between salinity and yield and discovered that when salinity exceeded a threshold of 1.3 dS/m, root production decreased by 13% per dS/m. However, with high humidity (90%), the salt tolerance threshold increased to roughly 5.2 dS/m with no change in the slope. Scialabba and Melati (1990) showed that *NaCl* salinity resulted in a loss of coordination between cellular growth and differentiation in radish seedlings. Wall thickening and metabolic aggregates inside parenchyma cells were visible as salinity rose, indicating structural and cellular changes. An ontogenetic analysis of xylem components can help pinpoint the developmental stage at which seedlings become salt stressed. Salinity had previously been found to variably impede the growth of various root types in radish (Waisel and Breckle 1987). Growth of new laterals was more tolerant than lateral root growth, which was more susceptible. According to studies conducted in NaCl solutions (Shadded and Zidan 1989; Scialabba and Melati 1990), the G50 for radish can range from 14 to 30 dS/m.

7.2.2.7 Chenopodiaceae

Since the Arabs brought it to Spain in the eleventh century, spinach (*Spinacia oleracea* L.), which has its origins in Persia, has been popular throughout Europe. One of the glycophytic chenopods is spinach, a green vegetable with a modest sensitivity to salt. The slope is 7.6%, and the tolerance level for spinach is 2.0 dS/m (Langdale et al. 1971). However, in Israel, irrigation with saline water with an EC of 4 dS/m on sandy soils did not reduce yields, and the harvestable product was of higher quality (Pasternak and De Malach 1994). Moreover, Tomemori et al. (1996)

discovered that seawater diluted to 1000 mg/L salt increased spinach growth in sandy soil. Speer and Kaiser (1991) revealed that spinach showed little growth impairment throughout a 17-day period following addition of 100 mM *NaCl* to hydroponic cultures. Research in solution cultures have demonstrated that spinach is less sensitive to *NaCl* salt than other single-salt formulations on an osmotic basis and that there is no appreciable growth inhibition up to an osmotic potential of around 0.3 MPa, or about 8 dS/m, for the plant (Nieman 1962; Osawa 1963). Spinach was utilized by Chow et al. (1990) to show that high-salinity circumstances demand more K^+ for shoot growth than low-salinity conditions. Reductions in shoot biomass brought on by rising salinity can be mitigated by raising substrate K^+ levels. Since spinach has a higher leaf K^+ content than other leafy vegetables, it is plausible that there is a K^+ requirement that may be important for spinach's apparent sensitivity to salinity. If this theory is correct, selection for K^+/Na^+ selectivity could be a helpful screening criterion to increase salt tolerance.

The widespread seaside plant known as wild sea beet (*Beta maritima*), which may be found throughout the whole coastline of western Asia and Europe, is thought to be the ancestor of both leaf and root beets. In the Mediterranean region, where beets are native, chard has been consumed by people since prehistoric times. The branched, stringy roots are typically removed, while the leaves with their white, green, or red midribs are consumed. Studies to determine the phytoavailability of potentially harmful ions like *Se* (Gutemann et al. 1993) and *Cd* have employed Swiss chard (*Beta vulgaris*) as a test species (Bingham et al. 1983, 1984; Smolders and McLaughlin 1996b). Chloride salinity enhanced *Cd* uptake by chard in every study, and several different mechanisms have been put forth to account for the increased *Cd* availability. 120 mM *Cl* had no impact on growth (Smolders and McLaughlin 1996a). Swiss chard was salinized after the emergence of the first genuine leaves in studies carried out in outdoor sand cultures at the US Salinity Laboratory in 1997 with six concentrations of simulated drainage waters predominately made of Na_2SO_4 salts. Dry weights rose up to a maximum of 11 dS/m, after which they began to fall at a rate of roughly 5.7% per dS/m. The irrigation solution's EC, which was used to calculate conductivity level of the yield, was 19.8 dS/m. Starting at the cotyledon stage of growth, Osawa (1966) cultivated Swiss chard and three other vegetables in sand cultures and watered them with strong solutions of nutritional salts. In this study, the chard yield's conductivity was calculated to be 17.5 dS/m.

About 300 BC, beets were recognized as vegetables. The bloated hypocotyl is consumed. Although no trustworthy research on soils has been done, it is considered to be fairly salt tolerant. Total yield of top plus roots rose in sand cultures salinized with *NaCl* and $CaCl_2$ salts up to salinities equivalent to about 0.2 MPa osmotic potential (EC 5.2 dS/m) and dropped at about 0.3 MPa (Bernstein et al. 1974). Hoffman and Rawlins (1971) discovered that osmotic potentials of 0.5, 1.0, and 1.5 MPa decreased beet yields by 40%, 72%, and 91%, respectively, in gravel cultures that were watered with nutrient solutions and *NaCl*. This information, along with prior data from the US Salinity Laboratory (Magistad et al. 1943), can be used to determine the threshold EC for beet, which is 4.0 dS/m, and the slope, which is 9%.

Orach (*Atriplex hortensis L.*) is a plant that is native to western Asia and southeast Europe. Since ancient times, these regions have grown orach for their young, tasty leaves. It was cultivated in kitchen gardens in western Europe up to the eighteenth century and is still farmed to a lesser amount in France and central Europe, despite being substantially replaced by spinach. In solution cultures with modest salinities of 50 mM *NaCl* or Na_2SO_4, Jeschke and Stelter (1983) investigated the growth and ion relationships of orach. At 50 mM *Na* salts, there was a considerable stimulation of growth, dry matter production, and leaf size. The existence of bladder hairs, which virtually completely remove Na^+ from young leaf lamina, and the recirculation of K^+ from leaves to roots in orach plants establish K^+/Na^+ selectivity. Regardless of whether the anion was Cl^-, Br^-, or SO_4^{2-}, Na^+ and K^+ were able to successfully promote the leaf succulence of orach. Ca^{2+} and Mg^{2+} had no impact on succulence. After 45 and 54 days of development, *A. hortensis*' growth was inhibited by around 9% and 35% at 100 mM *NaCl* in solution cultures (or about 10.1 dS/m), respectively (Handley and Jennings 1977). After 54 days of development, the predicted conductivity was around 300 mM *NaCl* (30 dS/m). Red orach was salinized 19 days after seeding in an experiment at the US Salinity Laboratory using eight levels of simulated drainage waters primarily made of Na_2SO_4 salts with EC ranging from 3 to 24 dS/m. The highest plant dry weights were taken from plots irrigated with simulated drainage water at 10 dS/m 70 days after sowing. Using drainage water at 24 dS/m, the dry weights of plants were lowered by 50%.

7.2.2.8 Convolvulaceae

Native to South America, sweet potatoes (*Ipomoea batatas*) were already being grown there by at least 2500 BC, according to archeological findings. It is a component of Polynesia's old agricultural complex as well, and because of its sporadic distribution, numerous theories concerning early migrations across the Pacific have been developed (DeRougemont 1989). In 1493, during his return voyage, Columbus brought these plants to Spain and Portugal. The swelling storage root is what is edible.

This plant ranks as the seventh most crucial dietary staple in the world. On the other hand, it is delicate to salinity, aluminum toxicity at low pH, and low fertility (Horton 1989). Vine growth is less vulnerable to salinity than root growth is (Greig and Smith 1962). According to reports, the EC for salinity in sweet potatoes is either 11.0 or 4.0 dS/m in irrigation water (Maas and Hoffman 1977). Using salinized in vitro cultures, Ekanayake and Dodds (1993) thoroughly evaluated the salt tolerance of sweet potato germplasm. They evaluated the growth and survival of seedlings in 38 farmed clones and 17 salt-resistant clones, the latter of which had been chosen from field locations with a wide range of salinity (Horton 1989). According to reports, the 55 sweet potato genotypes represented both different geographical areas in Peru and the International Potato Center's germplasm collection for sweet potatoes. *NaCl* concentrations as low as 0.5–1.0 mM in liquid media greatly decreased the number of nodes developing roots, the number of roots per node, and the dry weight of plantlets. Despite the fact that considerable clonal differences were discovered, field observations did not match up with the findings.

As a result, the screening process was changed to apply a 16 mM *NaCl* stress for a shorter amount of time. In these circumstances, a statistically significant association between field observations and the approach was discovered; nonetheless, the method was only thought to be useful for early vegetative survival and hence early generation testing.

7.2.2.9 Euphorbiaceae

Of all root crops, cassava (*Manihot esculenta Crantz*) is the one that is grown the most. Tapioca is produced from starchy tubers. The crop most likely originated in tropical Brazil and spread thousands of years ago to other regions of South America (Yamaguchi 1983). Although screening and selection have shown that there is room for improvement, cassava is a crop that is only moderately sensitive to salt.

When irrigated with solutions containing between 30 and 50 mM *NaCl*, the weight of the cassava's tuber was cut in half (Hawker and Smith 1982). Even putatively tolerant cultivars in Colombia's long-term yield studies showed a 50% yield drop at a salt level of just 0.7 dS/m (Anon. 1976, cited in Hawker and Smith 1982). Using soil cultures salinized with 1500 ppm *NaCl*, Indira and Ramanujam (1982) employed the leaf K^+:Na^+ ratio to screen and select 14 cassava genotypes for salt resistance. With a soil EC of 3 dS/m, six selections with leaf K^+:Na^+ values greater than 15 were produced in the field (pH 8.65). None of the genotypes displayed problems linked to salinity, and all genotypes stabilized properly.

7.2.2.10 Portulacaceae

The Mediterranean region and central Europe have supposedly been growing purslane (*Portulaca oleracea*) since ancient times. It is believed to be a native of western Asia. Archaeological sites in southern Canada and the USA contain seeds (Gorske et al. 1979). In Mediterranean areas, the crop is grown commercially, and the fleshy stems and leaves are utilized in salads. With a salinity threshold of 6.3 dS/m and a slope of 9.6%, purslane is classified as being moderately tolerant (Kumamoto et al. 1990). Purslane's halophytic character, however, manifests itself after the first cutting, and with each consecutive harvest, its tolerance to salt rises (Grieve and Suarez 1997).

7.2.2.11 Solanaceae

Prior to the Spanish discovery, the potato was grown for more than 2000 years by the Incas at altitudes of over 2000 m in the Andes. In 1537, these adventurers brought the potato to Europe. The tuber, a fleshy stem with buds in the axils of leaf scars, is the sole portion of the potato plant that may be eaten.

Potato has been categorized as being somewhat tolerant to salinity, but it is more susceptible when the tuber bud is first forming. Shortly after, it becomes more tolerant when the proportion of extra-large tubers decreases in favor of smaller, more marketable tubers due to salinity. Due to a decrease in average tuber size, potato is once more sensitive when salt duration and/or concentration increase.

Potato is extremely vulnerable to both drought and a calcium deficiency (Abdullah and Ahmad 1982; Bilski et al. 1988; van Hoorn et al. 1993), which increases the difficulties in conducting research on salt tolerance.

Salinity was observed to lower tuber size and number as well as hasten maturation in the "White Rose" potato in field plot research carried out at the US Salinity Laboratory in 1951 (Bernstein 1959). These trials used frequent irrigations supplemented with $NaCl\text{-}CaCl_2$ (1:1 by weight) to achieve average root zone salinities of 0.85, 3.37, 4.85, and 6.46 dS/m, with a 50% yield reduction at 6.2 dS/m. It was shown that salt had no discernible effects on potato quality as determined by specific gravity or the proportions of reducing sugar, sucrose, and starch in tubers, nor did it result in any signs of harm on the leaves. Across the treatment range, the amounts of Ca^{2+}, Cl^-, and Na^+ in the leaves and stems increased three- to fourfold, but the amounts of Na^+ in the leaves remained low. Treatment had little to no effect on K^+ and Mg^{2+} levels in leaves and stems (Maas 1986).

In the Negev desert at three salinities, Levy (1992) investigated the salt tolerance of 14 potato cultivars in field settings. The greatest saline solution's EC was 6.1–6.9 dS/m with a Na^+:Ca^{2+} ratio of roughly 2:1 by weight, and irrigations were applied often by drippers. Salinity sped up maturity, slowed down growth of both haulms (shoots) and tubers, and delayed plant emergence. Nadler and Heuer (1995) further supported earlier findings in a study also done in the Negev that salt had the favorable impact of lowering the proportion of extra-large tubers and increasing the yield of large tubers (Bernstein et al. 1951; Paliwal and Yadav 1980). It was found in the later investigation that tuber size, but not tuber number, decreased with increased salinity, supporting the initial finding by Bernstein and colleagues.

It has been shown that potato cultivars differ in how well they tolerate salt, although no correlation has been found between salt tolerance and physiological or morphological traits. Levy (1992) discovered that among 14 cultivars, the connection between maturity time and salt tolerance was inconsistent. However, the early-maturing cultivars Atica, Desiree, and clone LT4 were the least sensitive to mild salinity. Notwithstanding the pessimism of this researcher, a growing body of research on potato and other species suggests that some salt tolerance may be related to earlier maturation (salinity escape), provided that this early maturity is not linked to a decrease in yield. This theory is in line with common data that show how a plant's ability to mitigate the impacts of ions that build up in its tissues due to high salinity depends on its ability to develop at a faster rate.

Despite exposing seven cultivars to $NaCl$ concentrations as high as 51.3 mM, Levy et al. (1988) found no correlation between high proline and salt tolerance. *S. kurzianum*, a species of wild potato, has been discovered to be more salt tolerant due to lesser growth reductions with increasing salinity than the cultivars Alpha and Russet Burbank (Sabbah and Tal 1995). The accumulation of Na^+ in the shoot was shown to be higher in wild species than in cultivated species; however, the buildup of Cl^- and the presence of Ca^{2+} may have a substantial impact on salt tolerance. Moreover, field experiments have demonstrated the significance of gypsum applications on potatoes growing in salt (Abdullah and Ahmad 1982).

7.2.3 Soil Salinity and Salinity Resistance Mechanisms

7.2.3.1 Salt Stress and Nutrient Uptake

Although Na^+, Ca^{2+}, and Mg^{2+} are the most frequent cations found in SAS, along with anions Cl^-, SO_4, and HCO_3, these soils are primarily dominated by Na^+ and Cl^- (Brady and Weil 2005; Hardie and Doyle 2012; Qadir et al. 2005). When Na^+ and Cl^- replace the critical physiological roles that nitrogen, phosphorus, potassium, calcium, and magnesium play in plants, nutritional imbalances may result (Chen et al. 2010). An appropriate and balanced supply of mineral nutrients is crucial for the best growth and yield of crops (Sairam and Tyagi 2004). By the processes of fixation, adsorption, and transformation in soil (Maksimovic and Ilin 2012), high Na^+ and Cl^- levels in the rhizosphere might obstruct the uptake of critical elements, resulting in their shortfalls or imbalances (Hu and Schmidhalter 2005; Silva et al. 2003).

Calcium and Magnesium

With around 3% and 2.6% of the earth's crust, respectively, calcium makes up the fifth most abundant element and sodium makes up the sixth most abundant element. Ca^{2+} has a larger hydrated ionic radius of 0.44 nm due to its higher charge density than Na^+, which also draws more water (Cramer 2002). Due to its capacity to create intermolecular bonds, calcium, an essential inorganic nutrient, is crucial in preserving the structural and functional integrity of cell walls and membranes (Rameeh 2012; Tuna et al. 2007).

Many studies on *Na*:*Ca* interactions have been conducted in light of the significance of external *Ca* in maintaining *K* transport and *K*:*Na* selectivity in plants under *Na* stress conditions. A common sign of Na toxicity in plants is a *Ca* deficit (Davenport et al. 1997). The ability of a plant to withstand salt stress can be expressed by the *Na*:*Ca* ratio of plants (Hadi and Karimi 2012). By inhibiting *Ca* and *Mg* inflow through roots, decreasing the extracellular binding sites for *Ca* and *Mg* in the plasma membranes (Hadi and Karimi 2012; Lynch and Läuchli 1998), and decreasing the osmotic potential (Sahin et al. 2018) under saline conditions, a high *Na* concentration can reduce the uptake of *Ca* and *Mg*.

In saltmarsh grass (*Spartina alterniflora*), higher *Na* levels were found to dramatically lower *Ca* and *Mg* concentrations (Brown et al. 2006). The ratios of *Ca*:*Na* and *Mg*:*Na* in the leaves and roots of cabbage, *Brassica oleracea*, decreased with the increasing salinity levels, and there were also significant negative associations between salt stress and contents of *Ca* and *Mg* (Sahin et al. 2018). Similar findings were reported by other scientists, who found that plants under salt stress accumulated more *Na* and *Cl* and contained less *Ca* and *Mg* (Maksimovic and Ilin 2012; Parida and Das 2005. Nevertheless, Chen et al. (2010) discovered higher *Na* and *Ca* levels in cotton leaves as salt stress increased. *Fragaria ananassa*, a strawberry plant, has lower *Ca* and *Mg* contents in the shoot than in the root, according to Rahimi and Biglarifard's (2011) observations. Several studies have shown that adding additional *Ca* in conditions of high salinity enhanced the *Ca* concentration in rapeseed (*Brassica napus*) and tomato (*Lycopersicon esculentum*) plants. Salinity, however, has been reported to have more negative impacts in environments with varying degrees of drought stress (Sahin et al. 2018; Brown et al. 2006; Saud et al. 2013).

The amount of *Na* as well as genotypes can affect how much *Ca* and *Mg* enters the root and is transported to the shoot. In *Triticum aestivum*, a salt-sensitive cultivar of wheat, *Ca* was found to be more readily absorbed at low *Na* concentrations and less readily transported at higher concentrations (Davenport et al. 1997). Moreover, in comparison to sensitive accessions grown in high salinity, tolerant forms of the green chiretta (*Andrographis paniculata*) had higher *Ca* and *Mg* concentrations and lower *Na* concentrations (Talei et al. 2012).

Micronutrients

Micronutrients have a role in a number of crucial physiological and biochemical processes in plants, including the activation of enzymes, production of chlorophyll, protein synthesis, synthesis of carbohydrates, and metabolism of lipids and nucleic acids (Zaman et al. 2018). In salt-stressed conditions, various writers observed varying amounts of micronutrient uptake and accumulation, including *Fe*, *Mn*, *Zn*, and *Cu*. Several soil, plant, and ambient factors are thought to affect the uptake and accumulation of micronutrients in different plant sections during salt stress, while the precise mechanism is not well understood in the literature.

In earlier investigations, it was discovered that the response of micronutrient concentrations in plants under salt stress was variable. Because of their poor solubility and availability, micronutrients as *Fe*, *Zn*, *Mn*, and *Cu* frequently showed deficiencies in plants growing in salty soils (Chen et al. 2012). In the tissues of *Avicennia marina* (Forssk.), Patel et al. (2010) discovered a negative correlation between the salt content and the concentrations of *Cu*, *Mn*, and *Fe*. When exposed to high salt concentrations, Chakraborty et al. (2015) discovered that different sections of *Brassica* spp. accumulated less *Fe*, *Mn*, and *Zn* during the flowering and post-flowering stages. The contents of *Fe* and *Zn* in the roots and leaves of the cabbage (*Brassica oleracea*) were also found to decrease under salt stress (Sahin et al. 2018). While the *Mn* content of the green chiretta (*Andrographis paniculata*) dropped under high salinity, the concentrations of *Fe*, *Zn*, and *Cu* dramatically rose (Talei et al. 2012). When strawberry plants (*Fragaria ananassa*) were subjected to salt stress, the concentrations of *Fe*, *Mn*, and *Zn* increased in the aboveground portion of the plant while their contents did not alter in the root. Contrarily, salt stress enhanced the concentration of copper in the root while the content did not alter in the aerial portion (Turhan and Eris 2005). During salt stress, it was found that the Fe content of strawberry roots and shoots had significantly decreased, but that *Zn*, *Cu*, and *Mn* concentrations had not changed (Rahimi and Biglarifard 2011). *Triticum aestivum*, a type of wheat, did not appear to be constrained by changes in the concentrations of *Fe*, *Zn*, or *Mn*, and salt stress had no effect on the contents of plants containing these elements (Hu and Schmidhalter 2001). Several types of the green chiretta (*Andrographis paniculata*), the *Brassica* spp. (Chakraborty et al. 2015), and the strawberry (*Fragaria ananassa*) (Turhan and Eris 2005) responded differently to salt stress in terms of micronutrient contents.

Nitrogen

All micro- and macroorganisms have a significant amount of nitrogen in their proteins and nucleotides (Khan et al. 2017; Elser et al. 2000). When compared to other necessary nutrients, it is one of the limiting variables and is needed in much higher amounts for crop development, production, and quality (Blumenthal et al. 2008; Chen et al. 2016; Hammad et al. 2017; Hofman and Cleemput 2004). The average total *N* content in mineral soils is between 0.05% and 0.2%, or roughly 1750 and 7000 kg of *N* per hectare, while in plants, it is between 1% and 4% on a dry matter basis (Hofman and Cleemput 2004). More than 90% of the total *N* comes from soil organic matter, and nitrogen is only sometimes found in the lithosphere in rock deposits (Brady and Weil 2005). NH_4^+ and NO_3^+ are the main forms in which it is primarily absorbed by plants (Dai et al. 2015). The uptake and assimilation of NH_4^+ and NO_3 are inhibited by excessive concentrations of *Na* and *Cl* as a result of high salinity. As the presence of *N* in the growth medium encourages the uptake and assimilation of other important nutrients, the suppression of *N* metabolism may affect the uptake of other essential nutrients (Hofman and Cleemput 2004). Plants under salt stress may exhibit a physiological response by reducing *N* absorption (Bybordi and Ebrahimian 2011).

Salt stress was found to strongly inhibit the uptake and accumulation of *N* in different plant parts of green chiretta (*Andrographis paniculata*) (Talei et al. 2012), cabbage (*Brassica oleracea*) (Sahin et al. 2018), canola (*Brassica napus*) (Bybordi and Ebrahimian 2011), cotton (*Gossypium hirsutum*) (Dai etal. 2015), gray poplar (*Populus canescens*) [20], saltmarsh grass (*Spartina alterniflora*) (Brown et al. 2006), sesame (*Sesamum indicum*) (Kanagaraj and Desingh 2017), and wheat (*Triticum aestivum*) (Botella et al. 1997). The uptake and buildup of *N* in response to salt stress, however, depend on several soil and plant variables [80]. The biggest reduction was reported at the highest dose of NaCl, according to Kanagaraj and Desingh's (2017) observations of differences in foliar *N* across several sesame cultivars.

The increased uptake and buildup of Cl^- may be the cause of the decrease in *N* uptake by plants. *Oryza sativa* (Abdelgadir et al. 2005), a type of rice; *Brassica oleracea*, a type of cabbage; and *Hordeum vulgare*, a type of barley, were discovered to have an antagonistic association between the uptake of NO_3 and Cl^-. It was discovered that high soil salinity increased the concentration of Cl^- in the various plant sections of the Poaceae, *Catapodium rigidum* (Zribi et al. 2018); cowpea, *Vigna unguiculata* (Silva et al. 2003); and cotton, *Gossypium hirsutum* (Chen et al. 2010). However, both the length of the salt stress and the stage of the plant's life cycle affect the amount of Cl^- that accumulates (Lacerda et al. 2006). High salinity can also have an adverse effect on NH_4^+ uptake. A higher Na^+:NH_4^+ ratio in the growth medium was seen to be associated with a decrease in NH_4^+ uptake in plants (Dluzniewska et al. 2007).

During salt stress, several investigations found that nutrient buildup was hindered and water uptake decreased (Zribi et al. 2018) [64]. Depending on the types of *N* used, salt stress has different effects on N metabolism (Dluzniewska et al. 2007). When their compounds were incorporated as the source of *N*, Botella et al. (1997)

reported that the increased concentration of *NaCl* in the root medium decreased the net uptake of *N* more profoundly in the NO_3 form compared to the NH_4^+ form. This was assumed to be the cause of the greater affinity for NH_4^+ compared to NO_3 under the saline environment. Similar results were obtained by Saud et al. (2020) under abiotic stress, finding greater *N* content, *N* isotope abundance, and relative water content in the roots and leaves of Kentucky bluegrass, *Poa pratensis*, in the NH_4^+ treatment as compared to the NO_3 treatment. In contrast, Dluzniewska et al. (2007) found that poplars subjected to higher *NaCl* concentrations saw a decrease in the net uptake of NH_4^+, which led to a lower whole-plant *N* content than the control. In contrast to other forms, NH_4^+ is preferred by plants because it does not need to be reduced before incorporating into plant compounds, according to Hofman and Cleemput (2004). However, Dai et al. (2015) discovered that cotton seedlings fed with NO_3 had greater root growth and a lower *Na* content than those fed with NH_4, and they also noted that $NO_3{-}N$ was superior to $NH_4{-}N$ in terms of absorbing *N* when exposed to salt stress. The production of protein by plants may be hampered by the high salt concentration of the soil. Chakraborty et al. (2015) noted that the high salt content, which was reflected in the low protein levels in seeds, caused a decrease in the uptake of *N* by *Brassica* spp. The impaired metabolism of *N* (Maksimovic and Ilin 2012) that may result in the low protein content under saline conditions is blamed for the physiological drought of plants, which is brought on by the low osmotic potential of soil solution.

Phosphorus

For plants to flourish at their best, phosphorus, the second-most necessary nutrient, must make up 0.3–0.5% of the dry matter (Pessarakli et al. 2015). It is a crucial component of membrane lipids and nucleic acids (Elser et al. 2000). The primary sources of *P* in the lithosphere, as opposed to *N*, are rock deposits (Hofman and Cleemput 2004). Plants absorb *P* from the soil solution in the form of $H_2PO_4^-$ and HPO_4^{2-}, although $H_2PO_4^-$ is taken up to a greater amount (Menzies 2009; Syers et al. 2008). Phosphorus is present in the soil solution as orthophosphate ions such as $H_2PO_4^-$, HPO_4^{2-}, and PO_4^{3-} (Hutchins and Fu 2008).

Green chiretta, *Andrographis paniculata* (Talei et al. 2012); cabbage, *Brassica oleracea* (Sahin et al. 2018); canola, *Brassica napus* (Bybordi and Ebrahimian 2011); pistachio, *Pistacia vera* (2011); saltmarsh grass, *Spartina alterniflora* (Brown et al. 2006); and spinach, *Ipomoea aquatica* (Khan et al. 2018), all showed a significant drop in *P* levels after exposure to high salinity. When plants were subjected to both salinity and drought at the same time, it was discovered in multiple studies that the salt stress magnified the negative effects on the uptake of *P* by plants (Sahin et al. 2018; Brown et al. 2006). The struggle between H_2PO_4 and *Cl* ions may be the cause of the decrease in *P* concentration under high salinity (Maksimovic and Ilin 2012). The amount of *P* that builds up in plants during salt stress varies depending on the plant organs. According to Silva et al. (2003), leaves have more phosphorus than roots. On the other hand, Shahriaripour et al. (2011) observed that a high-salinity environment impeded the translocation of *P* from root to shoot.

Many investigations had also noted signs of increasing *P* under high salinity levels. The concentration of *P* in leaves rose when *NaCl* was included in the nutritional solution (Silva et al. 2003; Lacerda et al. 2006). Nevertheless, Zribi et al. (2018) showed that salinity had no appreciable impact on *P* concentration or acquisition efficiency (PAE), and they also discovered that salt stress plus *P* treatment increased the concentration of *Cl* in the shoot compared to salt stress alone. Temperature, moisture, soil pH, and the current soil *P* level are all factors that affect how much phosphorus plants absorb when they are under salt stress (Grattan and Grieves 1999; Lacerda et al. 2006; Syers et al. 2008). Due to their high sorption capacities, divalent cations like Ca^{2+}, Mg^{2+}, and Fe^{3+} at various pH ranges may also have an impact on *P* solubility (Akram et al. 2018b; Hutchins and Fu 2008; Marschner 1995). These cations' high sorption capacities regulate the absorption by plants.

Potassium

Potassium is a significant and plentiful mineral nutrient that makes up between 1% and 10% of plant tissue and 2.6% of the earth's crust on a dry weight basis (Maathuis and Podar 2011; Epstein and Bloom 2004). It supports crucial physiological, biochemical, and biophysical functions that control plant growth and development, including photosynthesis, osmotic adjustment, and turgor maintenance (Epstein and Bloom 2004; Greenway and Munns 1980; Reggiani et al. 1995). The net uptake of *K* is dependent on the *K* concentration in the growth media (Al-Karaki 2000). The relatively high concentration and increased mobility of *K* are thought to be the causes of the seldom deficit found in most soils (Huq and Jum 2013; Mahmud and Roy 2020). Yet, in a salt-stressed environment, the plants may experience *K* insufficiency due to *Na* toxicity. *Na* and *K* are both monovalent cations with comparable physical and chemical characteristics. *Na* and *K* ions have hydrated ion radius of 0.358 nm and 0.331 nm, respectively (Wakeel 2013). High *Na* concentrations can have a negative impact on the uptake and accumulation of *K* in plants because of similar ionic radius and cationic competition for entry into the plant cells.

Salt-affected soils had a considerable impact on *K* concentration in plant tissues, either raising or lowering it. One of the main responses of plants to excessive *Na*, which ultimately results in nutrient imbalances, is a decrease in *K* content in plant tissues (Rameeh 2012). High salt stress with increased concentrations of *NaCl* was found to decrease the total *K* and cause an increase in the *Na* content in Aloe vera (*Aloe vera*) (Jin et al. 2007), barley (*Hordeum vulgare*) (Jiang et al. 2006), bean (*Phaseolus vulgaris* and *P. acutifolius*) (Bayuelo-Jiménez et al. 2003), cabbage (*Brassica oleracea*) (Sahin et al. 2018), cotton (*Gossypium hirsutum*) (Chen et al. 2010), green chiretta (*Andrographis paniculata*) (Talei et al. 2012), maize (*Zea mays*) (Fahad and Bano 2012), pistachio (*Pistacia vera*) (Shahriaripour et al. 2011), Poaceae (*Catapodium rigidum*) (Zribi et al. 2018), rice (*Oryza sativa*) (Saleque et al. 2005), saltmarsh grass (*Spartina alterniflora*) (Brown et al. 2006), strawberry (*Fragaria ananassa*) (Rahimi and Biglarifard 2011), and tomato (*Lycopersicon esculentum*) (Adams and Ho 1989; Kaya et al. 2001). In saltbush (*Atriplex canescens*), salt tolerance ability was found to be positively correlated with

Na and negatively correlated with *K* (Glenn et al. 1996). Moreover, there is proof that as salinity grew, so did the amounts of *Na* and *K* in plants. In a greenhouse experiment, it was discovered that the concentrations of *Na* and *K* in the aboveground and belowground parts of the Jerusalem artichoke (*Helianthus tuberosus*) increased with the application of increasing amounts of seawater (Long et al. 2009). Similar to how the concentrations of *Na* and *K* in cowpea (*Vigna unguiculata*) leaves increased under salt stress, the responses varied depending on the length of the stress and the age of the leaves (Lacerda et al. 2006). With an increase in *K* concentration in the medium, Al-Karaki (2000) discovered that tomato plants translocated *K* from the roots to the shoots to a greater extent in saline environments compared to nonsaline ones. While AKT1 (a hyperpolarization-activated inward-rectifying *K* channel) is used as a pathway for *K* uptake, the increased concentration of *Na* inhibits *K* transport and lowers *K* uptake (Fuchs et al. 2005). The phase of *K* uptake from the soil solution is probably more essential for the inhibitory action of *Na* on the transport of *K* through channels in membranes than the phase of *K* transport to the xylem (Qi and Spalding 2004).

Moreover, the concentration of K in solution affects the inhibitory effect of *Na* on K translocation. Low Na and high K levels in the medium have a negative impact on the outcome (Al-Karaki 2000; Idowu and Aduayi 2006; Krishnasamy et al. 2014). However, the effects of *Na* on the uptake and accumulation of *K* can vary among species and seven between varieties within such same species of Aloe vera, *Aloe vera* (Jin et al. 2007); bean, *Phaseolus vulgaris* (Bayuelo-Jiménez et al. 2012); cotton, *Gossypium hirsutum* (Ali et al. 2009); green chiretta, *Andrographis paniculata* (Talei et al. 2012); mustard, *Brassica nigra* (Purty et al. 2008); tomato, *Lycopersicon esculentum* (Kaya et al. 2001; Tuna et al. 2007); and wheat, *Triticum aestivum* (Krishnasamy et al. 2014). It was discovered that salt-tolerant species retain a high *K* and low *Na* content. An excellent biomarker for salinity tolerance in plants is a high concentration of *K* and a low concentration of *Na* in a saline environment (Rameeh et al. 2004). The various plant sections also react to salt stress in different ways. In general, a larger concentration of *Na* was found in the roots compared to the leaves of different plant species, and this was attributed to the roots' high tolerance and the limited translocation of *Na* from the roots to the leaves (Alam 1999). Various authors observed that stems had decreased *K* content due to salt stress to a higher extent than leaves and roots (Jin et al. 2007; Bayuelo-Jiménez et al. 2012). To ensure consistent photosynthesis and leaf stomatal conductance, leaves must have a high concentration of *K* (Bayuelo-Jiménez et al. 2003).

7.2.3.2 Research About Salinity Resistance Mechanisms

To increase productivity in saline environments, salts must first be removed from the root zone. Nevertheless, the leaching method alone cannot recover saline-sodic and sodic soils. So, the dangerous concentration of salts must first be washed out of the root zone, which can be accomplished by leaching, the most efficient method for getting rid of too many soluble salts. This will improve the growth and yield performance of crops in salty soils. The starting salt concentration, the kind of soluble salts, the intended EC of the soil after leaching, the characteristics of the soil, the quality of the water to be used for leaching, and other variables all affect how effectively salts are removed from the soil profile during leaching (Abrol et al. 1988).

Leaching of soluble salts can be prevented by providing the right amount of water at the right time with sufficient drainage. To lessen the negative impacts of salinity by leaching, a trustworthy estimation of the ideal soil moisture content is necessary. In general, for continuous ponding, a depth of water equal to the depth of soil eliminates 70–80% of the soluble salts; hence, 15 cm of water is needed to reduce the salt content in the top 15 cm of soil by 70–80%. However, because continuous ponding reduces soil aeration and causes water to evaporate quickly in soils with a coarse texture, intermediate ponding or spray irrigation is favored for more effective salt leaching by lengthening the time that salts are in contact with water (Sommerfeldt et al. 1988).

On the other hand, extended drying may raise the salt concentration in the soil solution, lowering the soil solution's osmotic potential. With coarse-textured soils, a significant amount of water can be leached over a shorter period of time due to high permeability and reduced workability. To eliminate the salts from the soil profile, fine-textured soils with high CEC and organic matter need more water (Gaines and Gaines 1994). Leaching of soils can desalinate them, but it also depends on how well the soil drains. While subsurface drainage can maintain the depth of the groundwater and stop further salinization, the application of leaching with surface drainage at shallow groundwater levels may worsen salinity issues (Gelaye et al. 2019). In the section on organic amendments that follows, it is discussed how organic amendments can help SAS be recovered by enhancing its physical characteristics as described in the literature. Leaching is favored by soils with a sound structure and internal drainage (Provin and Pitt 2001). Thus, it is crucial to apply water with good drainage in the right amounts and at the right times to avoid irrigated lands from getting salinized.

Leaching is carried out by allowing irrigation water with a low salt content to drain out or by using natural precipitation or artificial irrigation water. Because saline water contains significant amounts of cations such as Na^+, Ca^{2+}, Mg^{2+}, and K^+ as well as anions like Cl^-, HCO_3^-, and SO_4^{2-}, the salinity issue of agricultural land may worsen if saline water is used for leaching purposes (Spark 2002; Roy et al. 2020).

By minimizing the negative impacts, appropriate nutritional management is frequently seen as practicable and affordable, which can improve the performance of plants cultivated in various conditions (Nasim et al. 2018; Roy et al. 2018; Saud et al. 2017). It has become a common practice over the past few years to apply organic manures as amendments to recover SAS, which is thought to be extremely sustainable (Leogrande and Vitti 2019). Farmers all over the world have accepted the use of organic amendments as a replacement for chemical fertilizers or in combination with chemical fertilizer due to the high cost and rapid release of nutrients involved with chemical fertilizers (Roy and Kashem 2014). Also, the widespread use of inorganic fertilizer increases the risk of groundwater contamination owing to leaching loss, and the production of powerful greenhouse gases like nitrous oxide from agricultural fields is a major factor in global climate change (Akram et al. 2018b; Khan et al. 2017).

Despite the fact that inorganic fertilizers make nutrients freely accessible to plants, they are quickly lost from the soil. On the other hand, organic manure increases soil's physical and biological properties while also releasing nutrients gradually. An efficient and long-lasting strategy for improving crops' tolerance to abiotic stress is the inclusion of organic manures alongside chemical fertilizers (Bhatt et al. 2019; Fahad et al. 2015b, 2016a; Liu et al. 2006). To increase the organic matter content and nutrient status, several sources of organic materials derived from plant and animal residues are used, such as green manure, cattle manure, chicken manure, and food processing wastes. The physical properties of soil are greatly impacted by organic additions.

According to a number of studies, applying organic manures boosted aggregate stability, total porosity, hydraulic conductivity, and permeability while decreasing bulk density and penetration resistance (Mahmood et al. 2017; Obi and Ebo 1995; Tester 1990), as well as aggregate stability (Barzegar et al. 2002; Gümüş and Şeker 2017). The type and texture of the soil, as well as agronomic aspects including management, inputs, and composition of the organic matter, all have an impact on soil organic matter, which is a crucial component of soil quality and aggregate stability (Akhtaruzzaman et al. 2020; Tobiašová 2011). Increased soil porosity and decreased bulk density are associated with improvements in aggregate stability. Good aggregation leads to an increase in porosity and a decrease in bulk density, which facilitates the leaching of soluble salts from the soil and maintains the root zone's appropriate oxygen supply, both of which are essential for crop productivity in SAS.

Increases in soil organic matter (Gümüş and Şeker 2017; Li et al. 2014), organic carbon (Roy and Kashem 2014; Mahmood et al. 2017; Gümüş and Şeker 2017; Kamal et al. 2009; Bai et al. 2017; Bakayoko et al. 2009; Shehzadi et al. 2017; Zhang et al. 2020), CEC (Kamal et al. 2009; Bakayoko et al. 2009), total nutrients (Oo et al. 2015; Gümüş and Şeker 2017; Li et al. 2014), and available nutrients (Oo et al. 2015; Roy and Kashem 2014; Mahmood et al. 2017; Tester 1990; Xiaohou et al. 2008; Bai et al. 2017; Bakayoko et al. 2009; Zhang et al. 2020; Adnan et al. 2017) have all been linked to the incorporation of organic amendments. The acidic nature of the amendments can be used to explain the drop in soil pH caused by their incorporation that has been noted in numerous investigations. The inclusion of organic materials causes microbial activity, which releases carbon dioxide that combines with water to generate carbonic acid and lowers the pH of the soil (Norton and Strom 2012). Several studies (Wang et al. 2014; Tejada et al. 2006; Ouni et al. 2013; Wu et al. 2011; Prapagar et al. 2012; Mkhabela and Warman 2005) also indicated that the addition of organic amendments raised the pH of soils. According to Roy and Kashem (2014), the pH of the soil was first only marginally raised by the addition of organic amendments before rapidly declining over time. However, the pH shift brought about by the addition of organic amendments varies on the original soil's pH, the type of organic materials used, and the rate at which they are applied (Oo et al. 2015; Ouni et al. 2013). The creation of OH^- ions by ligand exchange and the subsequent release of such basic ions as *K*, *Ca*, and *Mg* are responsible for the increase in pH, which is explained by the mineralization of *C* (Mkhabela and Warman 2005). The displacement of *Na* from the exchange sites and washing out with soluble salts during the leaching process may be the cause of the drop in EC brought on by the application of organic amendments. *Ca*, *Mg*, and *K* are largely present in organic amendments (Wang et al. 2014; Leogrande and Vitti 2019). Due to increased *Na* exchange by *Ca* at the cation exchange sites of soil particles, which allows for greater leaching of exchanged *Na* with percolating water, the presence of *Ca* can contribute to the low ESP of SAS (Oo et al. 2015; Qadir and Oster 2004). Moreover, the entry of *Na* into the exchange complex may be restricted by the presence of soluble and exchangeable *K* due to a similar ionic balance and lower ESP (Walker and Bernal 2008). *Ca* also enhances soil aggregation by forming cationic bridges between clay particles and soil organic matter, increasing soil porosity. The leaching of the soluble and exchangeable *Na* ions is inversely proportional to the total porosity, and the reduction in soil sodicity and salinity as measured by the ESP and EC values, respectively, is proportional to the total porosity (Wang et al. 2014).

The ability of organic additions to absorb salt is also known to be taken into account when reducing the EC of soils (Gunarathne et al. 2020). Depending on the rate and length of assimilation, various organic soil amendments have frequently led to a small rise in EC (Ding et al. 2020; Yu et al. 2015). High concentrations of *K* and *Ca* may be the cause of the increase in EC following the application of organic manures (Ouni et al. 2013). It is already known that organic materials help to raise the CEC.

Hue (1992) reported that higher CEC brought on by the assimilation of organic matter caused a decrease in *Na* in soil solution. It has also been shown that the removal of organic matter raises the CEC of a particular soil. This finding may be related to the exposure of the permanent charge of montmorillonite clay, which was previously prevented by the interaction of the organic matter with the clay (Tan and Dowling 1984).

According to numerous studies, adding organic matter significantly increased the microbiological and enzymatic activities of the soil. Enzymatic activity (Liang et al. 2003; Tejada et al. 2006; Gunarathne et al. 2020; Li et al. 2014; Zhang et al. 2020; Alharby and Fahad 2020), microbial biomass (Oo et al. 2015; Wu et al. 2018; Li et al. 2014; Zhang et al. 2020; Wichern et al. 2020), microbial biomass *N* (Oo et al. 2015; Li et al. 2014), basal respiration (Oo et al. 2015; Bouajila and Sanaa 2011), and nematode abundance (Wu et al. 2018) were all improved by the addition of organic supplements. However, depending on the types of amendments, rates of assimilation, types of crops cultivated, etc., the response of microbial and enzymatic activity to organic amendments varies. Comparing soils treated with reed, composted straw, or straw, it was discovered that soils amended with poultry manure had the highest rate of microbial respiration. All of the amended soils caused a rapid increase in respiration rate at the start of incorporation that gradually decreased over time (Shehzadi et al. 2017). The largest soil microbial biomass *C* and cumulative C-CO_2 were found by Tejada et al. (2006) in soils treated with a maximum dose of poultry manure, which was 37% greater than soils amended with cotton gin crushed compost. The activities of the enzymes urease, protease, b-glucosidase, phosphatase, arylsulfatase, and dehydrogenase were also found to be 34%, 18%, 37%, 39%, 40%, and 30% higher in soils amended with poultry manure than in soils treated with cotton gin broken compost.

In a different study, Liang et al. (2003) discovered that, compared to rice straw treatment alone, the urease activity rose by 62.3% and 117.4%, respectively, in soils modified with pig dung and rice straw plus pig manure. The addition of rice straw, poultry manure, or both enhanced the activity of the enzyme urease in rice by 21%, 96%, and 163%, respectively, while in barley, it increased it by 57.4%, 93.1%, and 152.5%, respectively, in comparison to the control. The application of vermicompost enhanced the activities of dehydrogenase, urease, and phosphatases by 37–68%, 22–107%, and 3.4–56%, respectively, but the addition of vermicompost decreased the activities of β-1,4-glucosidase and β-1,4-*N*-acetylglucosaminidase by 17–53% and 24–42%, respectively. While recalcitrant, lignin-rich amendments like woody biomass have a smaller but longer lasting impact on these soil properties, easily decomposable organic materials like biosolids, swine manure, and chicken manure may likely be retained in the soil over brief periods, resulting in an intense and short effect (Leogrande and Vitti 2019; Larney and Angers 2012).

According to several studies, phytohormones and rhizobacteria that promote plant growth can alter the physiological and metabolic reactions of plants to salt stress, improving their tolerance as well as growth and yield (Fahad et al. 2015c, d; Kamran et al. 2018; Egamberdieva 2009). Under salinity-induced conditions, Egamberdieva (2009) discovered that the bacterial strains that produce indoleacetic acid greatly improved the seedling root growth of wheat by up to 52% in comparison to controlled environment. Yet, when many plant growth regulators were administered in combination as opposed to their single dose, it was discovered in various research that the morphological and physiological growth and yield features of crops were boosted under abiotic stress (Fahad et al. 2016b, c).

The growth, nutrient uptake, and accumulation of plants under salt stress are significantly influenced by the favorable impacts of the organic amendments on the physical, chemical, and biological characteristics of SAS. Application of organic amendments in SAS is regarded as a practical and efficient technique to improve crop development and soil fertility (Liang et al. 2003; Šimanský and Zaujec 2009). The application of organic amendments in SAS increased the biomass yield of alfalfa, *Medicago sativa* (Mahdy 2011); barley, *Hordeum vulgare* (Šimanský and Zaujec 2009); cotton, *Gossypium hirsutum* (Wu et al. 2018); maize, *Zea mays* (Oo et al. 2015; Bai et al. 2017; Lashari et al. 2015; Prapagar et al. 2012); onion, *Allium cepa* (Prapagar et al. 2012); rice, *Oryza sativa* (Liang et al. 2003; Kamal et al. 2009; Cha-um et al. 2011); seepweed, *Suaeda salsa* (Sun et al. 2016); sweet fennel, *Foeniculum vulgare* (El-Magd et al. 2008); tomato, *Solanum lycopersicum* (Esteban et al. 2016); and wheat, *Triticum aestivum* (Ding et al. 2020). The addition of organic manure to the SAS also raised the *N*, *P*, and *K* contents of rice (*Oryza sativa*), barley (*Hordeum vulgare*), and sweet fennel (*Foeniculum vulgare*) as well as the *K* and *Ca* contents of tomato (*Solanum lycopersicum*) while decreasing the *Na* uptake (Liang et al. 2003; Cha-um et al. 2011; El-Magd et al. 2008; Esteban et al. 2016). Crops under salt stress develop and produce more effectively when organic amendments are incorporated into the soil because they improve the soil's physical characteristics, make macro- and micronutrients more available, and increase microbial activity (Wang et al. 2014; Larney and Angers 2012). A crucial defense mechanism for plants to withstand the negative effects of salts and perform better growth is the maintenance of a high *K*:*Na* ratio as a result of the incorporation of organic manure (Liang et al. 2003; El-Magd et al. 2008). *Na* absorption may have decreased as a result of organic matter's role as salt-ion chelating agents that detoxify harmful ions, particularly *Na* and *Cl* (Cha-um et al. 2011). By affecting the availability of nutrients, particularly *N*, the *C*:*N* ratio also affects plant growth. Using organic amendments with a lower *C*:*N* ratio results in more readily available *N* (Mahmood et al. 2017).

In most circumstances, the introduction of inorganic amendments is not necessary because saline soils typically have strong structural qualities and can be leached to remove excess salts. Exchangeable *Na* must first be taken out of the exchange sites of soil particles in saline-sodic and sodic soils, and then it must be leached to wash out of the root zone. In order to improve the soil structure and speed up the leaching process in sodic soils, which are known for their poor soil structure and low infiltration rates, various inorganic supplements are utilized in addition to organic materials. As previously noted in various research (Prapagar et al. 2012; Cha-um et al. 2011), the application of *Ca*-containing salt, particularly gypsum coupled with organic amendments in SAS, to replace exchangeable *Na*, improves the physical state of the soil, promotes salt leaching, and boosts crop production. Gypsum application followed by soil leaching improved reclamation and reduced salinity as well as sodicity levels (Abdel-Fattah et al. 2015). When gypsum was applied to salt-affected soil, Khattak et al. (2007) noticed that the pH, EC, and SAR of the leached soils decreased and the yields of rice and wheat increased by 9.8–25.3% and 10–80%, respectively. According to Khosla et al. (1979), applying gypsum can lessen the need for increased water usage while still reducing the SAR value to a higher level. Gypsum is needed to recover saline-sodic and sodic soils, and the amount depends on the exchangeable sodium content, soil texture, leaching rate, crop to be planted, solubility, and reaction rates of the amendments (Provin and Pitt 2001; Abrol et al. 1988). When Abdel-Fattah et al. (2015) investigated how different gypsum size fractions (0.5, 0.5–1, and 1.0–2.0 mm) affected the effectiveness of the reclamation of SAS, it was discovered that the salinity and sodicity reduced with increasing gypsum fineness. Gypsum should typically be placed evenly over the field and should be worked into the top 10–15 cm of soil by 2–3 shallow plowings at least 10–15 days prior to planting (Abrol et al. 1988).

With the intention of recovering SAS and promoting plant development, zeolite ($CaAl_2Si_4O_{12}{\cdot}nH_2O$), an aluminosilicate, has also been explored as an inorganic amendment.

Zeolite can reduce salt stress and improve plant growth and nutrient uptake. Zeolite application led to a considerable rise in the biomass of barley (*Hordeum vulgare*), as well as higher concentrations of *Ca*, *Mg*, and *K* in postharvest soils, according to an experiment by Al-Busaidi et al. (2008). Zeolite was added to the soil to improve its availability of *N*, *P*, *K*, *Ca*, and *Mg*. It also raised the fresh and dry weights of shoots and roots, fruit weight, and quantity of achenes on the strawberry (*Fragaria ananassa*) (Abdi et al. 2006). By using zeolite in conjunction with bovine dung and inorganic fertilizer, Milosevic and Milosevic (2009) discovered greater levels of humus, total *N*, and accessible *P* and *K* in the soil in addition to a considerable increase in the shoot length and trunk cross-sectional area. High sorption and ion exchange capability define zeolite. It has a significant role in the mobilization of both micronutrients and macronutrients as a sorbent (Badora 2016). The adsorption of *Na* under salt-stress conditions can balance the negative charges created in the structure of zeolite by the replacement of quadruply charged silicon cations with triply charged aluminum. In addition, the $[SiO_4]^{4-}$ and $[AlO_4]^{5-}$ tetrahedra that make up the three-dimensional framework of zeolite are joined by sharing the oxygen atoms that are located at each tetrahedron's corner, which causes the framework to develop pores or voids in the shape of cages and channels between the tetrahedra (Jha and Singh 2016). Hence, the addition of zeolite to SAS may result in the adsorption of *Na* on surfaces or its trapping in voids, which would reduce plant uptake. Furthermore, zeolite contributes significantly to the soil-plant system's *Ca* (CaO, 16%) needs. By reducing *Na* and promoting *Ca* absorption, the release of *Ca* in the root medium from the *Ca*-type zeolite can maintain a high *Ca*:*Na* ratio in the shoot and root (Song and Fujiyama 1996). The concentration of Na should be considered while employing zeolite as a modification to recover SAS. Zeolite spraying resulted in a significant rise in Na in soil and plant (Al-Busaidi et al. 2008). Moreover, in low-pH soils, zeolite disintegration and simultaneous addition of Al^{3+} and Mn^{2+} ions to the sorption complex may result in increased *Mg* and *Ca* leaching, root injury, a lack of *Mg* and *P*, toxicity of *Mn* and *Fe*, and reduced plant growth (Badora 2016).

The majority of salinity research is done on model systems. The application of salinity tolerance research findings to crop plants is extremely rare. One example is the use of AtNHX1 to enhance salt compartmentation in the tomato vegetative tissue vacuoles, which increased yield without causing the tomato fruit's salt content to increase (Bassil and Blumwald 2014). Recently, it was discovered that barley expression of AtCIPK16, an SNF1-related kinase/CBL-interacting protein kinase underpinning a quantitative trait locus for Na^+ exclusion in the *Arabidopsis thaliana* Bay-0 X Shahdara mapping population, improved Na^+ exclusion and biomass in a saline region (Roy et al. 2013).

Reactive oxygen species (ROS) serve as a warning indication under salt stress, but they can also cause harm to the root and shoot tissue of plants by disrupting the function of enzymes, cell walls, and membranes. SR3 wheat, a hybrid with a high level of salt tolerance, has allowed researchers to clone several genes involved in ROS detoxification (Dong et al. 2013). Reduced cellular damage, maintenance of photosynthetic energy absorption, and improved shoot and root growth under salinity conditions have all been attributed to the overexpression of genes involved in ROS scavenging (Roy et al. 2014). It is crucial to assess the energetics of ROS detoxification as well as the effects of ROS detoxification on final grain production because many of these transgenics have decreased growth under nonsaline conditions.

Most plant respiratory expenditures may be accounted for via ion transport (Van der Werf et al. 1988). Plant salinity tolerance is supported by ion transporters and their location in important cell types (Osakabe et al. 2014). For shoot NaCl exclusion, root xylem parenchyma cells serve as the "gatekeeper" cell types because of their physical position and special protein circuitry (Henderson and Gilliham 2015). TaHKT1;5-D supports the Kna1 locus, is found on the plasma membrane (PM) of root xylem parenchyma cells, and lowers Na^+ load in the xylem before entering the shoot to maintain high cytosolic K^+/Na^+ ratios in bread wheat shoots (Byrt et al. 2014). Arabidopsis, durum wheat, and rice all contain orthologous proteins in terms of both sequence and function (Henderson and Gilliham 2015). The *Triticum monococcum* HKT1;5-A was introduced into durum wheat, which enhanced shoot Na^+ exclusion and increased grain production by 25% (Munns et al. 2012). The salt excessively sensitive (SOS) pathway genes and AtCIPK16 are additional salt tolerance components that are expressed in the root stele (Roy et al. 2013). Very little is known about the proteins involved, but root stelar cells also provide control shoot Cl^- buildup, which can cause salinity toxicity in crops (Henderson et al. 2014).

The huge multigenic family of aquaporin proteins, which controls a significant amount of water transport across membranes, is quickly affected by salt both transcriptionally and posttranslationally (Chaumont and Tyerman 2014). In a saline area, soybeans overexpressing a PM intrinsic protein produced more seeds and increased shoot Na^+ exclusion (Zhou et al. 2014). Wheat TIP2;2 and Arabidopsis HKT1 are both controlled by methylation after salt treatment (Xu et al. 2013; Sani et al. 2013).

Garcia de la Garma et al. (2015) revealed significant Na^+ vesicle trafficking between the PM and the Na^+-rich vacuolar compartment in salt-acclimated tobacco BY2 cells. Arabidopsis roots were used to demonstrate the role of RAB6 GTPase ARA6 and VAMP727-mediated endocytotic machinery in salt tolerance; ARA6 overexpression improved salt tolerance, but ara6/vamp727 mutant plants were salt hypersensitive (Ebine et al. 2011).

At high salt loads, it is frequently claimed that the CPA1 family of Na^+/H^+ antiporters, NHX1 (localized in the tonoplast) and NHX7/SOS1 (localized in the PM), confer Na^+ compartmentation or exclusion. Nevertheless, their function is less obvious under moderate salinities. The moderate external Na^+ concentrations do not affect double-nhx1/nhx2 knockouts, while the moderate external K^+ concentrations do (reviewed in Bassil and Blumwald 2014). The trans-Golgi network-localized NHX double knockouts, nhx5/nhx6, on the other hand, are hypersensitive to moderate salinity and cause vesicle trafficking to the vacuole to be disrupted (Bassil and Blumwald 2014). The cation/H^+ exchanger (CHX), GmSALT3, a member of the CPA family, increases soybean shoot Na^+ exclusion and salt tolerance (Guan et al. 2014). Using fluorescent protein fusions, CHX proteins, such as GmSALT3, have frequently been localized to the endoplasmic reticulum. If these membrane localizations are reliable, this is additional proof that endosomal localized transport proteins play important roles in salt tolerance.

Under stressful and non-stressed conditions, rhizospheric fungi and plant growth-promoting rhizobacteria (PGPR) can boost plant yield (De-la-Pena and Loyola-Vargas 2014; Nadeem et al. 2014). Salt-tolerant PGPR populations can improve plant biomass under salinity stress by inducing greater proline synthesis, increasing the expression of stress-responsive transcription factors, reducing shoot Na^+ content, and increasing ROS scavenging (De-la-Pena and Loyola-Vargas 2014; Nadeem et al. 2014). Arbuscular mycorrhizal fungi can colonize plant roots to increase water uptake and shoot K^+ while lowering shoot Na^+ content, which can improve plant salt tolerance (Auge et al. 2014).

Due to their potential to increase access to water and nutrients while reducing salt accumulation, root systems are essential to increasing crop salt tolerance (Jung and McCouch 2013). In Arabidopsis and wheat, salt apparently inhibits root epidermal cell division and elongation rates through its osmotic effects, which reduces main root growth but promotes lateral root development (Rahnama et al. 2011; Jung and McCouch 2013). This would enable plants to mine nonsaline areas for water and minerals prior to the need to utilize saline areas. The salinity of the soil is always variable in the field and often rises with depth. According to Jung and McCouch (2013), a complex network of interconnected hormone-mediated pathways regulates the structure of the Arabidopsis root system. However, these processes have not been extensively studied in agricultural systems (Rogers and Benfey 2015). In sterile culture, Arabidopsis roots exposed to a region of high $NaCl$ show negative halotropism, or aversion to salt. An external Na^+ gradient triggers this asymmetric root development response, which is carried out by the PIN-FORMED 2 (PIN2) auxin efflux carrier, which actively redistributes the auxin gradient to the side of the root that is exposed to the salt (Galvan-Ampudia et al. 2013).

According to Richards et al. (2014), elite cultivars that excel in ideal environments also frequently produce their highest yields in water-restricted environments. This rule may also apply to saline environments as long as the increased yields are the consequence of energy-efficient processes. Schilling et al. (2014) demonstrated that overexpressing the Arabidopsis vacuolar proton pumping pyrophosphatase (H^+-PPase) increases the growth and yield of transgenic barley under saline conditions in both glasshouse and field conditions, and it improves the salinity tolerance of various crop species under controlled conditions. Interestingly, the AtAVP1-overexpressing barley increased grain production and shoot biomass under nonsaline conditions. More study is needed to determine the mechanism underlying these advancements (Schilling et al. 2014).

Phenology is a major determinant of grain yield in semiarid areas. In some semiarid locations, climate change models indicate rising temperatures and falling rainfall (Anwar et al. 2015). Due to exceptionally high salt concentrations in the soil during grain filling, which limit grain size, crops may become vulnerable to terminal droughts. To maximize prospects for photosynthetic uptake and transfer of photosynthate to grain, planting and flowering times are essential. According to the species and salinity level, salinity impacts flowering time and can either accelerate or delay it (Munns and Rawson 1999; Kim et al. 2013). To fully comprehend and take advantage of the differences in genotype- and species-specific responses to salt-induced phenology, more knowledge of the molecular regulators of flowering time and their interactions with soil salinity is required.

Even though many cereals still have low salt tolerance, breeders continue to make small annual gains in grain output. It has been hypothesized that the possibility for significant increases in stress tolerance has been diminished due to the shrinking crop genetic diversity as a result of domestication and intense breeding (Munns et al. 2012). "Exotic" cereals exhibit useful natural variation; for instance, several Tibetan wild barley lines have better levels of salt tolerance in terms of biomass buildup than is typical (Wu et al. 2011).

7.3 Future of Seafood Production and the Path Forward

Pandemic impacted the production of seafood around the world causing significant decline in the production of this sector. In 2022, world population passed eight billion people, forcing world leaders to look for more efficient and sustainable ways to feed the growing population. Traditional farming methods and practices along with rapid environmental changes prevent the aquaculture sector from responding to the needs of the growing population.

As shown, deep tech can empower farmers and communities allowing them to increase the pace of food production. Additionally, live environmental monitoring and data gathered for training the algorithms can be a great source for modelling and forecasting the climate condition in each country and the region.

The increase in aquaculture production can only be achieved if sustainable urban farming is involved, and as shown, IAS can be a solution for involving families and communities in urban areas. However, the freshwater species grown in IAS are not high-value species, even though these species are good sources of proteins. On the other hand, RAS is expensive with many extra parts that prevent them from being installed in small areas such as rooftops and backyards in apartments and houses in cities.

Poseidon-AI® algorithms are trained for six species, which are three freshwater and three saltwater species. These algorithms are aimed to act as an expert in Poseidon-AI® IAS models allowing people in urban areas to contribute to global aquaculture production. Usually, IAS modules are made with freshwater gathered from rivers, lakes, and rainwaters; however, saltwater modules are not common. For this reason, Poseidon-AI® IAS modules use Poseidon-AI® algorithms, IoT devices, and AI/ML to allow culture of saltwater species. The module consists of three main parts: a tank with 500 liters of water with chemical components of ocean waters, containing Cl^-, Na^+, SO_4^{2-}, Mg^{2+}, Ca^{2+}, and K^+: the tank is a suitable living habitat for high-value saltwater species such as seabass (*Dicentrarchus labrax*), red snapper (*Lutjanus campechanus*), grouper (*Epinephelus morio*), and tiger prawns (*Penaeus monodon*); a tank containing 200 liters of hydro pebbles made of clay as biofilters; and a media bed which contains salt-tolerant or -resistant plants with high nutritional value for human consumption. The veggies can be samphire (*Crithmum maritimum*), agretti (*Salsola soda*), and several species of the genus Salicornia, while other halophytes are grown for grain production such as the quinoa. Additionally, many plants from the family Chenopodiaceae, such as chard (*Beta vulgaris var. Maritima*) and beet (*Beta vulgaris* var. *Cycles*), and other species such as the common tomato (*Lycopersicon esculentum*), cherry tomatoes (*Lycopersicon esculentum var. Cerasiforme*), and basil (*Ocimum basilicum*), can be grown in these media beds. Saline, sodic, and saline-sodic soils differ in their physical characteristics, which affects how plants are stressed by them and how to treat them. The detrimental effects of salt stress on a plant's ability to absorb vital nutrients depend on a variety of factors, including genotype, growth stage, salt concentration in the medium, etc. While salts can be leached by high-quality water and a sufficient drainage system to restore salinity, saline-sodic and sodic soils cannot be restored just through leaching. To reclaim the saline-sodic and sodic soils, it is frequently necessary to apply inorganic and organic amendments. Saline, saline-sodic, and sodic soils can all be recovered with the help of organic additions. Saline, saline-sodic, and sodic soils benefit from organic additions in physical, chemical, and biological ways that increase the extent of their reclamation. Moreover, organic amendments serve as crucial suppliers of vital plant nutrients. Hence, for sustainable crop production and food security, applying organic amendments in conjunction with the selective use of inorganic amendments may be a superior strategy to enhance the plant's response to salt stress.

It is proved that the use of irradiation techniques can help introduce new varieties of plants and vegetables which are resistant toward salinity. These plants are often used in coastal lands where high salinity causes reduced soil productivity and various difficulties for communities living in these areas. These plants are already provided for coastal communities suffering from the climate change impacts such as sea level rise. However, it is not yet clear if these species can survive in Poseidon-AI® IAS modules due to high fluctuation in salinity level with every water circulation. Furthermore, a drop in the salinity level and change in the pH might cause high mortality in the cultured tanks, and so, profound studies are required. Of course, AI/ML and intelligence algorithms can help in healthy and efficient culture of the species, but the modules need to be user friendly and easy to use for the communities in urban areas, and hence, before presenting the Poseidon-AI® saline IAS modules to the market, more improvement in the industrial design of the modules is required.

References

Abdel-Fattah MK, Fouda S, Schmidhalter U (2015) Effects of gypsum particle size on reclaiming saline-sodic soils in Egypt. Commun Soil Sci Plant Anal 46:1112–1122

Abdelgadir EM, Oka M, Fujiyama H (2005) Characteristics of nitrate uptake by plants under salinity. J Plant Nutr 28(1):33–46

Abdi G, Khosh-Khui M, Eshghi S (2006) Effects of natural zeolite on growth and flowering of strawberry (Fragaria x ananassa Duch.). Int J Agric Res 1:384–389

Abdullah Z, Ahmad R (1982) Salt tolerance of Solanum tuberosum L. growing on saline soils amended with gypsum. J Agron Crop Sci 151:409–416

Abrol IP, Yadav JSP, Massoud FI (1988) Salt affected soils and their management. FAO Soils Bulletin 39. Food and Agriculture Organization of the United Nations, Rome

Adams P, Ho LC (1989) Effects of constant and fluctuating salinity in the yield, quality and calcium status of tomatoes. J Hort Sci 64:725–732

Adane A, Gebrekidan H, Kibret K (2019) Effects of treatment application rates (fym and gypsum) on selected chemical properties of saline sodic soils under water limited condition in eastern lowlands Ethiopia. Forest Res Eng 3(3):106–113

Adnan M, Shah Z, Fahad S, Arif M, Alam M, Khan IA et al (2017) Phosphate-solubilizing bacteria nullify the antagonistic effect of soil calcification on bioavailability of phosphorus in alkaline soils. Sci Rep 7:16131

Akhtaruzzaman M, Roy S, Mahmud MS, Shormin T (2020) Soil properties under different vegetation types in Chittagong University campus, Bangladesh. J Forest Environ Sci 36(2): 133–142

Akram R, Turan V, Hammad HM, Ahmad S, Hussain S, Hasnain A et al (2018a) Fate of organic and inorganic pollutants in paddy soils. In: Hashmi MZ, Varma A (eds) Environmental pollution of paddy soils. Springer, Cham, pp 197–214

Akram R, Turan V, Wahid A, Ijaz M, Shahid MA, Kaleem S et al (2018b) Paddy land pollutants and their role in climate change. In: Hashmi MZ, Varma A (eds) Environmental pollution of paddy soils. Springer International Publishing AG, pp 113–124

Alam SM (1999) Nutrient uptake by plants under stress conditions. In: Pessarakli M (ed) Handbook of plant and crop stress, 2nd edn. Marcel Dekker, Inc, New York

Al-Busaidi A, Yamamoto T, Inoue M, Eneji AE, Mori Y, Irshad M (2008) Effects of zeolite on soil nutrients and growth of barley following irrigation with saline water. J Plant Nutr 31(7): 1159–1173

Alharby HF, Fahad S (2020) Melatonin application enhances biochar efficiency for drought tolerance in maize varieties: modifications in physio-biochemical machinery. Agron J 112(4): 2826–2847

Ali L, Rahmatullah MMA, Kanwal S, Ashraf M, Hannan A (2009) Potassium substitution by sodium in root medium influencing growth behavior and potassium efficiency in cotton genotypes. J Plant Nutr 32(10):1657–1673

Al-Karaki GN (2000) Growth, sodium, and potassium uptake and translocation in salt stressed tomato. J Plant Nutr 23(3):369–379

Aloni B, Pressman E (1987) The effects of salinity and gibberellic acid on blackheart disorder in celery (Apium graveolens L.). J Hort Sci 62:205–209

Amthor JS (2000) The McCree–de Wit-Penning de Vries-Thornley respiration paradigms: 30 years later. Ann Bot 86:1–20

Anwar MR, Liu DL, Farquharson R, Macadam I, Abadi A, Finlayson J, Wang B, Ramilan T (2015) Climate change impacts on phenology and yields of five broadacre crops at four climatologically distinct locations in Australia. Agric Syst 132:133–144

Ashraf M (1992) Effect of NaCl on growth, seed protein and oil contents of some cultivars/lines of brown mustard (Brassica juncea (L.) Czern. and Coss.). Agrochimica 36:137–147

Ashraf M (1994) Organic substances responsible for salt tolerance in Eruca sativa. Biol Plant 36: 255–259

Ashraf M, McNeilly T (1990) Responses of four Brassica species to sodium chloride. Environ Exp Bot 30:475–487

Ashraf M, Naqvi MI (1992) Growth and ion uptake of four Brassica species as affected by Na/Ca ratio in saline sand culture. Z PflanzenernaÈhr Bodenk 155:101–108

Ashraf M, Noor R (1993) Growth and pattern of ion uptake in Eruca sativa Mill. under salt stress. Angew Bot 67:17–21

Auge RM, Toler HD, Saxton AM (2014) Arbuscular mycorrhizal symbiosis and osmotic adjustment in response to NaCl stress: a meta-analysis. Front Plant Sci 5:562

Ayers AD, Wadleigh CH, Bernstein L (1951) Salt tolerance of six varieties of lettuce. Proc Am Soc Hort Sci 57:237–242

Badora A (2016) The influence of zeolites on quality indicators of soil plant connection and food safety. In: Belviso C (ed) Zeolites—useful minerals. IntechOpen, London, pp 203–215
Bai Y, Yan Y, Zuo W, Gu C, Xue W, Mei L et al (2017) Coastal mudflat saline soil amendment by dairy manure and green manuring. Int J Agron 2017:1–9
Bakayoko S, Soro D, Nindjin C, Dao D, Tschannen A, Girardin O et al (2009) Effects of cattle and poultry manures on organic matter content and adsorption complex of a sandy soil under cassava cultivation (Manihot esculenta, Crantz). Afr J Environ Sci Technol 3(8):190–197
Barzegar AR, Yousefi A, Daryashenas A (2002) The effect of addition of different amounts and types of organic materials on soil physical properties and yield of wheat. Plant Soil 247:295–301
Bassil E, Blumwald E (2014) The ins and outs of intracellular ion homeostasis: NHX-type cation/H + transporters. Curr Opin Plant Biol 22:1–6
Bayuelo-Jiménez JS, Debouck DG, Lynch JP (2003) Growth, gas exchange, water relations, and ion composition of Phaseolus species grown under saline conditions. Field Crop Res 80(3): 207–222
Bayuelo-Jiménez JS, Jasso-Plata N, Ochoa I (2012) Growth and physiological responses of Phaseolus species to salinity stress. Int J Agron 2012:1–13
Bernstein L (1959) Salt tolerance of vegetable crops in the West. USDA Information Bulletin 205
Bernstein L, Ayers AD (1949) Salt tolerance of cabbage and broccoli. United States Salinity Laboratory Report to Collaborators, Riverside, CA, 39 pp
Bernstein L, Ayers AD (1953a) Salt tolerance of five varieties of carrots. Proc Am Soc Hort Sci 61: 360–366
Bernstein L, Ayers AD (1953b) Salt tolerance of five varieties of onions. Proc Am Soc Hort Sci 62: 367–370
Bernstein L, Hayward HE (1958) Physiology of salt tolerance. Annu Rev Plant Physiol 9:25–46
Bernstein L, Ayers AD, Wadleigh CH (1951) The salt tolerance of White Rose potatoes. Proc Am Soc Hort Sci 57:231–236
Bernstein L, Francois LE, Clark RA (1974) Interactive effects of salinity and fertility on yields of grains and vegetables. Agron J 66:412–421
Bhatt MK, Labanya R, Joshi HC (2019) Influence of long-term chemical fertilizers and organic manures on soil fertility—a review. Univ J Agric Res 7(5):177–188
Bilski JJ, Nelson DC, Conlon RL (1988) The response of four potato cultivars to chloride salinity, sulfate salinity and calcium in pot experiments. Am Pot J 65:85–90
Bingham FT, Strong JE, Sposito G (1983) Influence of chloride salinity on cadmium uptake by Swiss chard. Soil Sci 135:160–165
Bingham FT, Sposito G, Strong JE (1984) The effect of chloride on the availability of cadmium. J Environ Qual 13:71–74
Blumenthal JM, Baltensperger DD, Cassman KG, Mason SC, Pavlista AD (2008) Importance and effect of nitrogen on crop quality and health. In: Hatfield JL, Follett RF (eds) Nitrogen in the environment: sources, problems, and management, 2nd edn. Elsevier Inc., pp 51–70
Botella MA, Martínez V, Nieves M, Cerdá A (1997) Effect of salinity on the growth and nitrogen uptake by wheat seedlings. J Plant Nutr 20(6):793–804
Bouajila K, Sanaa M (2011) Effects of organic amendments on soil physicochemical and biological properties. J Mater Environ Sci 2(S1):485–490
Brady NC, Weil RR (2005) The nature and properties of soils, 13th edn. Pearson-Prentice Hall, Delhi, p 960
Brouk B (1975) Plants consumed by man. Academic Press, London, p 479
Brown CE, Pezeshki SR, DeLaune RD (2006) The effects of salinity and soil drying on nutrient uptake and growth of Spartina alterniflora in a simulated tidal system. Environ Exp Bot 58:140–148
Bybordi A, Ebrahimian E (2011) Effect of salinity stress on activity of enzymes involved in nitrogen and phosphorous metabolism case study: Canola (Brassica napus L.). Asian J Agric Res 5:208–214

Byrt CS, Xu B, Krishnan M, Lightfoot DJ, Athman A, Jacobs AK, Watson-Haigh NS, Plett D, Munns R, Tester M et al (2014) The Na+ transporter, TaHKT1;5-D, limits shoot Na+ accumulation in bread wheat. Plant J 80:516–526

Chakraborty K, Sairam RK, Bhaduri D (2015) Effects of different levels of soil salinity on yield attributes, accumulation of nitrogen, and micronutrients in Brassica spp. J Plant Nutr 39(7): 1026–1037

Cha-um S, Pokasombat Y, Kirdmanee C (2011) Remediation of salt affected soil by gypsum and farmyard manure—importance for the production of Jasmine rice. Aust J Crop Sci 5(4):458–465

Chaumont F, Tyerman SD (2014) Aquaporins: highly regulated channels controlling plant water relations. Plant Physiol 164:1600–1618

Chen Z, Newman I, Zhou M, Mendham N, Zhang G, Shabala S (2005) Screening plants for salt tolerance by measuring K+ flux: a case study for barley. Plant Cell Environ 28:1230–1246

Chen W, Hou Z, Wu L, Liang Y, Wei C (2010) Effects of salinity and nitrogen on cotton growth in arid environment. Plant Soil 326:61–73

Chen WP, Chang AC, Page AL (2012) Deficiencies and toxicities of trace elements. In: Wallender WW, Tanji KK (eds) Agricultural salinity assessment and management, 2nd edn. American Society of Civil Engineers, Virginia, pp 207–233

Chen B, Yang H, Song W, Liu C, Xu J, Zhao W et al (2016) Effect of N fertilization rate on soil alkali hydrolyzable N, subtending leaf N concentration, fiber yield, and quality of cotton. Crop J 4(4):423–430

Chow WS, Ball MC, Anderson JM (1990) Growth and photosynthetic responses of spinach to salinity: implications of K^+ nutrition for salt tolerance. Aust J Plant Physiol 17:563–578

Christiansen MN, Lewis CF (1982) Breeding plants for less favorable environments. Wiley-Interscience, New York, p 459

Clark RB (1982) Nutrient solution growth of sorghum and corn in mineral nutrition studies. J Plant Nutr 5:1039–1057

Conn S, Gilliham M (2010) Comparative physiology of elemental distribution in plants. Ann Bot 105:1081–1102

Coons JM, Kuehl RO, Simon NR (1990) Tolerance of ten lettuce cultivars to high temperature combined with NaCl during germination. J Am Soc Hort Sci 115:1004–1007

Cornish PS (1992) Use of high electrical conductivity of nutrient solution to improve the quality of salad tomatoes (Lycopersicon esculentum) grown in hydroponic culture. Aust J Exp Agric 32: 513–520

Cramer GR (2002) Sodium-calcium interactions under salinity stress. In: Läuchli A, Lüttge U (eds) Salinity: environment-plants-molecules, 1st edn. Springer, Dordrecht, pp 205–227

Cramer GR, Spurr AR (1986a) Response of lettuce to salinity I. Effects of NaCl and Na_2SO_4 on growth. J Plant Nutr 9:115–130

Cramer GR, Spurr AR (1986b) Salt responses of lettuce to salinity II. Effect of calcium on growth and mineral status. J Plant Nutr 9:131–142

Dai J, Duan L, Dong H (2015) Comparative effect of nitrogen forms on nitrogen uptake and cotton growth under salinity stress. J Plant Nutr 38(10):1530–1543

Davenport RJ, Reid RJ, Smith FA (1997) Sodium-calcium interactions in two wheat species differing in salinity tolerance. Physiol Plant 99:323–327

De-la-Pena C, Loyola-Vargas VM (2014) Biotic interactions in the rhizosphere: a diverse cooperative enterprise for plant productivity. Plant Physiol 166:701–719

Deo C, Lal P (1982) Effect of water quality and moisture regime on soil properties and yield of mustard and taramira (Eruca sativa). J Indian Soc Soil Sci 30:411–414

DeRougemont GM (1989) A field side to the crops of Britain and Europe. William Collins, London, p 367

Diacono M, Montemurro F (2012) Effectiveness of organic wastes as fertilizers and amendments in salt affected soils. Agriculture 5:221–230

Ding Z, Kheir AMS, Ali MGM, Ali OAM, Abdelaal AIN, Lin X et al (2020) The integrated effect of salinity, organic amendments, phosphorus fertilizers, and deficit irrigation on soil properties, phosphorus fractionation and wheat productivity. Sci Rep 10:2736

Dluzniewska P, Gessler A, Dietrich H, Schnitzler JP, Teuber M, Rennenberg H (2007) Nitrogen uptake and metabolism in Populus x canescens as affected by salinity. New Phytol 173(2): 279–293

Dong W, Wang M, Xu F, Quan T, Peng K, Xiao L, Xia G (2013) Wheat oxophytodienoate reductase gene TaOPR1 confers salinity tolerance via enhancement of abscisic acid signaling and reactive oxygen species scavenging. Plant Physiol 161:1217–1228

Doula MK, Sarris A (2016) Soil environment. In: Poulopoulos SG, Inglezakis VJ (eds) Environment and development: basic principles, human activities, and environmental implications. Academic Press, London, pp 213–286

Ebine K, Fujimoto M, Okatani Y, Nishiyama T, Goh T, Ito E, Dainobu T, Nishitani A, Uemura T, Sato MH et al (2011) A membrane trafficking pathway regulated by the plant-specific RAB GTPase ARA6. Nat Cell Biol 13:853–859

Egamberdieva D (2009) Alleviation of salt stress by plant growth regulators and IAA producing bacteria in wheat. Acta Physiol Plant 31:861–864

Ekanayake IJ, Dodds JH (1993) In vitro testing for the effects of salt stress on growth and survival of sweet potato. Sci Hort 55:239–248

El-Magd MMA, Zaki MF, Abou-Hussein SD (2008) Effect of organic manure and different levels of saline irrigation water on growth, green yield and chemical content of sweet fennel. Aust J Basic Appl Sci 2(1):90–98

Elser JJ, Fagan WF, Denno RF, Dobberfuhl DR, Folarin A, Huberty A et al (2000) Nutritional constraints in terrestrial and freshwater food webs. Nature 408:578–580

Endo T, Yamamoto S, Larrinaga JA, Fujiyama H, Honna T (2011) Status and causes of soil salinization of irrigated agricultural lands in Southern Baja California, Mexico. Appl Environ Soil Sci 2011:1–12

Epstein E, Bloom AJ (2004) Mineral nutrition of plants: principles and perspectives, 2nd edn. Sinauer Associates Inc., Sunderland, MA, p 380

Epstein E, Jefferies RL (1964) The genetic basis of selective ion transport in plants. Annu Rev Plant Physiol 15:169–184

Esteban W, Pacheco P, Tapia L, Bastías E (2016) Remediation of salt and boron-affected soil by addition of organic matter: An investigation into improving tomato plant productivity. IDESIA (Chile) 34(3):25–32

Etesami H, Noori F (2019) Soil salinity as a challenge for sustainable agriculture and bacterial-mediated alleviation of salinity stress in crop plants. In: Manoj Kumar M, Etesami H, Kumar V (eds) Saline soil-based agriculture by halotolerant microorganisms. Springer, Singapore

Fahad S, Bano A (2012) Effect of salicylic acid on physiological and biochemical characterization of maize grown in saline area. Pak J Bot 4(4):1433–1438

Fahad S, Chen Y, Saud S, Wang K, Xiong D, Chen C et al (2013) Ultraviolet radiation effect on photosynthetic pigments, biochemical attributes, antioxidant enzyme activity and hormonal contents of wheat. J Food Agric Environ 11(3&4):1635–1641

Fahad S, Hussain S, Saud S, Tanveer M, Bajwa AA, Hassan S et al (2015a) A biochar application protects rice pollen from high-temperature stress. Plant Physiol Biochem 96:281–287

Fahad S, Hussain S, Bano A, Saud S, Hassan S, Shan D et al (2015b) Potential role of phytohormones and plant growth promoting rhizobacteria in abiotic stresses: consequences for changing environment. Environ Sci Pollut Res 22:4907–4921

Fahad S, Nie L, Chen Y, Wu C, Xiong D, Saud S et al (2015c) Crop plant hormones and environmental stress. In: Lichtfouse E (ed) Sustainable agriculture reviews, vol 15. Springer, Cham, pp 371–400

Fahad S, Hussain S, Matloob A, Khan FA, Khaliq A, Saud S et al (2015d) Phytohormones and plant responses to salinity stress: a review. Plant Growth Regul 75:391–404

Fahad S, Hussain S, Saud S, Hassan S, Tanveer M, Ihsan MZ et al (2016a) A combined application of biochar and phosphorus alleviates heat induced adversities on physiological, agronomical and quality attributes of rice. Plant Physiol Biochem 103:191–198

Fahad S, Hussain S, Saud S, Hassan S, Ihsan Z, Shah AN et al (2016b) Exogenously applied plant growth regulators enhance the morphophysiological growth and yield of rice under high temperature. Front Plant Sci 7:1250

Fahad S, Hussain S, Saud S, Hassan S, Chauhan BS, Khan F et al (2016c) Responses of rapid viscoanalyzer profile and other rice grain qualities to exogenously applied plant growth regulators under high day and high night temperatures. PLoS One 11(7):e0159590

Fahad S, Bajwa AA, Nazir U, Anjum SA, Farooq A, Zohaib A et al (2017) Crop production under drought and heat stress: plant responses and management options. Front Plant Sci 8:1147

Fahad S, Ihsan MZ, Khaliq A, Daur I, Saud S, Alzamanan S et al (2018) Consequences of high temperature under changing climate optima for rice pollen characteristics—concepts and perspectives. Arch Agron Soil Sci 64(11):1473–1488

Fahad S, Rehman A, Shahzad B, Tanveer M, Saud S, Kamran M et al (2019a) Rice responses and tolerance to metal/metalloid toxicity. In: Hasanuzzaman M, Fujita M, Nahar K, Biswas JK (eds) Advances in rice research for abiotic stress tolerance. Woodhead Publishing

Fahad S, Adnan M, Hassan S, Saud S, Hussain S, Wu C et al (2019b) Rice responses and tolerance to high temperature. In: Hasanuzzaman M, Fujita M, Nahar K, Biswas JK (eds) Advances in rice research for abiotic stress tolerance. Woodhead Publishing, pp 201–224

Feigin A, Pressman E, Imas P, Miltau O (1991) Combined effects of KNO_3 and salinity on yield and chemical composition of lettuce and Chinese cabbage. Irrig Sci 12:223–230

Flowers TJ, Yeo AR (1986) Ion relations of plants under drought and salinity. Aust J Plant Physiol 13:75–91

Flowers TJ, Munns R, Colmer TD (2015) Sodium chloride toxicity and the cellular basis of salt tolerance in halophytes. Ann Bot 115:419–431

Francois LE (1984) Salinity effects on germination, growth, and yield of turnips. HortScience 19: 82–84

Francois LE (1987) Salinity effects on asparagus yield and vegetative growth. J Am Soc Hort Sci 112:432–436

Francois LE (1994) Yield and quality response of salt-stressed garlic. HortScience 29:1314–1317

Francois LE (1995) Salinity effects on bud yield and vegetative growth of artichoke (Cynara scolymus L.). HortScience 30:69–71

Francois LE, Clark RA (1979) Accumulation of sodium and chloride in leaves of sprinkler-irrigated grapes. J Am Soc Hort Sci 104:11–13

Francois LE, West DW (1982) Reduction in yield and market quality of celery caused by soil salinity. J Am Soc Hort Sci 107:952–954

Francois LE, Donovan TJ, Maas EV (1991) Calcium deficiency of artichoke buds in relation to salinity. HortScience 26:549–553

Fuchs I, Stölzle S, Ivashikina N, Hedrich R (2005) Rice K+ uptake channel OsAKT1 is sensitive to salt stress. Planta 221(2):212–221

Gaines TP, Gaines ST (1994) Soil texture effect on nitrate leaching in soil percolates. Commun Soil Sci Plant Anal 25(13–14):2561–2570

Galvan-Ampudia CS, Julkowska MM, Darwish E, Gandullo J, Korver RA, Brunoud G, Haring MA, Munnik T, Vernoux T, Testerink C (2013) Halotropism is a response of plant roots to avoid a saline environment. Curr Biol 23:2044–2050

Garcia de la Garma J, Fernandez-Garcia N, Bardisi E, Pallol B, Salvador Asensio-Rubio J, Bru R, Olmos E (2015) New insights into plant salt acclimation: the roles of vesicle trafficking and reactive oxygen species signaling in mitochondria and the endomembrane system. New Phytol 205:216–239

Gelaye KK, Zehetner F, Loiskandl W, Klik A (2019) Effects of soil texture and groundwater level on leaching of salt from saline fields in Kesem irrigation scheme, Ethiopia. Soil Water Res 14: 221–228

Ghassemi F, Jakeman AJ, Nix HA (1995) Salinization of land and water resources: human causes, extent, management and case studies. CABI Publishing, Wallingford, p 526

Glenn E, Pfister R, Brown J, Thompson T, O'Leary J (1996) Na and K accumulation and salt tolerance of Atriplex canescens (Chenopodiaceae) genotypes. Am J Bot 83(8):997–1005

Gorske SF, Rhodes AM, Hopen HJ (1979) A numerical taxonomic study of Portulaca oleracea. Weed Sci 27:96–102

Graifenberg A, Lipucci di Paola M, Giustiniani L, Temperini O (1993) Yield and growth of globe artichoke under saline-sodic conditions. HortScience 28:791–793

Graifenberg A, Botrini L, Giustiniani L, Lipucci di Paola M (1996) Salinity affects growth, yield and elemental concentration of fennel. HortScience 31:1131–1134

Grattan SR, Grieves CM (1999) Salinity-mineral nutrient relations in horticultural crops. Sci Hortic 78:127–157

Greenway H, Munns R (1980) Mechanisms of salt tolerance in nonhalophytes. Annu Rev Plant Physiol 31:149–190

Greenway H, Munns R (1983) Interactions between growth, uptake of Cl- and Na+, and water relations of plants in saline environments. 2. Highly vacuolated cells. Plant Cell Environ 6:575–589

Greig JK, Smith FW (1962) Salinity effects on sweet potato growth. Agron J 54:309–313

Grieve CM, Suarez DL (1997) Purslane (Portulaca oleracea L.): a halophytic crop for drainage water reuse systems. Plant Soil 192:277–283

Guan R, Qu Y, Guo Y, Yu L, Liu Y, Jiang J, Chen J, Ren Y, Liu G, Tian L et al (2014) Salinity tolerance in soybean is modulated by natural variation in GmSALT3. Plant J 80:937–950

Gull A, Lone AA, Wani NUI (2019) Biotic and abiotic stresses in plants. In: de Oliveira AB (ed) Abiotic and biotic stress in plants. IntechOpen, London

Gümüş İ, Şeker C (2017) Effects of spent mushroom compost application on the physicochemical properties of a degraded soil. Solid Earth 8:1153–1160

Gunarathne V, Senadeera A, Gunarathne U, Biswas JK, Almaroai YA, Vithanage M (2020) Potential of biochar and organic amendments for reclamation of coastal acidic-salt affected soil. Biochar 2:107–120

Gutemann WH, Doss GJ, Lisk DJ (1993) Selenium in Swiss chard grown on sewage sludge-cement kiln dust amended soil. Chemosphere 27:1461–1463

Hadi MR, Karimi N (2012) The role of calcium in plants' salt tolerance. J Plant Nutr 35(13): 2037–2054

Hadi MR, Khiyam-Nekoie SM, Khavarinejad R, Sima NAKK, Yavari P (2008) Accumulation and role of ions (Ca^{2+}, Mg^{2+}, SO_4^{2-}) on salt tolerance in Triticum turgidum L. J Biol Sci 8(1): 143–148

Hammad HM, Farhad W, Abbas F, Fahad S, Saeed S, Nasim W et al (2017) Maize plant nitrogen uptake dynamics at limited irrigation water and nitrogen. Environ Sci Pollut Res 24:2549–2557

Handley JF, Jennings DH (1977) The effect of ions on growth and leaf succulence of Atriplex hortensis var. cupreata. Ann Bot 41:1109–1112

Hardie M, Doyle R (2012) Measuring soil salinity. In: Shabala S, Cuin TA (eds) Plant salt tolerance: methods and protocols, methods in molecular biology. Humana Press, Totowa, NJ, pp 415–425

Hawker JS, Smith GM (1982) Salt tolerance and regulation of enzymes of starch synthesis in cassava (Manihot esculenta Crantz). Aust J Plant Physiol 9:509–518

He T, Cramer GR (1992) Growth and mineral nutrition of six rapid-cycling Brassica species in response to seawater salinity. Plant Soil 139:285–294

He T, Cramer GR (1993a) Salt tolerance of rapid-cycling Brassica species in relation to potassium/ sodium ratio and selectivity at the whole plant and callus level. J Plant Nutr 16:1263–1277

He T, Cramer GR (1993b) Growth and ion accumulation of two rapid-cycling Brassica species differing in salt tolerance. Plant Soil 153:19–31

He T, Cramer GR (1993c) Cellular responses of two rapid-cycling Brassica species, B. napus and B. carinata, to seawater salinity. Physiol Plant 87:54–60

He T, Cramer GR (1996) Abscisic acid concentrations are correlated with leaf area reductions in two salt-stressed rapid-cycling Brassica species. Plant Soil 179:25–33

Henderson SW, Gilliham M (2015) The "Gatekeeper" concept: cell-type specific molecular mechanisms of plant adaptation to abiotic stress. In: Laitinen R (ed) Molecular mechanisms in plant adaptation. John Wiley & Sons, New York, pp 83–115

Henderson SW, Baumann U, Blackmore DH, Walker AR, Walker RR, Gilliham M (2014) Shoot chloride exclusion and salt tolerance in grapevine is associated with differential ion transporter expression in roots. BMC Plant Biol 14:273

Hoffman GJ, Rawlins SL (1971) Growth and water potential of root crops as influenced by salinity and relative humidity. Agron J 63:877–880

Hofman G, Cleemput OV (2004) Soil and plant nitrogen, 1st edn. International Fertilizer Industry Association, Paris, p 48

Horton D (1989) Constraints to sweet potato production and use. In: Improvement of sweet potato in Asia. CIP, Lima, pp 219–223

Hu Y, Schmidhalter U (2001) Effects of salinity and macronutrient levels on micronutrients in wheat. J Plant Nutr 24(2):273–281

Hu Y, Schmidhalter U (2005) Drought and salinity: a comparison of their effects on mineral nutrition of plants. J Plant Nutr Soil Sci 168:541–549

Hue NV (1992) Correcting soil acidity of a highly weathered ultisol with chicken manure and sewage sludge. Commun Soil Sci Plant Anal 23:241–264

Huq SMI, Jum S (2013) The soils of Bangladesh, 1st edn. Springer, Dordrecht, p 165

Hussain S, Mubeen M, Ahmad A, Akram W, Hammad HM, Ali M et al (2019) Using GIS tools to detect the land use/land cover changes during forty years in Lodhran District of Pakistan. Environ Sci Pollut Res 27:39676

Hutchins DA, Fu FX (2008) Linking the oceanic biogeochemistry of iron and phosphorus with the marine nitrogen Cycle. In: Capone DG, Bronk DA, Mulholland MR, Carpenter EJ (eds) Nitrogen in the marine environment, 2nd edn. London, Academic Press, pp 1627–1666

Idowu MK, Aduayi EA (2006) Effects of sodium and potassium application on water content and yield of tomato in southwestern Nigeria. J Plant Nutr 29(12):2131–2145

Indira P, Ramanujam T (1982) Note on potassium: sodium ratio in cassava grown under salt and moisture stress. Indian J Agric Sci 52:198–199

Jacoby B (1994) Mechanisms involved in salt tolerance by plants. In: Pessarakli M (ed) Handbook of plant and crop stress. Marcel Dekker, New York, pp 97–123

Jacoby RP, Taylor NL, Millar AH (2011) The role of mitochondrial respiration in salinity tolerance. Trends Plant Sci 16:614–623

Jain RK, Jain S, Nainawatee HS, Chowdhury JB (1990) Salt tolerance in Brassica juncea L. I. In vitro selection, agronomic evaluation and genetic stability. Euphytica 48:141–152

Jamil A, Riaz S, Ashraf M, Foolad MR (2011) Gene expression profiling of plants under salt stress. Crit Rev Plant Sci 30(5):435–458

Jeschke WD, Stelter W (1983) Ionic relations of garden orache, Atriplex hortensis L.: growth and ion distribution at moderate salinity and the function of bladder hairs. J Exp Bot 34:795–810

Jha B, Singh DN (2016) Fly ash zeolites: innovations, applications, and directions, 1st edn. Springer Science Business Media, Singapore, p 211

Jiang Q, Roche D, Monaco TA, Durham S (2006) Gas exchange, chlorophyll fluorescence parameters and carbon isotope discrimination of 14 barley genetic lines in response to salinity. Field Crop Res 96:269–278

Jin ZM, Wang CH, Liu ZP, Gong WJ (2007) Physiological and ecological characters studies on Aloe vera under soil salinity and seawater irrigation. Process Biochem 42(4):710–714

Jones RA, Qualset CO (1984) Breeding crops for environmental stress tolerance. In: Collins GB, Petolino JG (eds) Applications of genetic engineering to crop improvement. Martinus Nijhoff/W. Junk, Dordrecht, pp 305–340

Jouyban Z (2012) The effects of salt stress on plant growth. Tech J Eng Appl Sci 2(1):7–10

Jung GA (1978) Crops tolerance to suboptimal land conditions. ASA-CSSA-SSSA, Madison, WI, p 343

Jung JKH, McCouch S (2013) Getting to the roots of it: genetic and hormonal control of root architecture. Front Plant Sci 4:186

Kamal AM, Alam MA, Uddin MZ, Hossain MM, Islam MN (2009) Impact of organic fertilizer on physical and chemical properties of soil as well as yield and quality of mango. J Bang Soc Agric Sci Technol 1&2:167–170

Kamran M, Cui W, Ahmad I, Meng X, Zhang X, Su W et al (2018) Effect of paclobutrazol, a potential growth regulator on stalk mechanical strength, lignin accumulation and its relation with lodging resistance of maize. Plant Growth Regul 85:171–172

Kanagaraj A, Desingh R (2017) Salinity influences physiological traits of seven sesame (Sesamum indicum L.) varieties. J Sci Agric 1:188–196

Kaya C, Kirnak H, Higgs D (2001) Enhancement of growth and normal growth parameters by foliar application of potassium and phosphorus in tomato cultivars grown at high (NaCl) salinity. J Plant Nutr 24:357–367

Khan MA, Duke NC (2001) Halophytes—a resource for the future. Wetl Ecol Manag 6:455–456

Khan A, Tan DKY, Afridi MZ, Luo H, Tung SA, Ajab M et al (2017) Nitrogen fertility and abiotic stresses management in cotton crop: a review. Environ Sci Pollut Res 24:14551–14566

Khan MZ, Islam MA, Azom MG, Amin MS (2018) Short term influence of salinity on uptake of phosphorus by Ipomoea aquatica. Int J Plant Soil Sci 25(2):1–9

Khattak SG, Haq IU, Malik A, Khattak MJ (2007) Effect of various levels of gypsum application on the reclamation of salt affected soil grown under rice followed by wheat crop. Sarhad J Agric 23(3):675–680

Khatun M, Shuvo MAR, Salam MTB, Rahman SMH (2019) Effect of organic amendments on soil salinity and the growth of maize (Zea mays L.). Plant Sci Today 6(2):106–111

Khosla BK, Gupta RK, Abrol IP (1979) Salt leaching and the effect of gypsum application in a saline-sodic soil. Agric Water Manag 2(3):193–202

Kim W-Y, Ali Z, Park HJ, Park SJ, Cha J-Y, Perez-Hormaeche J, Javier Quintero F, Shin G, Kim MR, Qiang Z et al (2013) Release of SOS2 kinase from sequestration with GIGANTEA determines salt tolerance in Arabidopsis. Nat Commun 4:1820

Krishnasamy K, Bell R, Ma Q (2014) Wheat responses to sodium vary with potassium use efficiency of cultivars. Front Plant Sci 5:631

Kumamoto J, Scora RW, Clerx WA, Matsumura M, Layfield D, Grieve CM (1990) Purslane: a potential new vegetable crop rich in omega-3 fatty acid with a controllable sodium chloride content. In: Naqvi HH, Estilai A, Ting IP (eds) Proceedings of the first international conference on new industrial crops and products. University of Arizona, Tucson, Riverside, CA, pp 229–233

Kumar D (1984) The value of certain plant parameters as an index for salt tolerance in Indian mustard (Brassica juncea L.). Plant Soil 79:261–272

Lacerda CF, Assis Júnior JO, Filho LCAL, de Oliveira TS, Guimarães FVA, Gomes-Filho E et al (2006) Morpho-physiological responses of cowpea leaves to salt stress. Braz J Plant Physiol 18(4):455–465

Langdale GW, Thomas JR, Littleton TG (1971) Influence of soil salinity and nitrogen fertilizer on spinach growth. J Rio Grande Valley Hort Soc 25:61–66

Larney FJ, Angers DA (2012) The role of organic amendments in soil reclamation: a review. Can J Soil Sci 92:19–38

Läuchli, A., (1976). Genotypic variation in transport. In: Lüttge, U., Pitman, M.G. (Eds.), Transport in plants, II. Part B. Encyclopedia of plant physiology. Springer. Berlin, pp. 372-393

Lashari MS, Ye Y, Ji H, Li L, Kibue GW, Lu H et al (2015) Biochar-manure compost in conjunction with pyroligneous solution alleviated salt stress and improved leaf bioactivity of maize in a saline soil from Central China: a 2-year field experiment. J Sci Food Agric 95 (6):1321–1327

Lazof D, Läuchli A (1991) The nutritional status of the apical meristem of Lactuca sativa as affected by NaCl salinization: an electron-probe microanalytic study. Planta 184:334–342
Leogrande R, Vitti C (2019) Use of organic amendments to reclaim saline and sodic soils: a review. Arid Land Res Manag 33(1):1–21
Levitt J (1972) Responses of plants to environmental stresses. Academic Press, New York, p 697
Levy D (1992) The response of potatoes (Solanum tuberosum L.) to salinity: plant growth and tuber yields in the arid deserts of Israel. Ann Appl Biol 120:547–555
Levy D, Fogelman E, Itzhak Y (1988) The effect of water salinity on potatoes (Solanum tuberosum L.): physiological indices and yielding capacity. Potato Res 31:601–610
Li J, Liang LN, Li X, Yang HF (2014) Effects of compost amendment on soil chemical and biological properties in greenhouse soil. In: Proceedings of the 1st international symposium on organic matter management and compost use in horticulture, 4–7 Apr 2011, Adelaide, South Australia, Australia, pp 195–202
Liang YC, Chen Q, Liu Q, Zhang WH, Ding RX (2003) Exogenous silicon (Si) increases antioxidant enzyme activity and reduces lipid peroxidation in roots of salt-stressed barley (Hordeum vulgare L.). J Plant Physiol 160(10):1157–1164
Lingle JC, Carolus RL (1956) Mineral nutrition of celery varieties with special reference to sodium and boron. Proc Am Soc Hort Sci 68:309–317
Liu X, Herbert SJ, Hashemi AM, Zhang X, Ding G (2006) Effects of agricultural management on soil organic matter and carbon transformation—a review. Plant Soil Environ 52(12):531–543
Long XH, Chi JH, Liu L, Li Q, Liu ZP (2009) Effect of seawater stress on physiological and biochemical responses of five Jerusalem artichoke ecotypes. Pedosphere 19(2):208–216
Lunin J, Gallatin MH, Batchelder AR (1963) Saline irrigation of several vegetable crops at various growth stages I. Effect on yields. Agron J 55:107–114
Lynch J, Läuchli A (1998) Salinity affects intracellular calcium in corn root protoplasts. Plant Physiol 87:351–356
Maas EV (1986) Salt tolerance of plants. Appl Agric Res 1:12–26
Maas EV (1990) Crop salt tolerance. In: Tanji KK (ed) Agricultural salinity assessment and management. ASCE manuals and reports on engineering no. 71. New York, ASCE, pp 262–304
Maas EV, Grattan SR (1998) Crop yields as affected by salinity. ASA drainage monograph, in press
Maas EV, Hoffman GJ (1977) Crop salt tolerance—current assessment. J Irrig Drainage Div ASCE 103(IR2):115–134
Maathuis FJM, Amtmann A (1999) K+ nutrition and Na+ toxicity: the basis of cellular K+/Na+ ratios. Ann Bot 84(2):123–133
Maathuis FJM, Podar D (2011) Uptake, distribution, and physiological functions of potassium, calcium, and magnesium. In: Hawkesford MJ, Barraclough P (eds) The molecular and physiological basis of nutrient use efficiency in crops, 1st edn. John Wiley & Sons Inc., Chichester, pp 265–293
Machado RMA, Serralheiro RP (2017) Soil salinity: effect on vegetable crop growth. Management practices to prevent and mitigate soil salinization. Horticulturae 3:30
Magistad OC, Ayers AD, Wadleigh CH, Gauch HG (1943) Effect of salt concentration, kind of salt, and climate on plant growth in sand cultures. Plant Physiol 18:151–166
Mahdy AM (2011) Comparative effects of different soil amendments on amelioration of saline-sodic soils. Soil Water Res 6(4):205–216
Mahmood F, Khan I, Ashraf U, Shahzad T, Hussain S, Shahid M et al (2017) Effects of organic and inorganic manures on maize and their residual impact on soil physico-chemical properties. J Soil Sci Plant Nutr 17(1):22–32
Mahmud MS, Roy S (2020) Physicochemical properties of soils collected from different agricultural lands of Chittagong district of Bangladesh. Acad J Environ Sci 8(2):22–29
Maksimovic I, Ilin Ž (2012) Effects of salinity on vegetable growth and nutrients uptake. In: Lee TS (ed) Irrigation systems and practices in challenging environments. IntechOpen, London
Malcolm CV, Smith ST (1971) Growing plants with salty water. J Agric West Aust 12, Bulletin No. 3785, pp 3–6

Malik YS, Mehrotra N, Pandita ML, Jaiswal RC (1983) Effects of soil salinity levels on the yield and quality of radish, carrot and turnip. Indian J Agric Sci 53:861–862

Maria SG, Rubio F, Dubcovsky J, Rodriguez-Navarro A (1997) The HAK1 gene of barley is a member of a large gene family and encodes a high-affinity potassium transporter. Plant Cell 9: 2281–2289

Marschner H (1995) Mineral nutrition of higher plants, 2nd edn. Academic Press, London, p 889

Menzies N (2009) The science of phosphorus nutrition: forms in the soil, plant uptake, and plant response. Available from: https://grdc.com.au/resources-andpublications/grdc-update-apers/tabcontent/grdc-update-papers/2009/02/the-science-of-phosphorus-nutritionforms-in-the-soil-plant-uptake-andplant-response

Mills D (1989) Differential response of various tissues of Asparagus officinalis to sodium chloride. J Exp Bot 40:485–491

Milosevic T, Milosevic N (2009) The effect of zeolite, organic and inorganic fertilizers on soil chemical properties, growth and biomass yield of apple trees. Plant Soil Environ 55(12): 528–535

Mizrahi Y, Pasternak D (1985) Effect of salinity on quality of various agricultural crops. Plant Soil 89:301–307

Mizrahi Y, Taleisnik E, Kagan-Zur V, Zohar Y, Offenbach R, Matan E, Golan R (1988) A saline irrigation regime for improving tomato fruit quality without reducing yield. J Am Soc Hort Sci 113:202–205

Mkhabela M, Warman PR (2005) The influence of municipal solid waste compost on yield, soil phosphorus availability and uptake by two vegetable crops, grown in a Pugwash sandy loam soil in Nova Scotia. Agric Ecosyst Environ 106:57–67

Munns R (2005) Genes and salt tolerance: bringing them together. New Phytol 167:645–663

Munns R, Rawson HM (1999) Effect of salinity on salt accumulation and reproductive development in the apical meristem of wheat and barley. Aust J Plant Physiol 26:459–464

Munns R, Termaat A (1986) Whole-plant responses to salinity. Aust J Plant Physiol 13:143–160

Munns R, Tester M (2008) Mechanisms of salinity tolerance. Annu Rev Plant Biol 59:651–681

Munns R, James RA, Xu B, Athman A, Conn SJ, Jordans C, Byrt CS, Hare RA, Tyerman SD, Tester M et al (2012) Wheat grain yield on saline soils is improved by an ancestral Na+ transporter gene. Nat Biotechnol 30:360–364

Nadeem SM, Ahmad M, Zahir ZA, Javaid A, Ashraf M (2014) The role of mycorrhizae and plant growth promoting rhizobacteria (PGPR) in improving crop productivity under stressful environments. Biotechnol Adv 32:429–448

Nadler A, Heuer B (1995) Effect of saline irrigation and water deficit on tuber quality. Potato Res 38:119–123

Nasim W, Ahmad A, Amin A, Tariq M, Awais M, Saqib M et al (2018) Radiation efficiency and nitrogen fertilizer impacts on sunflower crop in contrasting environments of Punjab, Pakistan. Environ Sci Pollut Res 25:1822–1836

Newton PJ, Myers BA, West DW (1991) Reduction in growth and yield of Jerusalem artichoke caused by soil salinity. Irrig Sci 12:213–221

Nieman RH (1962) Some effects of sodium chloride on growth, photosynthesis, and respiration of twelve crop plants. Bot Gaz 123:279–285

Norton J, Strom C (2012) Successful restoration of severely disturbed Wyoming lands: reclamation on salt/sodium-affected soils, bulletin-1231. University of Wyoming Extension and Wyoming Reclamation and Restoration Center (WRRC), Laramie

Nyman LP, Gonzales CJ, Arditti J (1983) In-vitro selection for salt tolerance of taro (Colocasia esculenta var. antiquorum). Ann Bot 51:229–236

Obi ME, Ebo POC (1995) The effect of organic and inorganic amendments on soil physical properties and maize production in severely degraded sandy soil in southern Nigeria. Bioresour Technol 51(2-3):117–123

Odegbaro OA, Smith OE (1969) Effect of kinetin, salt concentration, and temperature on germination and early seedling growth of Lactuca sativa L. J Am Soc Hort Sci 94:167–170

Oo AN, Iwai CB, Saenjan P (2015) Soil properties and maize growth in saline and nonsaline soils using cassava-industrial waste compost and vermicompost with or without earthworms. Land Degrad Dev 26(3):300–310

Osakabe Y, Yamaguchi-Shinozaki K, Shinozaki K, Lam-Son Phan T (2014) ABA control of plant macroelement membrane transport systems in response to water deficit and high salinity. New Phytol 202:35–49

Osawa T (1955) Effects of different forms of available nitrogen on Pak-choi (Brassica chinensis L.) injured by salinity. Bull Univ Osaka Pref Ser B 5:15–18

Osawa T (1961) Studies on the salt tolerance of vegetable crops in sand cultures. II. Leafy vegetables. J Jpn Soc Hort Sci 30:48–56

Osawa T (1963) Studies on the salt tolerance of vegetable crops with special reference to osmotic effects and specific ion effects. J Jpn Soc Hort Sci 32:211–223

Osawa T (1965) Studies on the salt tolerance of vegetable crops with special reference to mineral nutrition. Bull Univ Osaka Pref Ser B 16:13–57

Osawa T (1966) Effect of the stage of growth on the salt tolerance of some vegetable crops. J Jpn Soc Hort Sci 35:291–296

Osman KT (2018) Management of soil problems, 1st edn. Springer International Publishing AG, Cham, p 474

Ouni Y, Lakhdar A, Scelza R, Scotti R, Abdelly C, Barhoumi Z et al (2013) Effects of two composts and two grasses on microbial biomass and biological activity in a salt-affected soil. Ecol Eng 60: 363–369

Paek KY, Chandler SF, Thorpe TA (1988) Physiological effects of Na2SO4 and NaCl on callus cultures of Brassica campestris (Chinese cabbage). Physiol Plant 72:160–166

Paliwal KV, Yadav BR (1980) Effect of saline irrigation water on the yield of potato. Indian J Agric Sci 50:31–33

Parida AK, Das AB (2005) Salt tolerance and salinity effects on plants: a review. Ecotoxicol Environ Saf 60(3):324–349

Pasternak D, De Malach Y (1994) Crop irrigation with saline water. In: Pessarakli M (ed) Handbook of plant and crop stress. Marcel Dekker, New York, pp 599–622

Pasternak D, Twersky M, De Malach Y (1979) Salt resistance in agricultural crops. In: Mussel HW, Staples RC (eds) Stress physiology of crop plants. Wiley, New York, pp 107–142

Pasternak D, De Malach Y, Borovic I (1984) Irrigation with brackish water under desert conditions I. Problems and solutions in production of onions (Allium cepa L.). Agric Water Manag 9:225–235

Pasternak D, De Malach Y, Borovic I, Shram M, Aviram C (1986) Irrigation with brackish water under desert conditions. IV. Salt tolerance studies with lettuce (Lactuca sativa L.). Agric Water Manag 11:303–311

Patel NT, Gupta A, Pandey AN (2010) Salinity tolerance of Avicennia marina (Forssk.) Vierh. from Gujarat coasts of India. Aquat Bot 93(1):9–16

Pessarakli M, Haghighi M, Sheibanirad A (2015) Plant responses under environmental stress conditions. Adv Plants Agric Res 2(6):276–286

Prapagar K, Indraratne SP, Premanandharajah P (2012) Effect of soil amendments on reclamation of saline-sodic soil. Trop Agric Res 23(2):168–176

Provin T, Pitt JL (2001) Managing soil salinity. Available from: http://oaktrust.library.tamu.edu/bitstream/handle/1969.1/86985/pdf_1397.pdf.

Purty RS, Kumar G, Singla-Pareek SL, Pareek A (2008) Towards salinity tolerance in Brassica: an overview. Physiol Mol Biol Plants 14:39–49

Qadir M, Oster JD (2004) Crop and irrigation management strategies for saline-sodic soils and waters aimed at environmentally sustainable agriculture. Sci Total Environ 323:1–19

Qadir M, Schubert S, Steffens D (2005) Phytotoxic substances in soils. In: Hillel D (ed) Encyclopedia of soils in the environment. Academic Press, Oxford, pp 216–222

Qi Z, Spalding EP (2004) Protection of plasma membrane K+ transport by the salt overly sensitive Na+-H+ antiporter during salinity stress. Plant Physiol 136:2548–2555

Rahimi A, Biglarifard A (2011) (2011). Influence of NaCl salinity and different substrates on plant growth, mineral nutrient assimilation and fruit yield of strawberry. Notulae Botanicae Horti Agrobotanici Cluj-Napoca 39(2):219–226

Rahnama A, Munns R, Poustini K, Watt M (2011) A screening method to identify genetic variation in root growth response to a salinity gradient. J Exp Bot 62:69–77

Rameeh V (2012) Ions uptake, yield and yield attributes of rapeseed exposed to salinity stress. J Soil Sci Plant Nutr 12(4):851–861

Rameeh V, Rezai A, Saeidi G (2004) Study of salinity tolerance in rapeseed. Commun Soil Sci Plant Anal 35:2849–2866

Rana RS (1985) Breeding for salt resistance: concept and strategy. Int J Trop Agric 3:236–254

Raven JA (1985) Regulation of pH and generation of osmolarity in vascular plants–a cost-benefit analysis in relation to efficiency of use of energy, nitrogen and water. New Phytol 101:25–77

Reggiani R, Bonzo S, Bertani A (1995) The effect of salinity on early seedling growth of seeds of three wheat (Triticum aestivum L.) cultivars. Can J Plant Sci 75:175–177

Rhoades JD (1976) Measuring, mapping and monitoring field salinity and water table depths with soil resistance measurements. FAO Soils Bull 31:159–186

Rhoades JD (1979) Inexpensive four-electrode probe for monitoring soil salinity. Soil Sci Soc Am J 43:817–818

Rhoades JD (1993) Electrical conductivity methods for measuring and mapping soil salinity. In: Sparks DL (ed) Advances in agronomy, vol 49, pp 201–251

Richards RA, Hunt JR, Kirkegaard JA, Passioura JB (2014) Yield improvement and adaptation of wheat to water-limited environments in Australia—a case study. Crop Pasture Sci 65:676–689

Rogers ED, Benfey PN (2015) Regulation of plant root system architecture: implications for crop advancement. Curr Opin Biotechnol 32:93–98

Roy S, Kashem MA (2014) Effects of organic manures in changes of some soil properties at different incubation periods. Open J Soil Sci 4:81–86

Roy SJ, Huang W, Wang XJ, Evrard A, Schmoeckel SM, Zafar ZU, Tester M (2013) A novel protein kinase involved in Na+ exclusion revealed from positional cloning. Plant Cell Environ 36:553–568

Roy SJ, Negrao S, Tester M (2014) Salt resistant crop plants. Curr Opin Biotechnol 26:115–124

Roy S, Kashem MA, Osman KT (2018) The uptake of phosphorous and potassium of rice as affected by different water and organic manure management. J Plant Sci 6(2):31–40

Roy S, Akhtaruzzaman M, Nath B (2020) Spatio-seasonal variations of salinity and associated chemical properties in the middle section of Karnaphuli river water, Chittagong, Bangladesh using laboratory analysis and GIS technique. Int J Environ Sci Dev 11(8):372–382

Russo D (1987) Lettuce yield-irrigation water quality and quantity relationships in a gypsiferous desert soil. Agron J 79:8–14

Sabbah S, Tal M (1995) Salt tolerance in Solanum kurzianum and S. tuberosum cvs Alpha and Russet Burbank. Potato Res 38:319–330

Sahin U, Ekincib M, Orsa S, Turanc M, Yildizb S, Yildirimb E (2018) Effects of individual and combined effects of salinity and drought on physiological, nutritional and biochemical properties of cabbage (Brassica oleracea var. capitata). Sci Hortic 240:196–204

Sairam RK, Tyagi A (2004) Physiology and molecular biology of salinity stress tolerance in plants. Curr Sci 86:407–421

Saleque MA, Choudhury NN, Rezaul-Karim SM, Panaullah GM (2005) Mineral nutrient and yield of four rice genotypes in the farmer's fields of salt affected soils. J Plant Nutr 28(5):865–875

Sani E, Herzyk P, Perrella G, Colot V, Amtmann A (2013) Hyperosmotic priming of Arabidopsis seedlings establishes a long-term somatic memory accompanied by specific changes of the epigenome. Genome Biol 14:R59

Saud S, Yajun C, Long B, Fahad S, Sadiq A (2013) The different impact on the growth of cool season turf grass under the various conditions on salinity and draught stress. Int J Agric Sci Res 3(4):77–84

Saud S, Li X, Chen Y, Zhang L, Fahad S, Hussain S et al (2014) Silicon application increases drought tolerance of Kentucky bluegrass by improving plant water relations and morph physiological functions. Sci World J 2014:1–10

Saud S, Yajun C, Fahad S, Hussain S, Na L, Xin L et al (2016) Silicate application increases the photosynthesis and its associated metabolic activities in Kentucky bluegrass under drought stress and post-drought recovery. Environ Sci Pollut Res 23:17647–17655

Saud S, Fahad S, Yajun C, Ihsan MZ, Hammad HM, Nasim W et al (2017) Effects of nitrogen supply on water stress and recovery mechanisms in Kentucky bluegrass plants. Front Plant Sci 8:983

Saud S, Fahad S, Cui G, Yajun C, Anwar S (2020) Determining nitrogen isotopes discrimination under drought stress on enzymatic activities, nitrogen isotope abundance and water contents of Kentucky bluegrass. Sci Rep 10:6415

Schilling RK, Marschner P, Shavrukov Y, Berger B, Testerm M, Roy SJ, Plett DC (2014) Expression of the Arabidopsis vacuolar H+-pyrophosphatase gene (AVP1) improves the shoot biomass of transgenic barley and increases grain yield in a saline field. Plant Biotechnol J 12:378–386

Schmidhalter U, Oertli JJ (1991) Germination and seedling growth of carrots under salinity and moisture stress. Plant Soil 132:243–251

Schmidt C, Tie H, Cramer GR (1993) Supplemental calcium does not improve growth of salt stressed Brassicas. Plant Soil 155/156:415–418

Schroeder JI, Delhaize E, Frommer WB, Guerinot ML, Harrison MJ, Herrera- Estrella L, Horie T, Kochian LV, Munns R, Nishizawa NK et al (2013) Using membrane transporters to improve crops for sustainable food production. Nature 497:60–66

Scialabba A, Melati MR (1990) The effect of NaCl on growth and xylem differentiation of radish seedlings. Bot Gaz 151:516–521

Shaaban MM, El-Fouly MM (2002) Nutrient content and salt removal potential of wild plants grown in salt affected soils. Acta Hortic 573:377–385

Shabala S (2013) Learning from halophytes: physiological basis and strategies to improve abiotic stress tolerance in crops. Ann Bot 112:1209–1221

Shadded MA, Zidan MA (1989) Effect of NaCl salinity on the rate of germination, seedling growth, and some metabolic changes in Raphanus sativum L. and Trigonella foenum-graecum L. Beitr Trop Landwirtsch Vet Med 27:187–194

Shah F, Nie L, Cui K, Shah T, Wu W, Chen C et al (2014) Rice grain yield and component responses to near 2°C of warming. Field Crop Res 157:98–110

Shahriaripour R, Pour AT, Mozaffari V (2011) Effects of salinity and soil phosphorus application on growth and chemical composition of pistachio seedlings. Commun Soil Sci Plant Anal 42: 144–158

Shannon MC (1980) Differences in salt tolerance within 'empire' lettuce. J Am Soc Hort Sci 105: 944–947

Shannon MC (1996) New insights in plant breeding efforts for improved salt tolerance. HortTechnology 6:96–99

Shannon MC, Francois LE (1978) Salt tolerance of three muskmelon cultivars. J Am Soc Hort Sci 103:127–130

Shannon MC, Grieve CM (1999) Tolerance of vegetable crops to salinity. Sci Hort 78:5–38

Shannon MC, McCreight JD (1984) Salt tolerance of lettuce introductions. HortScience 19:673–675

Shannon MC, Noble CL (1990) Genetic approaches for developing economic salt-tolerant crops. In: Tanji K (ed) Agricultural salinity assessment and management. ASCE manuals and reports on engineering no. 71. ASCE, New York, pp 161–185

Shannon MC, McCreight JD, Draper JH (1983) Screening tests for salt tolerance in lettuce. J Am Soc Hort Sci 108:225–230

Shannon MC, Grieve CM, Francois LE (1994) Whole-plant response to salinity. In: Wilkinson RE (ed) Plant-environment interactions. Marcel Dekker, New York, pp 199–244

Sharma PC, Gill KS (1994) Salinity-induced effect on biomass, yield, yield-attributing characters and ionic contents in genotypes of Indian mustard (Brassica juncea). Indian J Agric Sci 64:785–788

Shehzadi S, Shah Z, Mohammad W (2017) Impact of organic amendments on soil carbon sequestration, water use efficiency and yield of irrigated wheat. BASE 21(1):36–49

Silva JV, Lacerda CF, Costa PHA, Enéas Filho J, Filho EG, Prisco JT (2003) Physiological responses of NaCl stressed cowpea plants grown in nutrient solution supplemented with CaCl2. Braz J Plant Physiol 15:99–105

Šimanský V, Zaujec A (2009) Suitable parameters for soil organic matter changes evaluation in agro-ecosystems. Folia Oecol 36(1):50–57

Smolders E, McLaughlin MJ (1996a) Effects of Cl on Cd uptake by Swiss chard in nutrient solutions. Plant Soil 179:57–64

Smolders E, McLaughlin MJ (1996b) Chloride increases cadmium uptake in Swiss chard in a resin-buffered nutrient solution. Soil Sci Soc Am J 60:1443–1447

Sommerfeldt TG, Rapp E, Chang C, Janzen HH (1988) Management of saline soils under irrigation. Publication 1624/E. Agriculture Canada, Ottawa, ON

Song JQ, Fujiyama H (1996) Differences in response of rice and tomato subjected to sodium salinization to the addition of calcium. Soil Sci Plant Nutr 42(3):503–510

Sonneveld C (1988) The salt tolerance of greenhouse crops. Neth J Agric Sci 36:63–73

Spark DL (2002) Environmental soil chemistry, 2nd edn. Academic Press, London, p 352

Speer M, Kaiser WM (1991) Ion relations of symplastic and apoplastic space in leaves from Spinacia oleracea L. and Pisum sativum L. under salinity. Plant Physiol 97:990–997

Sun J, He F, Shao H, Zhang Z, Xu G (2016) Effects of biochar application on Suaeda salsa growth and saline soil properties. Environ Earth Sci 75:630

Syers JK, Johnston AE, Curtin D (2008) Efficiency of soil and fertilizer phosphorus use: reconciling hanging concepts of soil phosphorus behaviour with agronomic information. FAO Fertilizer and Plant Nutrition Bulletin 18. Food and Agriculture Organization of the United Nations, Rome

Szabolcs I (1974) Salt-affected soil in Europe, 1st edn. Springer, Dordrecht, p 57

Szabolcs I (1994) Salt affected soils as the ecosystem for halophytes. In: Squires VR, Ayoub AT (eds) Halophytes as a resource for livestock and for rehabilitation of degraded lands, 1st edn. Springer, Dordrecht, pp 19–24

Takatori FH, Lorenz OA, Cannell GH (1961) Strontium and calcium for the control of blackheart of celery. Proc Am Soc Hort Sci 77:406–414

Talei D, Kadir MA, Yusop MK, Valdiani A, Abdullah MP (2012) Salinity effects on macro and micronutrients uptake in medicinal plant king of bitters (Andrographis paniculata Nees.). Plant Omics J 5(3):271–278

Tan KH, Dowling PS (1984) Effect of organic matter on CEC due to permanent and variable charges in selected temperate region soils. Geoderma 32(2):89–101

Tejada M, Garcia C, Gonzalez JL, Hernandez MT (2006) Use of organic amendment as a strategy for saline soil remediation: Influence on the physical, chemical and biological properties of soil. Soil Biol Biochem 38(6):1413–1421

Tester CF (1990) Organic amendment effects on physical and chemical properties of a sandy soil. Soil Sci Soc Am J 54(3):827–831

Tobiašová E (2011) The effect of organic matter on the structure of soils of different land uses. Soil Tillage Res 114:183–192

Tomemori H, Toyama M, Takeuchi Y (1996) Effects of irrigation by diluted sea water on the cultivation of spinach. Sand Dune Res 43:15–18

Tuna AL, Kaya C, Ashraf M, Altunlu H, Yokas I, Yagmur B (2007) The effects of calcium sulfate on growth, membrane stability and nutrient uptake of tomato plants grown under salt stress. Environ Exp Bot 59(2):173–178

Turhan E, Eris A (2005) Changes of micronutrients, dry weight, and chlorophyll contents in strawberry plants under salt stress conditions. Commun Soil Sci Plant Anal 36:1021–1028

United States Department of Agriculture (1954) Diagnosis and improvement of saline and alkali soils. Agriculture handbook No. 60, 160 pp

Uno Y, Kanechi M, Inagaki N, Sugimoto M, Maekawa S (1996) The evaluation of salt tolerance during germination and vegetative growth of asparagus, table beet and sea aster. J Jpn Soc Hort Sci 65:579–585

Van der Werf A, Kooijman A, Welschen R, Lambers H (1988) Respiratory energy costs for the maintenance of biomass, for growth and for ion uptake in roots of Carex diandra and Carex acutiformis. Physiol Plant 72:483–491

van Genuchten MT, Hoffman GJ (1984) Analysis of crop salt tolerance data. In: Shainberg I, Shalhevet J (eds) Soil salinity under irrigation—process and management. Springer, New York, pp 258–271

van Hoorn JW, Katerji N, Hamdy A, Mastrorilli M (1993) Effect of saline water on soil salinity and on water stress, growth, and yield of wheat and potatoes. Agric Water Manag 23:247–265

Vose PB (1963) Varietal differences in plant nutrition. Herbage Abstr 33:1–13

Waisel Y, Breckle SW (1987) Differences in responses of various radish roots to salinity. Plant Soil 104:191–194

Wakeel A (2013) Potassium-sodium interactions in soil and plant under saline-sodic conditions. J Plant Nutr Soil Sci 176:344–354

Walker DJ, Bernal PM (2008) (2008). The effects of olive mill waste compost and poultry manure on the availability and plant uptake of nutrients in a highly saline soil. Bioresour Technol 99(2): 396–403

Wang L, Sun X, Li S, Zhang T, Zhang W, Zhai P (2014) Application of organic amendments to a coastal saline soil in North China: effects on soil physical and chemical properties and tree growth. PLoS One 9(2):e89185

Wannamaker MJ, Pike LM (1987) Onion responses to various salinity levels. J Am Soc Hort Sci 112:49–52

West DW (1978) Water use and sodium chloride uptake by apple trees. II. The response to soil oxygen deficiency. Plant Soil 50:51–56

West DW, Taylor JA (1984) Response of six grape cultivars to the combined effects of high salinity and rootzone waterlogging. J Am Soc Hort Sci 109:844–851

Wichern F, Islam MR, Hemkemeyer M, Watson C, Joergensen RG (2020) Organic amendments alleviate salinity effects on soil microorganisms and mineralization processes in aerobic and anaerobic paddy rice soils. Front Sustain Food Syst 4:30

Wright MJ (1976) Plant adaptation to mineral stress in problem soils. A Special Publication of Cornell University Agricultural Experiment Station, Cornell University, Ithaca, NY, 420 pp

Wu D, Qiu L, Xu L, Ye L, Chen M, Sun D, Chen Z, Zhang H, Jin X, Dai F et al (2011) Genetic variation of HvCBF genes and their association with salinity tolerance in Tibetan annual wild barley. PLoS One 6:e22938

Wu Y, Li Y, Zhang Y, Bi Y, Sun Z (2018) Response of saline soil properties and cotton growth to different organic amendments. Pedosphere 28(3):521–529

Xiaohou S, Min T, Ping J, Weiling C (2008) Effect of EM Bokashi application on control of secondary soil salinization. Water Sci Eng 1(4):99–106

Xu C, Wang M, Zhou L, Quan T, Xia G (2013) Heterologous expression of the wheat aquaporin gene TaTIP2;2 compromises the abiotic stress tolerance of Arabidopsis thaliana. PLoS One 8: e79618

Yamaguchi M (1983) World vegetables: principles, production and nutritive values. AVI, Westport, CT

Yang Z, Zhang Z, Zhang T, Fahad S, Cui K, Nie L et al (2017) The effect of season-long temperature increases on rice cultivars grown in the central and southern regions of China. Front Plant Sci 8:1908

Yeo AR (1983) Salinity resistance—physiologies and prices. Physiol Plant 58:214–222

Yu Y, Liu J, Liu C, Zong S, Lu Z (2015) Effect of organic materials on the chemical properties of saline soil in the Yellow River Delta of China. Front Earth Sci 9:259–267

Zaman QU, Aslam Z, Yaseen M, Ihsan MZ, Khaliq A, Fahad S et al (2018) Zinc biofortification in rice: Leveraging agriculture to moderate hidden hunger in developing countries. Arch Agron Soil Sci 64(2):147–161

Zhang ZJ, Wang XZ, Wang H, Huang E, Sheng JL, Zhou LQ et al (2020) Housefly larvae (Musca domestica) vermicompost on soil biochemical features for a chrysanthemum (Chrysanthemum morifolium) farm. Commun Soil Sci Plant Anal 51(10):1315–1330

Zhao D, Wang Z, Yang F, Zhu W, An F, Ma H et al (2020) Amendments to saline-sodic soils showed long-term effects on improving growth and yield of rice (Oryza sativa L.). PeerJ 8: e8726

Zhou L, Wang C, Liu R, Han Q, Vandeleur RK, Du J, Tyerman S, Shou H (2014) Constitutive overexpression of soybean plasma membrane intrinsic protein GmPIP1;6 confers salt tolerance. BMC Plant Biol 14:188

Zia Z, Bakhat HF, Saqib ZA, Shah GM, Fahad S, Ashraf MR et al (2017) Effect of water management and silicon on germination, growth, phosphorus and arsenic uptake in rice. Ecotoxicol Environ Saf 144:11–18

Zidan MA, Elewa MA (1995) Effect of salinity on germination, seedling growth and some metabolic changes in four plant species (Umbelliferae). Indian J Plant Physiol 38:57–61

Zribi OT, Slama I, Trabelsi N, Hamdi A, Smaoui A, Abdelly C (2018) Combined effects of salinity and phosphorus availability on growth, gas exchange, and nutrient status of Catapodium rigidum. Arid Land Res Manag 32(3):277–290